AF616204

BD 034445685 X

Alignment Technologies and Applications of Liquid Crystal Devices

THE LIQUID CRYSTALS BOOK SERIES

Edited by

G. W. GRAY, J. W. GOODBY & A. FUKUDA

The Liquid Crystals book series publishes authoritative accounts of all aspects of the field, ranging from the basic fundamentals to the forefront of research; from the physics of liquid crystals to their chemical and biological properties; and, from their self-assembling structures to their applications in devices. The series will provide readers new to liquid crystals with a firm grounding in the subject, while experienced scientists and liquid crystallographers will find that the series is an indispensable resource.

Introduction to Liquid Crystals
Chemistry and Physics
P. Collings and M. Hird

The Static and Dynamic Continuum Theory of Liquid Crystals
I. W. Stewart

Crystals That Flow Classic Papers from the History of Liquid Crystals
T. J. Sluckin, D. A. Dunmur and H. Stegemeyer

Alignment Technologies and Applications of Liquid Crystal Devices

Kohki Takatoh, Masaki Hasegawa, Mitsuhiro Koden, Nobuyuki Itoh, Ray Hasegawa and Masanori Sakamoto

LONDON AND NEW YORK

First published 2005
by Taylor & Francis
2 Park Square, Milton Park, Abingdon, Oxon OX14 4RN

Simultaneously published in the USA and Canada
by Taylor & Francis Inc,
270 Madison Ave, New York, NY 10016

Taylor & Francis is an imprint of the Taylor & Francis Group

Typeset in Century Schoolbook by
Integra Software Services Pvt. Ltd, Pondicherry, India
Printed and bound in Great Britain by
MPG Books Ltd, Bodmin

British Library Cataloguing in Publication Data
A catalogue record for this book is available
from the British Library

Library of Congress Cataloging in Publication Data
A catalog record for this book has been requested

ISBN 0–748–40902–5

Contents

Preface

The aim of this book is to present a systematic discussion of the alignment technologies of liquid crystalline materials for liquid crystal displays (LCDs). These technologies determine the molecular orientation and conformation of liquid crystal phases on surfaces, and by understanding the technologies, both the structure and the properties of the LCDs can be better understood. The reader will obtain essential knowledge of LCDs through the discussion of the alignment technologies given in the text, which as far as the authors know is the first to be written from this standpoint.

Originally, the book was planned to be in two parts, theoretical and practical. However, as writing progressed, we came to the conclusion that the chapters based upon discussions and descriptions of actual devices could provide the reader with the most useful and comprehensive information to cover most requirements. Although discussions from the theoretical point of view may be a little limited, each chapter does provide indispensable theoretical explanations, supplemented and supported by references and the bibliography.

The first half of the book deals with alignment technologies for nematic liquid crystals, and in the second half those for smectic liquid crystals are covered. Almost all commercially available LCDs use nematic liquid crystals, and so the first chapters provide information that is practical and useful. However, alignment of smectic liquid crystals with all their variations is a very fruitful area of study, and with developments that are possible in the future, a knowledge of smectic liquid crystal alignment could be even more important. It is therefore one of the key features of the book that the alignments for both smectic and nematic liquid crystals are presented.

The book provides useful information and references for researchers and engineers working on developments and production processes for LCDs. It also provides a very appropriate introduction for students just starting to study liquid crystals or beginning to work in the LCD industry. It also provides a guide to LCD science and the related industries, and as such, establishes a good introduction to the field, with references and bibliography, for researchers and engineers from other fields.

The authors have all worked in the LCD industry from the 1980s to the time of writing, in the following companies: Sharp, IBM Japan and Toshiba. Their projects during these years were concerned with managing alignment technologies for LCDs. During these years the LCD industry has developed rapidly in Japan. From the first proposal for this book in 1996, its accomplishment with appropriate translation into English has taken time. Some of the very latest topics may not be discussed as fully as would be desired, but the basic concepts for these topics have all been introduced. The authors will be delighted if this book offers not only information, but also clues and ideas for solving the problems of those studying and working in the field of LC technologies.

We would like to record our sincere thanks to Professor Hiroshi Yokoyama and Professor Yukio Ouchi for their useful suggestions and discussions at the initial planning stage of this book. We also wish to thank the editors of the series, Professor George Gray, FRS, Professor John Goodby and Professor Atsuo Fukuda, for the opportunity and invitation to write this book and for their helpful suggestions and guidance. We are also grateful to Grant Soanes, Frances Horrocks and Janie Wardle for their editorial assistance during the different stages of preparation of this book. In addition, we would like to extend our heartfelt thanks to our families for their constant support.

Acknowledgements

The authors and publishers are pleased to acknowledge the following organizations for permission to reproduce selected papers and illustrations. Every effort has been made to contact and acknowledge copyright holders. If any errors or omissions have been made we would be pleased to correct them at a later printing.

Taylor & Francis, as publishers, and the authors of this book are most grateful to all who have given permission for their published work to be reproduced here.

Control and modification of nematic liquid crystal pretilt angles on polyimides, Figure 2, K.-W. Lee, A. Lien, J. Stahis and S.-H. Paek, *Japanese Journal of Applied Physics*, Part 1, **36**, p. 3591 (1997). Reproduced by permission of Institute of Pure and Applied Physics.

A mechanistic picture of the effects of rubbing on polyimide surfaces and liquid crystal pretilt angles, Figure 3, S.-H. Paek, C.J. Duming, K.-W. Lee and A. Lien, *Japanese Journal of Applied Physics*, **83**, p. 1270 (1998). Reproduced by permission of Institute of Pure and Applied Physics.

Figure 2, W. Chen, M.B. Feller and Y.R. Shen, *Physical Review Letters*, **63**, p. 2665 (1989). Copyright American Physical Society. Reproduced by permission of the American Physical Society.

Effect of rubbing on the molecular orientation within polyimide orienting layers of liquid-crystal displays, Figures 5 and 12, N.A.J.M. van Aerle, M. Barmentlo and R.W.J. Hollering, *Journal of Applied Physics*, **74**, p. 3111 (1993). Reproduced by permission of the American Institute of Physics.

Surface orientation of polyimide alignment layer studied by optical second-harmonic generation, Figure 5, K. Shirota, K. Ishikawa, H. Takezoe, A. Fukuda and T. Shiibashi, *Japanese Journal of Applied Physics*, **34**, p. L316 (1995). Reproduced by permission of Institute of Pure and Applied Physics.

Molecular orientation of polyimide films determined by an optical retardation method, Figures 1 and 2, K. Sakamoto, R. Arafune, S. Ushioda, Y. Suzuki and S. Morokawa, *Applied Surface Science*, **100 (101)**, p. 124 (1996). Reproduced with permission from Elsevier Science.

Determination of molecular orientation of very thin rubbed and unrubbed polyimide films, Figures 4 and 9, K. Sakamoto, R. Arafune, N. Ito and S. Ushioda, *Journal of Applied Physics*, **80**, p. 431 (1996). Reproduced by permission of American Institute of Physics.

A study of the relation between surface alignment of polymers and liquid-crystal pretilt angle, Figures 12 and 14, K.-Y. Han and T. Uchida, *Journal of the Society for Information Display*, **3 (1)**, p. 18 (1995). Permission for Reprint, courtesy of the Society for Information Display.

Polarized XANES studies on the rubbed polyimide, Figure 2, Y. Ouchi, I. Mori, M. Sei, E. Ito, T. Araki, H. Ishii, K. Seki, and K. Kondo, *Physica B*, **308 (209)**, p. 407 (1995). Reproduced with permission from Elsevier Science.

NEXAFS studies on the rubbing effects of the surface structure of polyimides, Figures 2, 5 and 6, I. Mori, T. Araki, H. Ishii, Y. Ouchi, K. Seki and K. Kondo, *J. Electron Spectrosc. Relat. Phenom.*, **78**, p. 371 (1996). Reproduced with permission from Elsevier Science.

Correlation between the pretilt angle of liquid crystal and the inclination angle of the polyimide backbone, Figure 4, R. Arafune, K. Sakamoto and S. Ushioda, *Applied Physics Letters*, **71**, p. 2755 (1997). Reproduced by permission of the American Institute of Physics.

Table 1 and Figure 6, R. Arafune, K. Sakamoto, S. Ushioda, S. Tanioka and S. Muraya, *Physical Review E*, **58**, p. 5914 (1998). Reproduced by permission of the American Physical Society.

Director orientation of a ferroelectric liquid, Figures 6 and 7, T. Uchida, M. Hirano and H. Sasaki, *Liquid Crystals*, **5**, p. 1127 (1989). Reproduced by permission of Taylor & Francis.

The mechanism of polymer alignment of liquid-crystal materials, Figure 1, J.M. Geary, J.W. Goodby, A.R. Kmetz and J.S. Patel, *Journal of Applied Physics*, **62**, p. 4100 (1987). Reproduced by permission of American Institute of Physics.

Characteristics of polyimide liquid crystal alignment films for active matrix-LCD use, Figure 9, M. Nishikawa, Y. Matsuki, N. Bessho, Y. Iimura and S. Kobayashi, *Journal of Photopolymer Science and Technology*, **8**, p. 233 (1995). Reproduced by permission of Technical Association of Photopolymers, Japan.

Figure 1, H. Mori, Y. Itoh, Y. Nishiura, T. Nakamura and Y. Shinagawa, *Digest of Society for Information Display 97*, p. 941 (1997). Permission for use courtesy of the Society for Information Display.

Mechanism of fast response and recover in bend alignment cell, Figures 3 and 5, S. Onda, T. Miyashira and T. Uchida, *Molecular Crystals and Liquid Crystals*, **331**, p. 383 (1999). Reproduced by permission of Taylor & Francis (http://www.tandf.co.uk).

MVA, Figures 5, 6 and 7, A. Takedo, *Ekisho*, **3 (2)**, p. 117 (1999). Reproduced by permission of Ekisho, Japan.

Tilted alignment of MBBA induced by short-chain surfants, Figure 1, G. Porte, *Le Journal de Physique*, **37**, p. 1246 (1976). Reproduced by permission of EDP Sciences.

Anisotropic interactions between MBBA and surface-treated substrates, Figure 5, S. Naemura, *Journal de Physique*, C3-517 (1979). Reproduced by permission of EDP Sciences.

High surface ordering of nematic liquid crystal using periodicity grating, Figure 2, A. Sugiura, N. Yamamoto and T. Kawamura, *Japanese Journal of Applied Physics*, **20 (7)**, pp. 1334–1343 (1981). Reproduced by permission of Institute of Pure and Applied Physics.

Consideration of alignment control on liquid crystal by grating surface, Figures 4 and 5, K. Sugawara, Y. Ishitaka, M. Abe, K. Seino, M. Kano and T. Nakamura, *Proceeding of Japan Display '92*, p. 815 (1992). Reproduced by permission of The Institute of Image Information and Television Engineers.

Alignment of nematic liquid crystals by polyimide Langmuir-Blodgett films, Table 11, M. Murata, H. Awaji, M. Isurugi, M. Uekita and Y. Tawada, *Japanese Journal of Applied Physics*, **31**, pp. L189–L191 (1992). Reproduced by permission of Institute of Pure and Applied Physics.

Charge-controlled phenomena in the surface-stabilized ferroelectric liquid crystal structure, Figure 4, W.J.A.M. Hartmann, *Journal of Applied Physics*, **66**, p. 1132 (1989). Reproduced by permission of American Institute of Physics.

FLC displays for TV application, Figures 9 and 15, W.J.A.M. Hartmann, *Ferroelectrics*, **122**, pp. 1–15 (1991). Reproduced by permission of Taylor & Francis.

Effect of alignment layer conductivity on the instability of surface stabilized ferroelectric liquid crystal devices, Figures 1 and 2, T.C. Chieu and K.H. Yang, *Applied Physics Letters*, **56 (14)**, p. 1326 (1990). Reproduced by permission of the American Institute of Physics.

Electrooptic bistability of a Ferroelectric Liquid Crystal Device prepared using charge transfer complex doped polyimide orientation films, Figures 1 and 2, K. Nakaya, B.Y. Zhang, M. Yoshida, I. Isa, S. Shindoh and S. Kobayashi, *Japanese Journal of Applied Physics*, **28 (1)**, p. L116 (1989). Reproduced by permission of Institute of Pure and Applied Physics.

5.5 inch video graphic array antiferroelectric display, Figure 2, E. Tajima, S. Kondoh and Y. Suzuki, *Ferroelectrics*, **149**, p. 255 (1993). Reproduced by permission of Taylor & Francis.

Polymeric alignment layers for AFLCD Cells, Figure 8, Y.S. Negi, I. Kawamura, Y. Suzuki, N. Yamamoto, Y. Yamada, M. Kakimoto and Y. Imai, *Molecular Crystals Liquid Crystals*, **239**, pp. 11–23 (1994). Reproduced by permission of Taylor & Francis.

Polyimide aligned ferroelectric smectic C liquid crystals, Figures 1 and 3, B.O. Myrvold, *Liquid Crystals*, **4 (6)**, p. 637 (1989). Reproduced by permission of Taylor & Francis.

Aliphatic polyimides as alignment layers for surface stabilized ferroelectric smectic C liquid crystal cells, Figure 1, B.O. Myrvold, *Liquid Crystals*, **7 (2)**, p. 262 (1990). Reproduced by permission of Taylor & Francis.

Effect of material constants on the orientation structure of ferroelectric liquid crystal cells, Figures 8 and 9, T.C. Chieu, *Journal of Applied Physics*, **64**, p. 6234 (1988). Reproduced by permission of the American Institute of Physics.

Essential factors in high-duty FLC Matrix display, Figure 3, Y. Inaba, K. Katagiri, H. Inoue, J. Kanbe, S. Yoshihara and S. Iijima, *Ferroelectrics*, **85**, pp. 255–257 (1988). Reproduced by permission of Taylor & Francis.

A reliable method of alignment for smectic liquid crystals, Tables 1 and 2, J.S. Patel, T.M. Leslie and J.W. Goodby, *Ferroelectrics*, **59**, pp. 137–139 (1984). Reproduced by permission of Taylor & Francis.

A complementary TN-LCD with wide-viewing-angle gray scale, Figure 7 (p. 34), K. Sumiyoshi, K. Takatori, Y. Hirai, and S. Kaneko, *Journal of the Society for Information Display*, 1/2, p. 31 (1994). Permission for Reprint, courtesy of the Society for Information Display.

Notations

duty ratio	$1/N$
director	$\mathbf{n}$
tilt angle	θ
azimuthal angle	Φ
layer tilt angle	δ
layer spacing	d
pretilt angle	θ_{p}
apparent tilt angle	$\theta_{\mathrm{app.}}$
spontaneous polarization	$\mathbf{p}$
memory angle	θ_{m}
director tilt angle	γ
director twist angle	ϕ

Chapter 1

Overview

Kohki Takatoh

1.1 Development of the LCD Market

This book discusses the alignment phenomena and technologies employed in LCD production. The LCD industry began in the early 70s with small segment-type displays for calculators and watches. These applications exploited the advantages of liquid crystals to the full, namely low voltage and low power consumption and compact size. Displays with matrix electrodes of low resolution also became available. These industries were incubated mostly in Japan. In the late 70s, the application of liquid crystals spread to a wide range of products, such as measuring instruments, domestic appliances and audio equipment, office equipment, game machines and vehicle instruments. From the early 80s onward, the technologies of active matrix displays with thin film transistors, TFT-LCDs, were developed, although mass production did not start until the late 80s. Application of active matrix devices such as thin film transistor devices led to the availability of LCDs offering high resolution. Because the problem of the initially low yields in TFT production were overcome, the market for TFT-LCDs larger than 10 inches, mainly for displays in portable computers, has been expanding. Furthermore, since 1997, TFT-LCDs larger than 15 inches have been produced, and this has enlarged the LCD market to include displays for monitors and television sets. TFT-LCDs larger than 10 inches constitute the main market of the LCD industry. In the market segment of displays larger than 30 inches, other flat panel displays, such as plasma display panels (PDPs), are more suitable. Therefore, the LCD market is thought to be limited to displays of not more than 30 inches. Moreover, the LCD market has been expanding to include small displays, such as those for car navigation systems, personal digital assistants (PDAs), handheld PCs, electronic organizers, AV monitors, video projectors, and digital cameras.

Since the start of the LCD industry in the early 70s, the twisted nematic (TN) mode has been adopted. For the TN mode, the alignment directions of the liquid crystal molecules are twisted through 90° on alignment layers on both glass plates. By adopting this structure, displays offering high contrast and low driving voltage can be realized. Furthermore, the combination of the TN mode and active matrix driving by thin film transistors led to the availability of high resolution and high contrast.

On the other hand, in the early 80s, the supertwisted nematic (STN) mode was proposed for passive matrix displays. In this mode, by using a concentration of optically active compound in the nematic liquid crystalline materials, the twisted structure is enhanced and the switching threshold becomes sharp. Because of the sharp threshold property, matrix driving of high resolution can be realized without any active matrix. With regard to liquid crystalline alignment technologies, an alignment layer for high pretilt angle is required for the STN mode. In spite of the drawback of low contrast ratio and slow switching speed, the market for STN mode expanded on the strength of its low price, reflecting the fact that it does not need TFT devices. Especially in the early stages of TFT-LCD production, the yields of TFT devices were low and the production cost was high. So, the market for the STN type of display increased, centreing on the personal notebook computers. However, in the late 90s, yields of TFT-LCDs improved and the market for TFT-LCDs overtook that for STN-LCDs, and now the main market for STN-LCDs is low-priced displays principally for game machines and portable telephones. It is still a large market.

1.2 Improvement of Viewing Angle Dependence of the Contrast Ratio

The TN mode is excellent in that a high contrast ratio can be easily realized. However, this mode shows a large viewing angle dependence of the contrast ratio, especially in grey scale expression. As the size of displays increased, improvement of the viewing angle dependence of the contrast ratio became an important theme for LCDs. Responding to the need for improvement of viewing angle dependence, the in-plane switching (IPS) mode, the multi-domain vertical alignment (MVA) mode, and the optically compensated birefringence (OCB) mode were developed. In the IPS mode, switching in one plane is achieved by the arrangement of electrodes in one plane. The other modes realize wide viewing angles by changes of initial alignment and switching mechanism of the liquid crystalline materials. The IPS mode and MVA mode have already been commercialized.

1.3 Ferroelectric and Antiferroelectric Liquid Crystals

For the LCDs described above, nematic type of liquid crystalline materials are used. Liquid crystalline materials can be categorized into two large groups: nematic liquid crystals and smectic liquid crystals. With the exception of the thermal effect mode, using the smectic A phase and cholesteric phase, smectic liquid crystals had not been investigated for practical device applications. In 1980, the application of ferroelectric liquid crystals using the chiral smectic C phase for optical switching devices was proposed. The proposed surface stabilized ferroelectric liquid crystal (SSFLC) possesses memory properties without an active matrix. Furthermore, its fast switching and wide viewing angle dependence were highly evaluated. Studies on the application of SSFLCs to high-resolution passive matrix displays have been carried out intensively. As a result, in the latter half of the 90s, large-size SSFLC displays for monitor usage were commercialized. However, production was suspended, partly because of the difficulty of grey scale expression.

In 1988, the antiferroelectric liquid crystal state was discovered and investigations started with a view to applying the materials to display devices. AFLC materials show a symmetrical response to positive and negative electric signals and a large hysteresis. For AFLC materials, a driving method in which an offset voltage is applied in the holding period of the line sequential driving scheme was proposed. By using this method, high-resolution passive matrix LCDs showing grey scale expression were realized. A 17-inch trial display using AFLC material was reported.

Besides the memory property and large hysteresis, SSFLC and AFLC possess excellent properties, such as fast switching speed and wide viewing angle. These properties are indispensable for high-quality displays, although the memory property and large hysteresis are inadequate for active matrix driving. Therefore, FLC or AFLC modes adequate for active matrix driving without memory property and large hysteresis have also been investigated.

The FLCs or AFLCs mentioned above are liquid crystalline materials showing a chiral smectic C phase or related phases. Although nematic liquid crystal possesses only directional order, smectic liquid crystals shows layer structures or periodic order of the liquid crystalline molecular centres. In this respect, smectic liquid crystalline materials have more in common with crystalline materials than with nematic liquid crystalline materials. As a result, the alignment of smectic liquid crystal is quite different from that of nematic liquid crystals. Smectic liquid crystals show a large variety of defects because they possesses highly ordered structures. Moreover, the layer structures are irreversibly destroyed by applying stress. This phenomenon poses a major problem for display applications. Therefore, several techniques to prevent the application of force to the liquid crystalline materials of FLC or AFLC devices have been proposed.

1.4 Development of Novel Alignment Method

From the early stages of LCD production in the 70s, only the rubbing method has been adopted for alignment technologies. The rubbing method is a simple, convenient, and low-cost, and it is considered unlikely that the rubbing method will be replaced by any other method without realizing a marked improvement. However, when the production of TFT-LCDs started in the 80s, defects of the rubbing method were highlighted. In the rubbing method, the surface of polyimide layers is rubbed with a cloth. Several problems were pointed out: The thin film transistors (TFTs) are destroyed by the static electricity caused by the friction between the surfaces of the cloth and the polyimide layers, the direct contact between the surfaces and the cloth causes defects and stains on the surface of alignment layer, and the clean room is polluted by fibre from the cloth.

Especially in the early stages of TFT device production, the destruction of thin film transistors by static electricity was a serious problem because the reliability of the transistors was quite low. This problem has been dealt with gradually. However, the occurrence of fibre dust has been a continuing problem, and has imposed restrictions on production lines. Moreover, the mechanism of the liquid crystalline alignment is not clearly understood.

In view of the problems mentioned above, studies on both the mechanism of the rubbing process and the development of other methods for liquid crystalline alignment have been carried out. Methods using microgroove structures and UV light radiation on UV curable resin or polyimide surfaces were investigated. On the other hand, the reliability of TFT devices has been improved and the problems associated with the rubbing method have been ameliorated. Thus, novel alignment technologies are required that not only achieve increase in production yields, but also offer additional advantages, such as the convenience of multi-domain alignment.

1.5 The Characteristics of this Book

This book discusses liquid crystalline alignment technologies investigated, mainly in industrial laboratories, from the early 80s to 2000. The authors are researchers who have been studying liquid crystalline alignment or developing LCDs during that period at the laboratories of their respective companies (Sharp, Toshiba, Japan IBM).

The investigation of liquid crystal devices can be divided into four categories: studies on driving methods, studies on active matrix devices such as those using thin film transistors, studies on materials such as liquid crystalline materials and polymers for alignment layers, and studies on LCD structures. The authors usually call the last one "studies on LCD cells". In studies concerning "the cell" structures, the orientation of the liquid crystalline molecules and the switching mechanism for new types of LCDs have been investigated. The most important theme in these studies concerns the orientation of the liquid crystalline molecules. The orientation of the liquid crystalline molecules is determined *only* by the selection of the liquid crystalline materials and the surface of the alignment layers. Therefore, the most important and exciting work in research in this field concerns alignment technologies; it concerns the selection of the surface on the alignment layers.

In this book, alignment technologies are divided into those for nematic liquid crystals and those for smectic liquid crystals. The alignment phenomena are discussed for each kind of liquid crystal. Regarding alignment technologies for nematic liquid crystals, typical alignment methods, which include the rubbing method, are discussed. In particular, rubbing technologies and the usage of UV radiation are discussed in detail. Moreover, liquid crystalline devices using nematic liquid crystalline materials are also explained.

In the sections on smectic liquid crystals, first the alignment and molecular orientation of surface stabilized ferroelectric liquid crystals (SSFLCs) are treated in detail. Next, the alignment technologies needed for the occurrence of bistability are detailed. Furthermore, liquid crystalline devices made of AFLC materials and the applications of FLC and AFLC materials to active matrix devices are discussed.

Chapter 2

Rubbing Technologies: Mechanisms and Applications

Masaki Hasegawa

2.1 Introduction

Rubbing is still the dominant alignment process, although it has been a long time since Mauguin reported this method to align liquid crystals eighty years ago [1]. The present rubbing process in the manufacture of liquid crystal displays (LCDs) is to rub an organic polymer coated substrate using a rotating drum which is covered by a cloth with short fibres. Figure 2.1 shows a typical rubbing machine consisting of a rotating drum and a substrate holding platform. Although this process has many problems, for example the generation of dust and static electricity, LCDs cannot be manufactured without it, because it is a short, simple LC alignment process, and can be applied to a large area at low cost.

Although the process is simple, it has large effects on the image quality and reliability of the LCDs. The rubbing controls both the azimuthal and polar angles of the liquid crystal alignment. Unevenness in the rubbing process causes alignment defects. These defects harm the optical characteristics of the LCDs, because the LCDs use the light modulation phenomena that depend on the LC's alignment. Not only the rubbing itself, but also the alignment materials affect the quality of the LCDs. In the early stages of liquid crystal manufacturing, polyvinyl alcohol, acrylic polymers, and vinyl polymers were tested as alignment materials for rubbing. Finally, a high temperature polyimide was used. The development of a low temperature polyimide then allowed a coloured liquid crystal display using colour filters. The reason why polyimide became the dominant alignment material is its stability and superior electric characteristics. In the case of display driven by thin film transistors (TFT), the electric characteristics are as important as the alignment quality.

What happens to the alignment material as a result of the rubbing, and why are the liquid crystals aligned in a uniform direction? There are many reports analyzing

Fig. 2.1 Photograph of a rubbing machine.

mechanisms of the rubbing process. Here, we will review recent studies of the rubbed surfaces, and describe our current understanding of the alignment mechanisms of rubbing. In addition to the rubbing mechanisms, several topics related to rubbing and the alignment of the LCs in the LCD manufacturing process will be discussed later, such as typical defects in LCDs, and the evaluation of the alignment materials.

2.2 Rubbing Mechanisms

2.2.1 Observations of rubbed surfaces

To examine the alignment mechanisms of the rubbing process, many studies have been made by using modern surface observation techniques such as AFM observations for the surface structure, polarized UV and IR absorption spectral measurements for molecular anisotropy, near edge X-ray absorption fine structure (NEXAFS) spectral measurements to determine the anisotropic distribution of molecules near the surface, retardation measurements for optical anisotropy, and water contact angle measurements for the polar functional group distribution. Table 2.2.1 summarizes the characteristics of each observation method. Here, we will review previously reported analytical studies of the rubbed surfaces intended to discover the alignment mechanisms of the rubbing process.

Table 2.2.1 Summary of surface observation methods

Method	Subject	Depth	Target of analysis
AFM	Surface	0.5–0.8 nm	Morphology
UV absorption	Volume	Whole	Molecular direction
IR absorption	Volume	Whole	Molecular direction
X-ray scattering	Surface	5 nm	Molecular direction
X-ray absorption	Surface	1 nm	Molecular direction
Second Harmonic Generation (SHG)	Surface	Several molecular layers	Molecular direction
Water contact angle	Surface		Polarity
Optical retardation	Volume		Anisotropy

2.2.1.1 AFM observations of microgrooves

In the early stages of studies of the rubbing mechanism, two mechanisms, grooves in the surface and reorientation of the alignment material, were considered as candidates for the alignment mechanism of the rubbing process. Therefore, many papers described the surface morphology of the rubbed surface. The development of the scanning tunnel microscope (STM) and the atomic force microscope (AFM) has supplied us with high-resolution observation techniques which create images that could not be created by using earlier techniques. Several groups have used the STM [12, 14] and the contact mode AFM [17, 23, 26, 30, 34, 35] to make observations of the rubbed alignment materials. To study the alignment mechanism of the rubbing process more deeply, Pidduck *et al.* studied the surface morphology dependency on the rubbing strength by using a tapping mode AFM [48]. Observations using an AFM, which has 0.5–0.8 nm of height resolution over 10 nm^2 area, revealed that the polyimide surface rubbed by three kinds of cloth each had unique textures [48]. They observed three kinds of structures in the background surface texture: isolated islands, grooves, and nanoscale undulations. Isolated islands might be caused by a redistribution of residual surface polyimide chains during rubbing. Shear force, caused by the friction between the rubbing fibres and the polymers, stretches the polymer chains, and orientation takes place. Two types of grooves were observed, deep and shallow scratches. Irregular deep scratches, 2–20 nm in depth, 100–400 nm wide, and several mm long were observed. Periodic shallow scratches, less than 1 nm depth and 20 nm wide, were also seen. Nanoscale undulations in the background texture showed that not only the contact area of the rubbing fibres, but also the background areas are modified by the rubbing process.

Ito *et al.* reported on grooves on a rubbed polyimide by observing RuO_4 sputtered on a surface using scanning electron microscopy (SEM). These grooves had a spacing period of 30 nm and a depth of 1 nm. They concluded that such grooves cause the LC alignment, because they found that only widely spaced grooves were observed on a weakly alignment polymer.

What effect does the structure of the rubbing fibre have on the rubbed surface? Mahajan reported that for a wide range of rubbing strengths, the microstructure of the grooves, as determined by their radii of curvatures, correlates well with the microscopic topography of the fibres [61]. The rubbing-induced topography depends not only on the rubbing strength, but also on the structure of the rubbing fibre as well [61]. Using a single fibre rubbing, and observing the microgrooves using the AFM, the friction force microscope (FFM), and SEM, Wako *et al.* found that the microgrooves are formed by the microstructure of the edges of the fibres [58]. The curvature of the microstructure of the fibre edge and that of the cross section of the microgroove agree well with each other, and their shapes are similar. Therefore, they concluded that the structure of the microgrooves from ordinary rubbing is formed by the microstructure of the fibre edge [58]. The anchoring strength is almost proportional to the ratio of the total area of each microgroove to the unit area [58].

Summarizing the AFM observations, fine grooves similar to the rubbing fibre structure, and nanoscale undulations in the background texture are observed on the rubbed polymer surfaces.

2.2.1.2 Chemical observations

To observe chemical changes in the rubbed surface, Lee and Paek measured the polarity of a rubbed surface by using water contact angle measurements, and discussed a correlation between the pretilt angle and the polarity of the rubbed surface [44, 62]. The measurement method is illustrated in Fig. 2.2.

Contact angles show the polarity of the surface. However, they are affected by surface roughness, although surface chemistry has the more dominant influence. As a polymer surface becomes microscopically rougher and consequently the surface area gets larger, the advancing contact angle increases while the receding contact angle decreases, especially in the case of a chemically inhomogeneous surface. Rubbing polyimides films with a cotton cloth decreases the receding water contact angle for polyimides containing polar groups or linear alkyl side chains in the repeating unit, due to microscopic reorientation to expose the polar groups to the surface or to move the non-polar side chains into the depth of the body.

In the regime of relatively weak rubbing where inhomogeneous, partial surface modification occurs, the surface polarity, the anchoring energy, and the pretilt angle all increase monotonically with rubbing strength. These increases correlate with an increase in the area fraction of the reoriented alignment layer surface [62]. In the strong rubbing regime, the surface chemistry and roughness have dominant effects on the pretilt angle and the anchoring energy. The increase in surface polarity and a larger reoriented surface area enhance the attractive interaction between the LC monolayer and the rubbed surface. Therefore, the anchoring energy increases and the pretilt angle decreases [62]. The relationship between water contact angles and the number of rubbings is shown in Fig. 2.3.

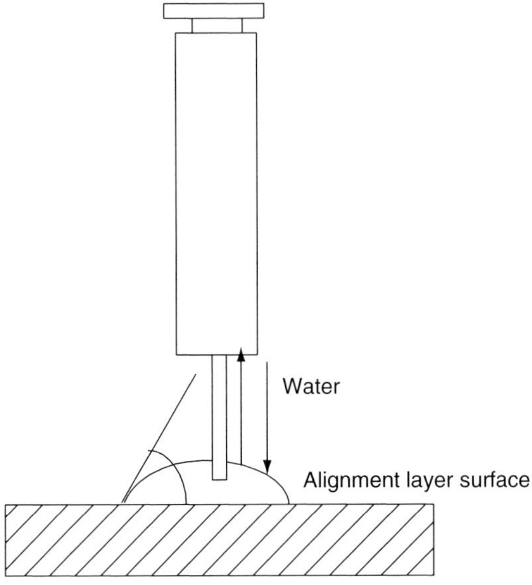

Fig. 2.2 Illustration of water contact angle measurements. Advancing and receding contact angles are those when water is added and removed through a needle, respectively. Reproduced by permission from [55].

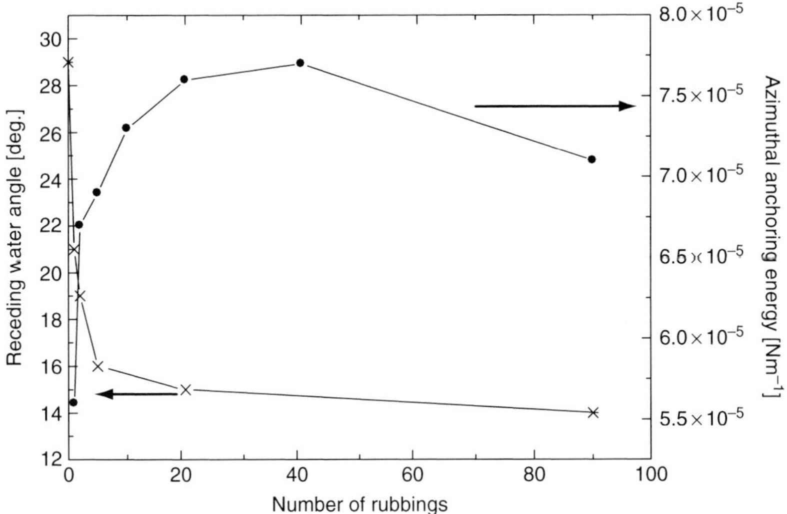

Fig. 2.3 Receding water contact angles on polyimide and the azimuthal anchoring energy in LC cells as a function of the number of rubbings and the pile impression. Reproduced by permission from [62].

2.2.1.3 SHG measurements

Only a surface or an interfacial monolayer, which consists of asymmetrically distributed second harmonic generation (SHG) active molecules, generates second harmonic waves. Therefore, the polar and azimuthal distributions of the surface and interfacial molecules can be measured by using the SHG technique. Due to this characteristic, SHG can be used both for measuring the surface distribution of the alignment materials, and for observing the alignment of the LC monolayer at the interface with the alignment layer [4, 6, 7, 8, 9, 10, 13, 22].

Chen *et al.* measured the alignment of the LC at the interface between the rubbed polyimide or the surfactant (methylaminopropyltrimethoxysilane, MAP) and the LC bulk, and reported that the LC monolayer absorbed by the polyimide controlled the alignment of the bulk LC (see Fig. 2.4). They also reported that the LCs at the interface of the rubbed MAP alignment layer do not align, and the elastic energy exerted by the grooved structure controls the alignment of the bulk LC [6, 13]. Feller *et al.* reported that LCs at the interface between the bulk LC and the obliquely evaporated SiO were isotropic, and they concluded that the elastic energy exerted by the grooves was the main mechanism of the LC alignment [13].

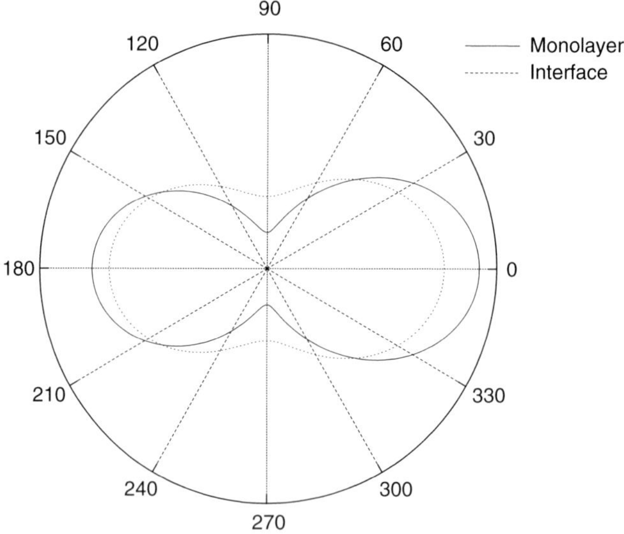

Fig. 2.4 Azimuthal orientational distribution functions of an 8CB monolayer on a rubbed polyimide-coated substrate (solid line), and an 8CB interfacial layer between a rubbed polyimide-coated surface and bulk 8CB (dashed line). Reproduction by permission from [6].

SHG measurements by Barmentlo showed that orientational asymmetry along the rubbing direction varied with the alignment layer polymer for identical rubbing conditions, while the average polar angle from the surface normal of the LC 8CB at the alignment layer surface was nearly constant (77°), regardless of the polymers [16]. Their studies found: (i) even a small orientational asymmetry along the rubbing direction is able to generate bulk alignment, (ii) the orientational asymmetry varies with the alignment layer polymers and increases with rubbing strength, (iii) the average polar angle is minimally affected by the polymers and rubbing strength, i.e., it is independent of the structure and treatment of the alignment layer surfaces, and (iv) the pretilt is linearly proportional to the orientational asymmetry along the rubbing direction. Thus, the pretilt angle is determined by the orientational asymmetry along the rubbing direction. Several studies have reported that this relationship is limited to weak rubbing [16, 19, 31, 43, 44, 55, 62].

Aerle *et al.* measured the azimuthal distribution function of a monolayer of 8CB on the alignment layer, and discussed the correlation between the distribution function and the optical retardation of the alignment layer. The order parameter of the azimuthal alignment increased with the optical retardation of the alignment layer, but it was saturated over some range of values of the retardation. The relationship between the in-plane orientation function and the rubbing induced retardation is shown in Fig. 2.5. Observing the order parameter of the bulk LC by using the same cells, it was found that the order parameter of the surface monolayer is directly related to that of the bulk LC [22].

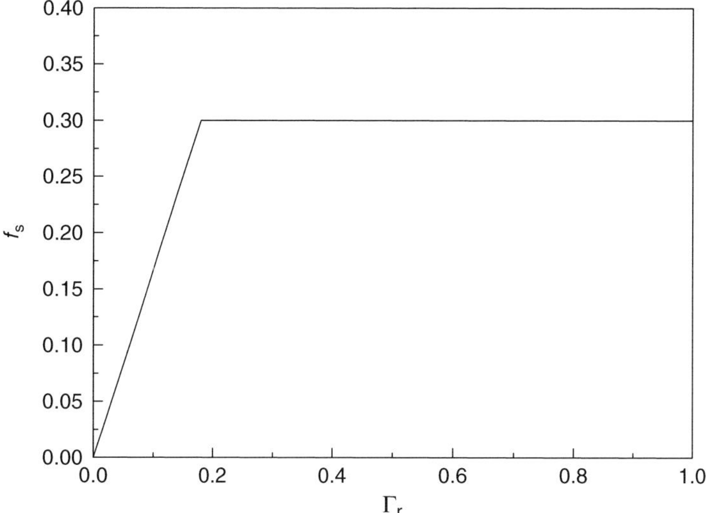

Fig. 2.5 In-plane orientation function f_s of an 8CB monolayer as a function of the rubbing-induced optical retardation Γ_r induced in the polyimide orientation layer. Reproduction by permission from [22].

Distributions of polyimide side chains were also studied by using SHG [38, 52, 56]. A distribution function of a polyimide side chain, which is a non-linear active trifluorocarbon-substituted side chain, was measured [38]. Though the polyimide side chains are distributed on a conical surface with some polar angle before rubbing, they were pulled down at the side parallel to the rubbing, and pushed up on the side opposite to the direction of the rubbing. The azimuthal distribution was biased to rubbing and anti-rubbing direction [38]. The rubbing effect on the polyimide side chains is schematically represented in Fig. 2.6. Although the polar angle of the side chain of the alignment layer in the rubbing direction is different from that in the opposite direction, that of the first monolayer of the LC on the alignment layer is same. This effect correlates only with the rubbing strength and the distribution of the azimuthal direction. There was a correlation between the alignment distribution functions of the side chains of the alignment layer and that of the LC monolayer and the pretilt angle of the LC. When the asymmetry between the rubbing direction and the opposite direction of the distribution functions decreases, the pretilt angle decreases. From these results, they concluded that the azimuthal asymmetric distribution caused the pretilt angle of the LC [52].

Sakai *et al.* separated the SHG signals from the two interfaces, air-polyimide and polyimide-substrate by making the alignment film thick [56], and analyzed the two interfaces independently. For as-coated films, the average tilt angle of the side chain at the air-PI interface is smaller than that of the PI-substrate interface. The rubbing process caused in-plane anisotropy only for the air-PI interface [56]. The polar angles determined were about 60° and 25° at the air-PI and at the substrate-PI interfaces, respectively. Characteristic anisotropic patterns are obtained at the air-PI interface. The polar angles of the side chains in the parallel and the antiparallel directions with respect to the rubbing direction are different:

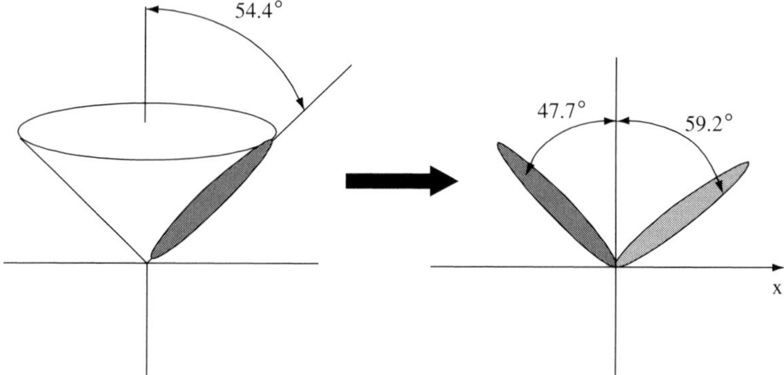

Fig. 2.6 The rubbing effect on the polyimide side chains is schematically represented. The rubbing direction corresponds to the x-direction. Reproduction by permission from [38].

67° and 45.3° from the surface normal, respectively. Before rubbing, the side chains were distributed randomly for azimuthal angles with an average polar angle of 60°. Because of rubbing, the side chain tends to align toward the rubbing direction, so that the side chain is pushed down to 67° in the rubbing direction, and is pulled up to 45.3° in the anti-rubbing direction [56].

2.2.1.4 Optical retardation measurements

Measurement of the infrared (IR) absorption cannot be used to evaluate the anisotropy of the alignment layer of actual display devices because their substrates are made of glass and are not transparent to IR radiation. However, optical retardation measurements using ellipsometry for visible light can evaluate the anisotropy of a thin membrane, and can be used to evaluate the alignment of the molecules in the alignment material after rubbing [5, 23, 24, 29, 50, 51].

Aerle *et al.* reported that the phase retardation increased steadily with pile impression. The influence of the pile impression of the rubbing cloth on the rubbing-induced optical retardation Γ_r of the orienting layer, and on the normal force F_N exerted by the rubbing cloth on the samples is shown in Fig. 2.7.

Geary *et al.* measured the optical retardation of several rubbed polymers, and studied the relationship between the optical retardation and the rubbing

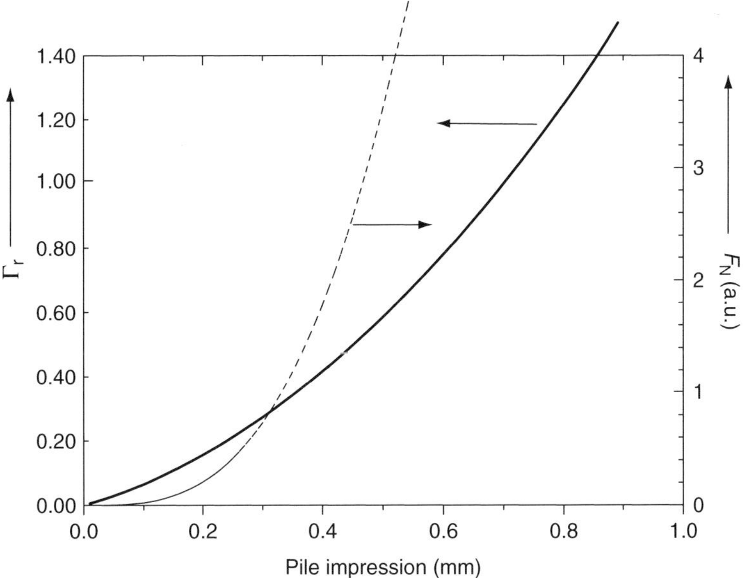

Fig. 2.7 Influence of the pile impression of the rubbing cloth on the rubbing-induced optical retardation Γ_r of the orienting layer, and on the normal force F_N exerted by the rubbing cloth on the samples. Reproduction by permission from [22].

strength. The retardation increased with rubbing strength, and became saturated. The slow axis of polystyrene was perpendicular to the rubbing direction, although the axis of the other polymers was parallel. They also studied the alignment of an LC aligned by films which are made of elongated bulk polymers, and showed that the crystallized polymers aligned the LC without rubbing [5].

The incident angle dependence of the optical retardation of polyimide films on quartz substrates is shown in Fig. 2.8. Measuring the azimuthal and polar angle

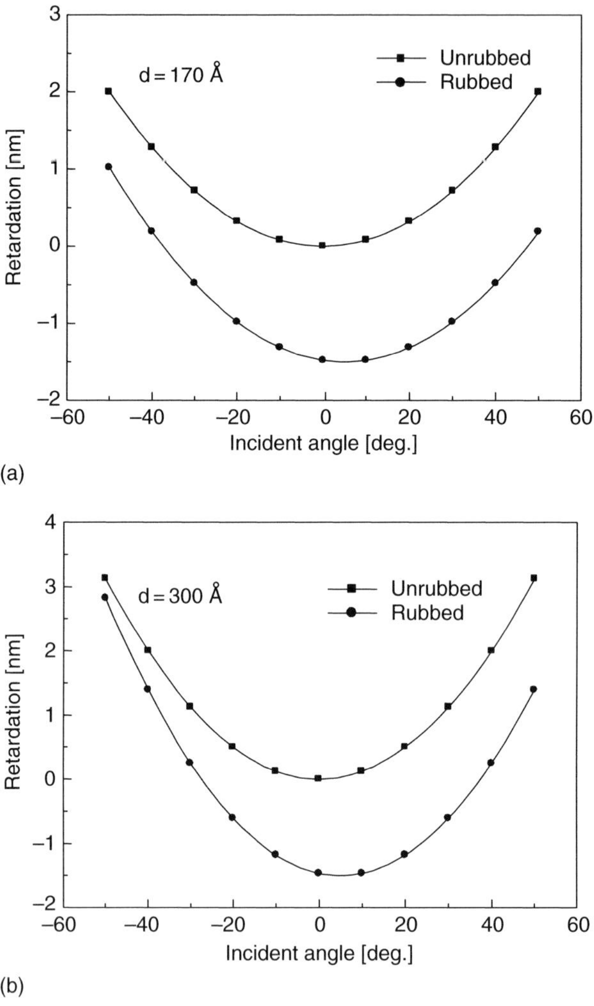

Fig. 2.8 Incident angle dependence of the optical retardation of polyimide films on quartz substrates: (a) for a 17 nm thick film, and (b) for a 30 nm thick film. Reproduction by permission from [50].

dependency of the optical retardation, and analyzing its results quantitatively, the three dimensional distributions of the molecules of the alignment material were calculated [32, 51, 57].

Sakamoto calculated the tilt angles of the molecules of the alignment material from the results of retardation measurements. He used a thin alignment layer, less than 13 nm, so that the entire thickness can be assumed to be processed uniformly. This assumption was also confirmed by the IR measurements. He calculated the distribution of the alignment molecules by assuming a three layer structure consisting of a vacuum, an alignment layer, and a substrate. The planar stack model of three dielectric layers used for molecular orientation calculations is shown in Fig. 2.9. The analysis showed that the main chain of the polymer is oriented in the rubbing direction, and in the case of PMDA-ODA, it generated a tilt angle of 9°. This result was in good agreement with the result from IR measurements [51].

Han and Seo *et al.* studied the relationship between the tilt angle of the main and side chains of the polymer and the rubbing strength [32, 57]. The relationship between the inclination angle of the polymer and the LC pretilt angle as a function of the rubbing strength is shown in Fig. 2.10. In the case of main chain polymers, their retardation before rubbing had no dependency on the incident angle of the light, because they are aligned randomly. The average tilt angle of the main chains increased with rubbing strength. The pretilt angle showed the same behaviour as the inclination angle of the polymer in an alignment layer. The pretilt angle and

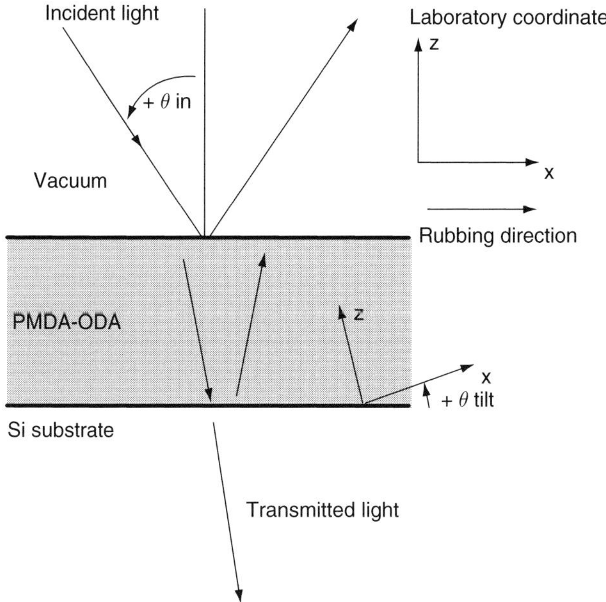

Fig. 2.9 Planar stack of three dielectric layers used for molecular orientation calculation. Reproduction by permission from [51].

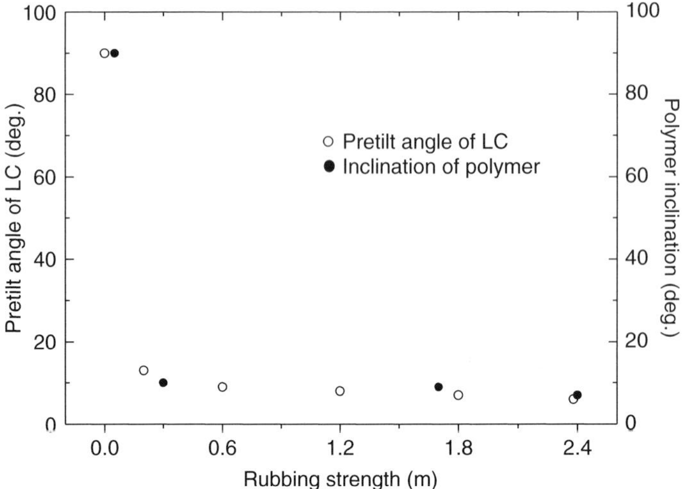

Fig. 2.10 Relation between the inclination angle of the polymer and the LC pretilt angle as a function of rubbing strength. Permission to print from [32], courtesy Society for Information Display.

the inclination angle of the polymer became saturated at a rubbing strength of approximately $L = 2\,\mathrm{m}$. In the case of side chain polymers used for high-pretilt alignment, the retardation had a maximum of 0 nm at normal incidence and decreased at large angles. This means that the average optical axis is perpendicular to the substrate. Since the polymer inclination and the LC pretilt angle agree well with each other, the LC tends to align almost parallel to the side chain of the polymer. For the case of polymers used for homeotropic alignment, the retardation before rubbing had an extremum of 0 nm at normal incidence. This can be modeled with an index ellipsoid processing an optical axis parallel to the direction normal to the surface. It can be concluded that the pretilt of the LC is mainly caused by the tilted orientation of the polymer molecules aligned by the rubbing treatment [32, 57]. An alignment model for a main chain polymer for low pretilt is shown in Fig. 2.11.

2.2.1.5 X-ray scattering and absorption

X-ray scattering and absorption spectrum measurement techniques have been used to study the structure of the rubbed surface. Toney *et al.* studied the surface structure by using an X-ray scattering technique [39]. For surface structure measurements, they used grazing-incidence X-ray scattering (GIXS), which is based on a critical angle that exists because the refractive index of the X-rays is lower than one. The X-rays which interact with the surface at the critical angle become evanescent radiation and pass through a surface depth less than 5 nm. The

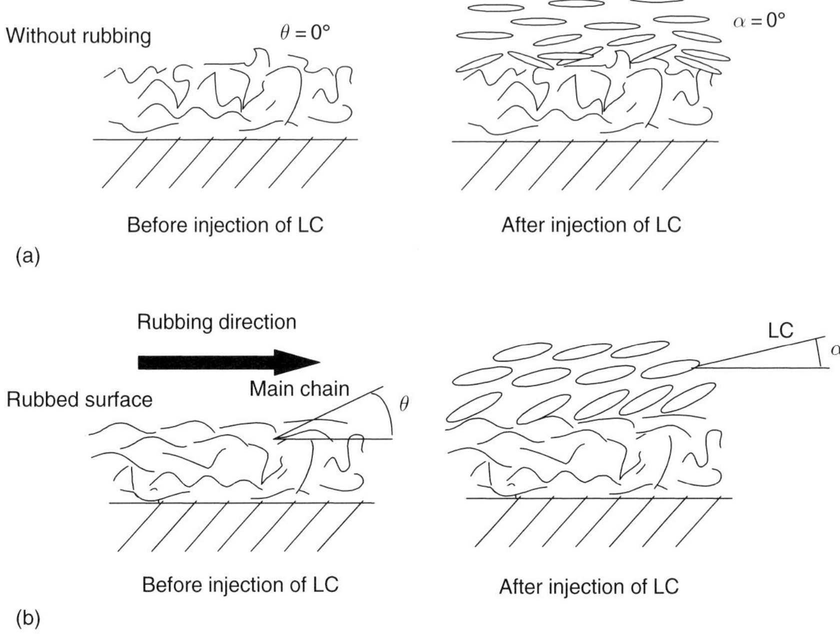

Fig. 2.11 Alignment model for a main chain polymer for low pretilt (a) before rubbing, (b) after rubbing. Permission to print from [32], courtesy Society for Information Display.

observed structure had a 3.1 nm periodicy perpendicular to the main chain direction caused by replicated molecular structures. The orientation of the molecules was confirmed based on this structure. Based on quantitative measurements of the Bragg diffraction by this structure, it was confirmed that the molecules in a surface at a depth of less than 6 nm were distributed within $\pm 10°$ of the rubbing direction, and the molecules within a 6 nm depth relative to the surface were better aligned than those within a depth of 20 nm. From this result, they estimated that the glass transition point and the yield strength of a thin layer are lower than those of a thick layer.

In addition to X-ray scattering, the absorption of soft X-rays can be used for surface analysis. A NEXAFS spectrum measurement can observe molecular orientation to a surface depth of about 1 nm, though the IR and UV absorption spectra and optical retardation measurements reveal the bulk anisotropy of the material. Therefore, these techniques can directly measure the reorientation of the alignment material caused by the rubbing. The LC pretilt angle with BPDA-type polyimide exhibits an odd–even effect; the odd number polyimide shows nearly $\pm 10°$ of the LC pretilt angle, while the even numbered polyimide induces a relatively high LC pretilt angle. The NEXAFS spectra have successfully revealed the surface structure differences between the odd- and even-numbered BPDA polyimides, a *trans-cisoid* type and a *trans-transoid* type, respectively [37]. The rigid biphenyl part of BPDA-C8 selectively tilts away from the surface by 15°, while for BPDA-C7 it does not. Similar behaviours were also observed in other numbered BPDA polyimides. Assuming that LC molecules are adsorbed onto the imide/biphenyl part, the asymmetric surface structure of BPDA-C8 may induce a larger biased pretilt angle of the LC molecules than the symmetric case of BPDA-C7 [37]. The angular dependences of the peak height for BPDA-C7 and C8 polyimides are shown in Fig. 2.12.

The relationship between the LC pretilt angle and the surface structure of BPDA-Cn and PMDA-Cn polyimide was studied by using the NEXAFS technique [46]. The BPDA type polyimide shows an odd–even effect. However, PMDA type polyimide does not show such an odd–even effect. The chain length dependence of the LC pretilt angle for BPDA and PMDA types of polyimide is shown in Fig. 2.13. The conformation of PMDA-C8 is determined as a *cis-transoid* type. As for PMDA-C7, a *trans-cisoid* type conformation of the polymer chain is a possible structure on the substrate surface. The incident angle dependence of the peak intensity for PMDA-C8 and PMDA-C7 polyimides is shown in Fig. 2.14. The inclination of the aromatic core part has been calculated to be 25°. In contrast to the case of PMDA-C8, the aromatic part of BPDA-C8 selectively tilts toward the rubbing direction. Experimental results strongly suggest that the all-*trans*-structure of BPDA-C8 is formed on the substrate. The incident angle dependence of the peak

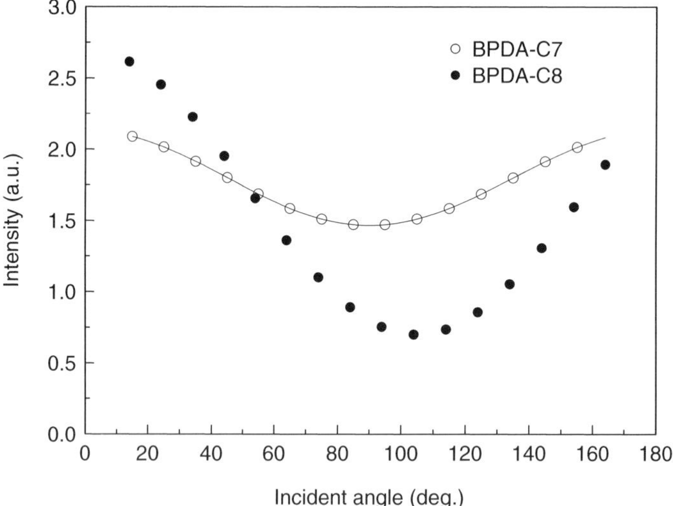

Fig. 2.12 Angular dependences of the peak height for BPDA-C7, C8 polyimides. Permission to print from [37], courtesy Elsevier Science.

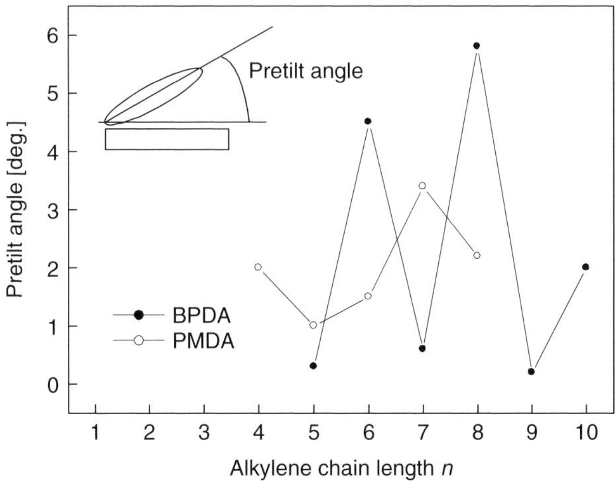

Fig. 2.13 Chain length dependence of the LC pretilt angle for BPDA and PMDA type polyimides. Reproduction by permission from [46].

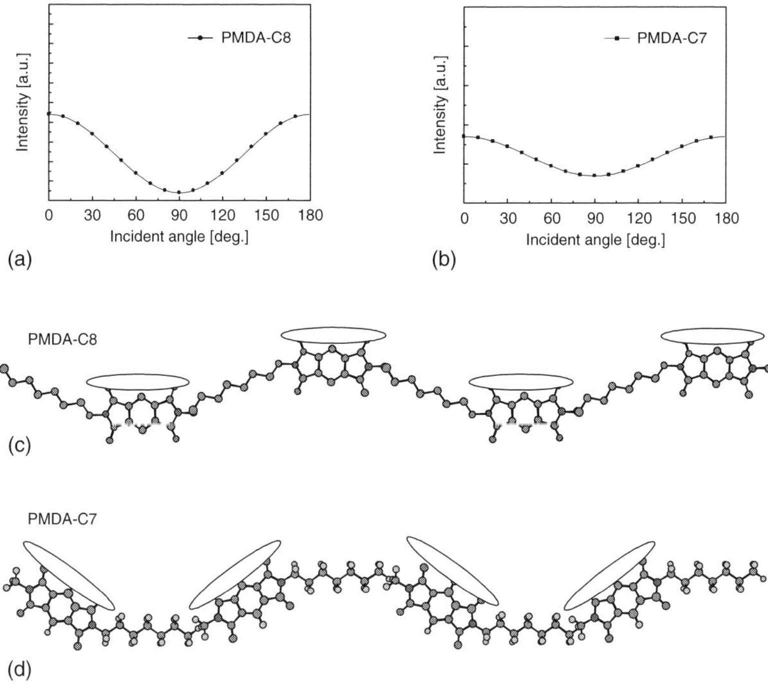

Fig. 2.14 Incident angle dependence of the peak intensity for (a) PMDA-C8 and (b) PMDA-C7 polyimides. Model structures of (c) PMDA-C8 (*cis-transoid*) and (d) PMDA-C7 (*trans-cisoid*). Reproduction by permission from [46].

intensity for (a) BPDA-C8 and (b) BPDA-C7 polyimides is shown in Fig. 2.15. The aromatic core part tilts from the surface by 13°. It should be first noted that PMDA-C8 and BPDA-C8 exhibit different conformations though they have similar molecular structures. This is probably due to the existence of the twisted structure of the biphenyl part of the BPDA polyimide. Only BPDA-C8 possesses an asymmetric conformation with respect to the surface normal. If it were assumed that the LC molecules are adsorbed only by the aromatic part of the polyimide, the LC alignment on BPDA-C8 also results in the LCs being asymmetric. It is highly probable that the high pretilt observed for BPDA-C8 is caused by the *trans-transoid* conformation on the polymer surface, which is produced by the rubbing process [46].

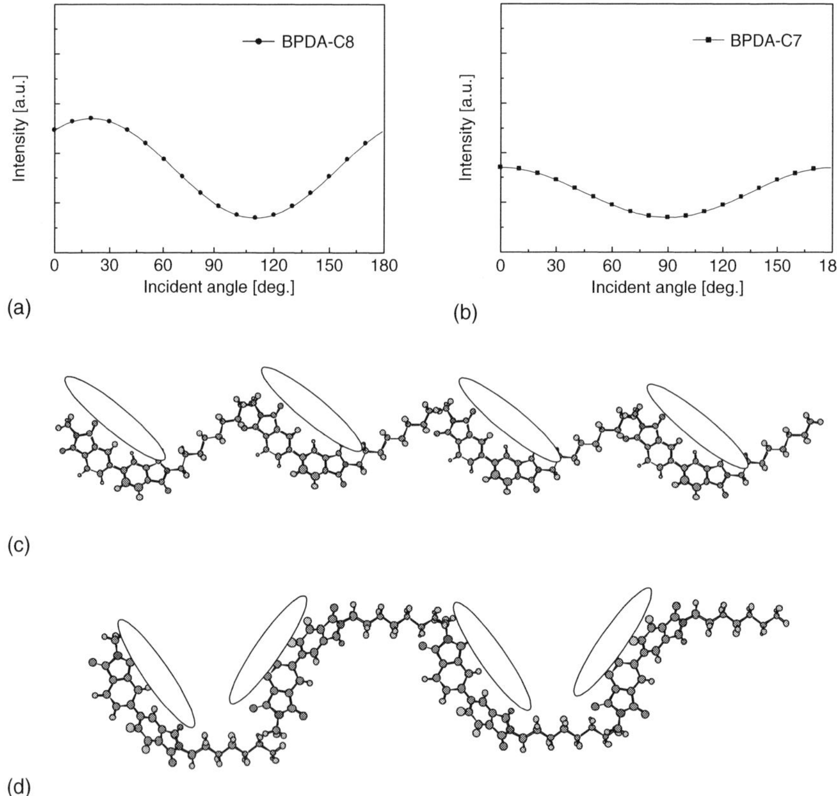

Fig. 2.15 Incident angle dependence of the peak intensity for (a) BPDA-C8 and (b) BPDA-C7 polyimides. Model structures of (c) BPDA-C8 (*trans-transoid*) and (d) BPDA-C7 (*trans-cisoid*). Reproduction by permission from [46].

2.2.1.6 Infrared (IR) absorption measurements

Although a transmittance IR absorption spectrum measurement evaluates the bulk of the alignment material, the distribution of the alignment of the molecules can be determined by a qualitative analysis by changing the polar and azimuthal angles of the sample. The relationship between the dichroism of the polyimide and its thickness was studied using polarized infrared absorption measurements, and it was found that the dichroism was consistently less than 17 nm and decreased when the polyimide becomes thicker [22]. From these results and retardation measurements, a molecular distribution model was proposed which shows uniform anisotropy to some depth, lower anisotropy in deeper areas, and isotropy in areas which are even deeper [22].

Sakamoto *et al.* quantitatively analyzed IR absorption spectra by using a three-layer model, vacuum, an alignment layer, and the substrate, in the same way to that used for the analysis of retardation, and estimated changes in the molecular distribution of PMDA-ODA polyimide on a Si substrate caused by rubbing. Figure 2.16 shows an experimental configuration for IR absorption measurement for quantitative analysis. From this analysis, it was found that the polyimide main chains were aligned parallel to the substrate without rubbing. They became reoriented in the rubbing direction with an 8.5° tilt angle after rubbing, as shown in Fig. 2.17. This tilt angle causes the pretilt angle of the LC [29]. The depth of the alignment material, which was affected by the rubbing, was estimated to be 12.5 nm from the results of the thickness dependency measurements of the IR dichroism. This depth is in good agreement with Aerle's estimate [22]. Based on this fact, the IR measurement for quantitative analysis was done on a thin alignment layer less than 12.5 nm in depth, and it was assumed that the alignment layer was reoriented uniformly by the rubbing in the analysis of the results [50]. The relationship between the pretilt angle of the LC and the tilt angle of the main chains of PMDA-ODA, with different lengths of alkyl chain, was studied, and the generation of the pretilt angle of the LC was discussed [41, 54, 59]. Figure 2.18 shows the relationship between the pretilt angle of the LC and the inclination angle of the polyimide main chain. Figure 2.19 shows the relationship between the water contact angle and the number of carbon atoms in the side chains of the polyimides. A larger contact angle corresponds to a more hydrophobic surface. Therefore, the rubbing process might reduce the number of alkyl side chains exposed on the surface. It was concluded that alkyl chains do not directly affect the pretilt angle generation, since it was confirmed by the contact angle measurements that alkyl side chains orient inwards [59].

2.2.2 Definition of rubbing strength

To describe the rubbing strength quantitatively, a definition of rubbing strength is required. The structure of a rubbing machine is shown as Fig. 2.20. It consists of a rotating drum covered by a cloth whose surface is made of short fibres, and

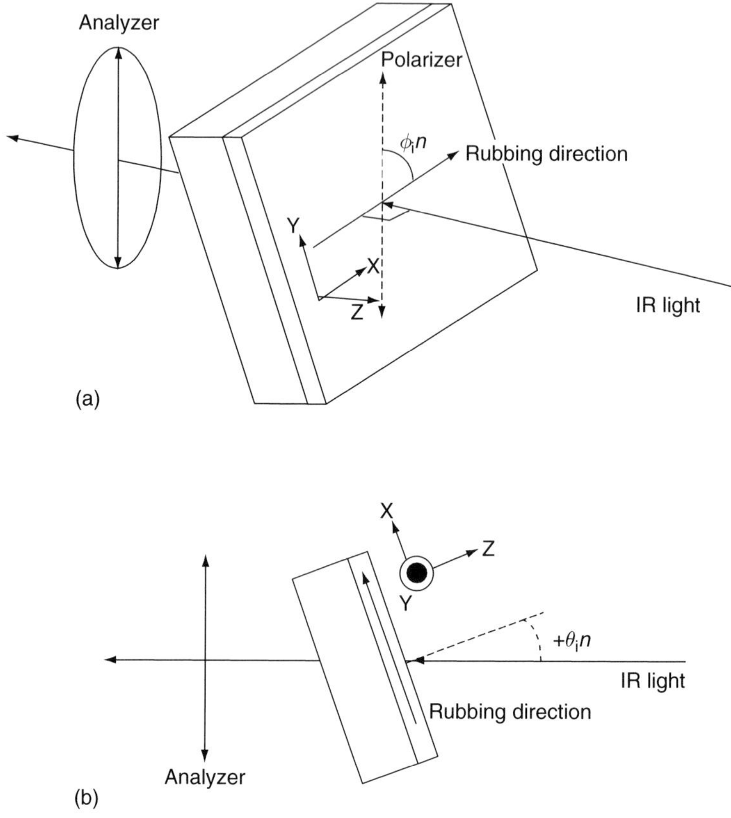

Fig. 2.16 Experimental geometries for the measurements of the polarization angle dependence (a) and the incident angle dependence (b) of IR absorption. $\phi_i n$ is the polarization angle, which is defined by the angle between the rubbing direction and the analyzer direction. $\theta_i n$ is the incident angle of the IR light. The polarization angle and the incident angle were varied by rotating the sample around the Z and the Y axes, respectively. Reproduction by permission from [50].

a uniaxially movable platform to hold the substrate. The drum moves on the polymer coated substrate while rotating and pressing against the substrate. The relative speed of movement between the buffing cloth and the substrate, the pressure of the buffing cloth on the substrate, and the number of times a specific substrate goes through the rubbing machine can be controlled easily, and these parameters affect the rubbing process. Therefore, the rubbing strength is defined by Uchida *et al.* as,

$$R_s = \gamma L, \tag{2.1}$$

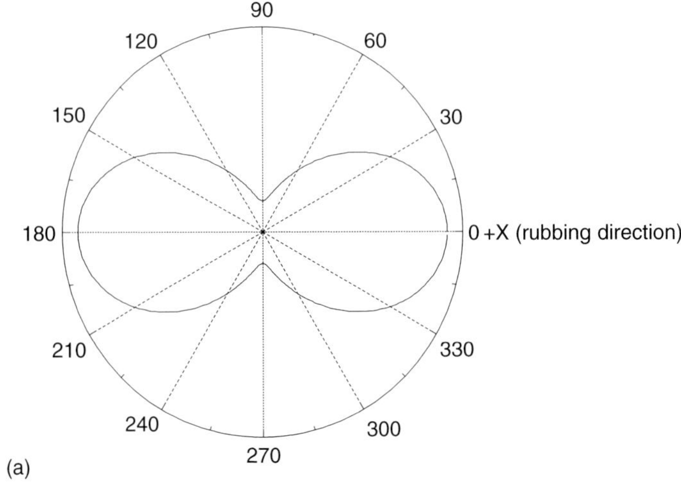

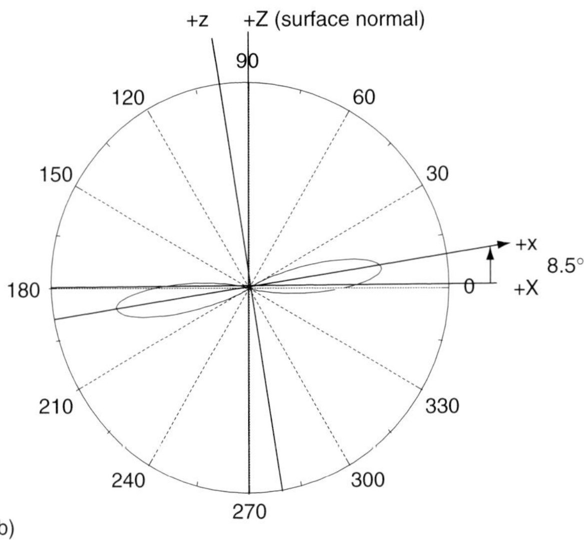

Fig. 2.17 Molecular orientation of the rubbed film in the azimuthal angle direction (a) and in the polar direction (b). Reproduction by permission from [51].

where γ is a coefficient related to the rubbing pressure, the fibre density of the rubbing cloth, the coefficient of friction, etc., and L is the total length of the rubbing cloth given by

$$L = Nl(1 + 2\pi rn/60\nu), \tag{2.2}$$

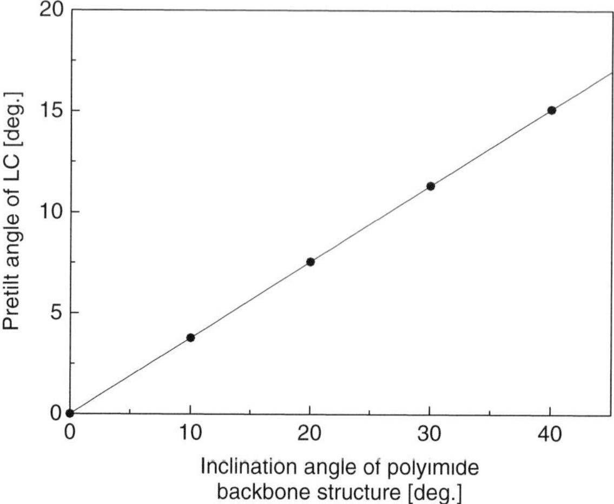

Fig. 2.18 Relation between the pretilt angle of the LC and the inclination angle of the polyimide backbone structure. Reproduction by permission from [54].

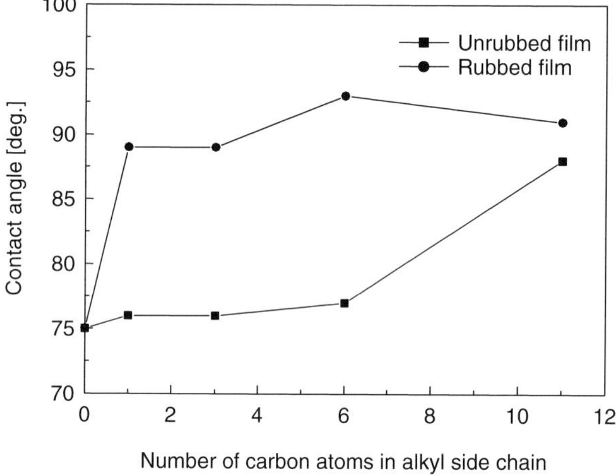

Fig. 2.19 The contact angle of water on An-PI and PMDA-ODA films. The horizontal axis is the number (n) of carbon atoms in the alkyl side chain. The data at $n=0$ are for the PMDA-ODA films. Reproduction by permission from [59].

where l is the contact length of the rubbing roller and the substrate, r is the radius of the roller, n is the number of rotations per second of the roller, and v is the speed of movement of the substrate stage, as shown in Fig. 2.21 [11, 20, 21]. In this equation, l increases with rubbing pressure as controlled by the gap between the roller and the substrate, and the rubbing pressure is included in this contact

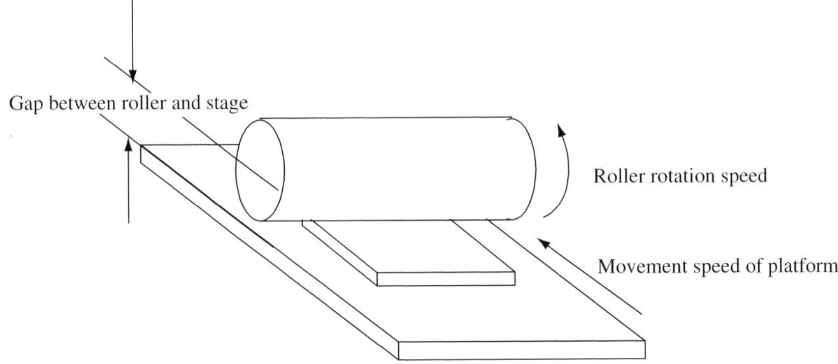

Fig. 2.20 Structural sketch of a rubbing machine. A rubbing machine consists of a roller covered by rubbing cloth, and a platform to hold the substrate. The roller or the substrate stage moves at a constant speed while the roller is rotating at several hundreds of rpm. The gap between the roller and the stage is adjusted to change the pile impression.

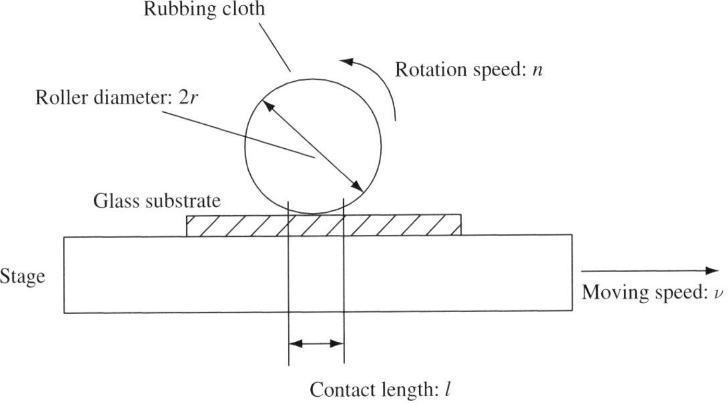

Fig. 2.21 Definition of the parameters associated with rubbing. Reproduction by permission from [11].

length. The utility of this equation was confirmed by demonstration that the effects of cylinder rotation speed and number of times a substrate is rubbed are independent of the anchoring strength of a ferroelectric LC as shown in Fig. 2.22 [11]. From this result, the linear relationship between l and the anchoring strength was confirmed.

Aerle *et al.* defined the rubbing strength as the rubbing density R_{d} described as [22],

$$R_{\mathrm{d}} = [(\pi C w + v)/\nu]N \qquad (2.3)$$

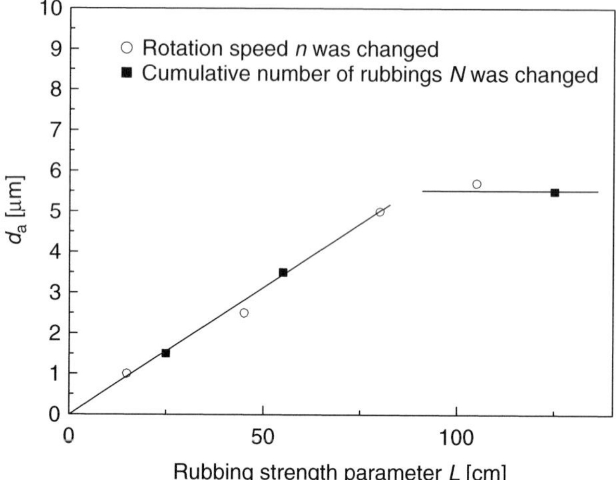

Fig. 2.22 Relation between the largest cell-gap to give homogeneous alignment, d_a, and the rubbing strength parameter, L, when n and N are changed independently. Reproduction by permission from [11].

Here, N, C, ω, and v are the number of times a location is rubbed, the diameter of the cylinder, the rotational speed of the cylinder, and the movement speed of the substrate stage, respectively. This equation is almost the same as Uchida's except that it does not include the contact length l. Since the rubbing pressure is constant, the rubbing density does not include this parameter. In these equations, the rubbing strength is proportional to the number of times a substrate is rubbed.

However, this relationship holds only in the weak rubbing regime. Actually, the effects of the number passes through rubbing and the rubbing pressure are not the same. An analysis of the rubbing strength in a weak rubbing regime has been reported [62]. The rubbing stress F can be expressed by the equation [25],

$$F = m_f a Y, \tag{2.4}$$

Here, m_f, a, and Y are the effective contact density between the alignment material and the rubbing cloth, the contact area, and the yield stress of the alignment material. If we assume that the contacts of the fibres of the rubbing cloth and the alignment material are a series of point contacts by a single fibre as shown in Fig. 2.23,

$$a = \pi w_f^2 / 4F = m_f \pi w_f^2 Y / 4 \tag{2.5}$$

Then,

$$w_f \propto \sqrt{F}. \tag{2.6}$$

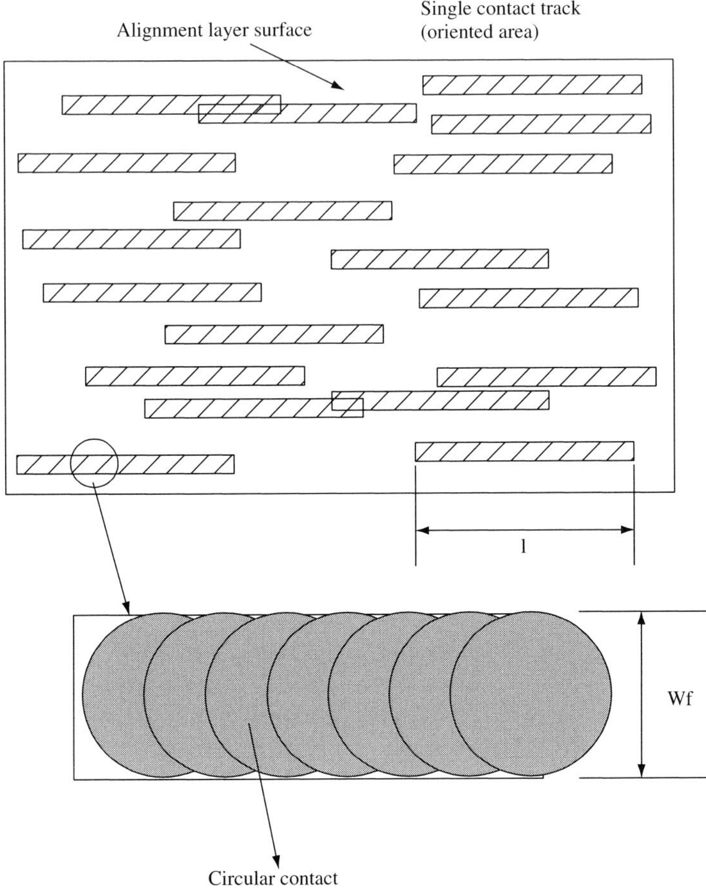

Fig. 2.23 Alignment layer structure partially modified by weak rubbing. A smooth surface was assumed. Reproduction by permission from [62].

The buffing length of a single fibre l can be expressed by the rubbing pressure M, and the cylinder diameter r as,

$$l \approx \sqrt{2}rM. \tag{2.7}$$

From this equation, the ratio of the area, affected by rubbing, A_f is proportional to $w_\mathrm{f}l$ and increases with increase in rubbing pressure, and

$$w_\mathrm{f}l \propto A_\mathrm{f} \approx F(M)M^{1/2}. \tag{2.8}$$

Here, *F(M)* is the rubbing stress related to the rubbing pressure. Increasing the number of rubbings increases A_f, the ratio of the area affected by rubbing, but it becomes saturated when all of the area has been affected by rubbing. In contrast to this, increasing the rubbing pressure also increases A_f, but it causes stronger interaction with the alignment layer for larger rubbing pressures and becomes saturated to A_f. This can be confirmed by considering the relationship between the number of rubbings, the rubbing pressure, and the anchoring energy.

2.2.3 Alignment mechanisms

Here, we summarize the alignment mechanisms based on our knowledge of the rubbing process. Several alignment mechanisms exist and align the LC comparably, such as alignment by reorientation of the rubbed polymer, and the grooves on the inorganic surfaces by rubbing, as described by Castellano [3]. The main alignment mechanism of the rubbing process depends on the alignment material. Here, we discuss the alignment mechanism of the rubbed polyimide, which is a standard material used for the mass production of LCD panels. Berreman calculated the anchoring energy by using the elastic energy based on the continuum theory [2]. Here, we review his calculations. The undulated surface defined by the depth A and the frequency $\lambda = 2\pi/q$ show the azimuthal anchoring energy W_ϕ calculated as,

$$W_\phi = \frac{1}{4} K_{11} (A_q)^2 q. \tag{2.9}$$

Here, K_{11} is a splay elastic constant of the LC. By using this equation, we can calculate the relationship between the groove frequency and the anchoring energy, which is shown in Fig. 2.24.

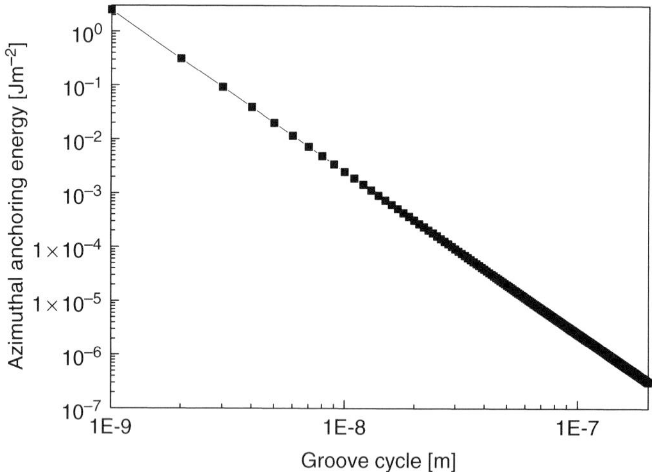

Fig. 2.24 Azimuthal anchoring energy as a function of the groove frequency. Calculated based on the elastic energy.

Here we assume the depth of the groove is 1 nm. The actual azimuthal anchoring energy of the rubbed polyimide is about 10^{-3} Jm^{-2}. If the main mechanism of the rubbing process is the elastic energy exerted by the grooves, then grooves which have a frequency less than several nm should exist. But what is the actual state of the rubbed surface of the polyimide? A frequency of several nm is the same approximate size as the polymer molecules, and the structure as observed by X-ray diffraction had a periodicy of 3.1 nm perpendicular to the main chain direction caused by the repeating molecular structure [39]. The finest structures observed by AFM had a depth of 2 nm and a frequency of 20 nm [48, 49]. The anchoring energy calculated from this structure is 2×10^{-5} Jm^{-2}, and it is two orders smaller than the actual anchoring energy. Therefore, the main alignment mechanism of the rubbed polyimide is the epitaxial growth of the LC due to the reorientation of polymer molecules [48, 49].

The reorientation of polymers which gives the anisotropic characteristics caused by rubbing was confirmed by optical retardation, IR absorption spectra, SHG, and NEXAFS spectral observations. The idea was suggested from results of the comparisons of the LC alignment on rubbed polymer and the elongated bulk polymer that the buffing of the polymer induces alignment of its molecular chains, and that the orientational order of the liquid crystal then grows epitaxially from the oriented polymer surface [5]. Figure 2.25 illustrates the surface deformation of a rubbed polymer.

Two alignment mechanisms, a short range interaction by the reorientation of the polymers and a long range interaction caused by the elastic energy induced in the rubbed surface, are suggested by the observations of the alignment by rubbed polyimide and MAP [6]. From these facts, the surface reorientation of the polymers rubbed by cloth might be the dominant mechanism of LC alignment by the rubbed polymer. However, alignment strength dependency of the rubbing cloth and groove structural differences caused by the cloths have been reported. Therefore, the surface grooves caused by rubbing probably play an important role in the alignment of the LC. It is still not clear why the surface reorientation of the polymer occurs. Pidduck *et al.* have estimated a local temperature increase of up to 180 °C. The T_g of polyimide is 300 °C, and thus locally the polyimide would experience a temperature comparable to T_g, facilitating flow and chain realignment [48]. Mada *et al.* suggested a model where the rubbing effect is local and not uniform, and the temperature of the hot spots caused by textile buffing was estimated to be 230 °C and its radius r was estimated at 2 μm [25]. The temperature at the hot spot exceeds the T_g of the polyimide, and it melts and reorients in the rubbing direction at the hot spot. The distance between each hot spot is estimated to be 20 μm, and this is shorter than the long range interaction. Therefore, the alignment of the LC by the hot spots affects the effective range of other areas of the surface, and aligns the bulk LC [25]. Toney suggested that the glass transition temperature of the surface is lower than that of the bulk, and the reorientation might have occurred at the lower temperatures [39]. However, direct observation of this local melting and reorientation has not been reported yet.

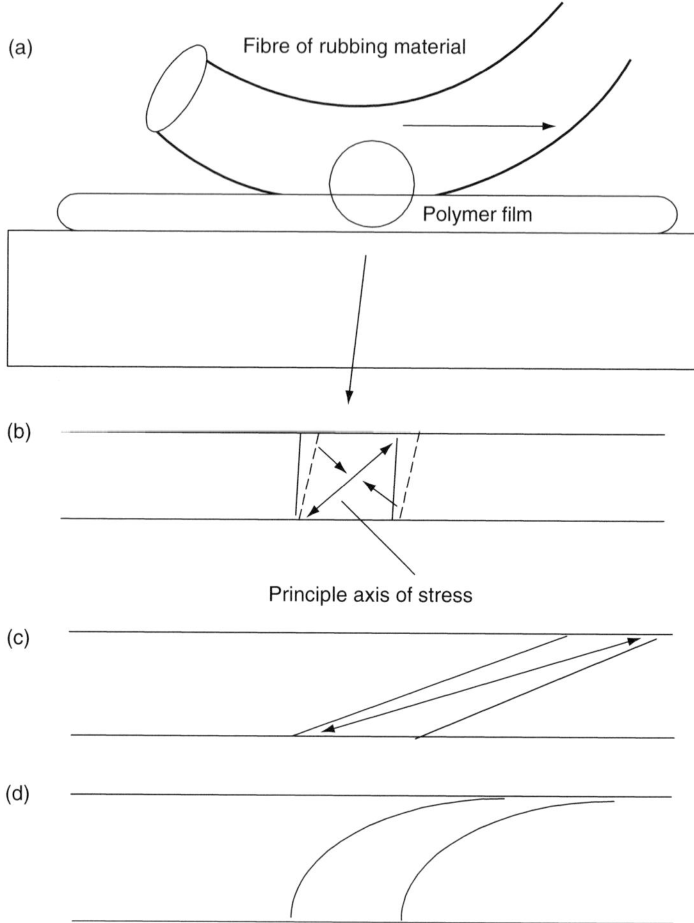

Fig. 2.25 (a) Cross sectional view of a polymer film in contact with a moving fibre of buffing material. (b) Magnified view of the film under shearing forces, showing the principle axes of stress. (c) Effect of large shear deformation, similar to that which would be produced by simple elongation parallel to the substrate. (d) Modified version of large shear deformation, taking into account a possible proximity effect exerted by the substrate surface on the polymer. Reproduction by permission from American Institute of Physics [5].

2.2.4 Pretilt mechanisms

The rubbing generates not only an azimuthal alignment, but also a polar directional alignment, called the pretilt angle, of the LC. SHG, optical retardation, and NEXAFS observations revealed that the main chains of the polymers in the alignment layer are aligned obliquely by the rubbing. In addition to this, a correlation between the polymer's tilt angle and the pretilt angle of the LC was confirmed, as shown in Fig. 2.18. From these results, it can be considered likely that the pretilt angle of the LC is caused by the tilt angle of the polymer main chains. Asymmetrical distribution of the polymer main chain between the parallel and antiparallel direction relative to rubbing was confirmed by the asymmetric absorption of the SHG and the IR spectrum. In the case of the main chain type polymers, the rubbing stress causes an asymmetric zigzag structure of the main chains. In the case of side chain type polymers, the tilt angle of the main chain increases with the length of the side chain. A larger tilt angle for the main chain was generated by a side chain that gets into the bulk below the surface.

Opinions on the mechanism of the pretilt angle of the LC in the case of polyimide with side chains are divided. One view is that it is directly caused by the side chains and another view is that it is caused by the main chains as shown in Fig. 2.26. An increase in the tilt angle of the main chain with increase in the side chain length was also observed as shown in Fig. 2.27 [41, 59]. From the results of water contact angle measurements, the side chains are oriented inward. Therefore, the side chains may not directly affect the pretilt angle generation of the LC. A linear relationship between the pretilt angle of the LC and the tilt angle of the main chain of the polymer was observed with or without side chains, independent of the length of the side chains, and an asymmetric distribution between the parallel and antiparallel directions relative to the rubbing direction was considered to be due to the asymmetric tilt angle of the main chains. Therefore, the dominant

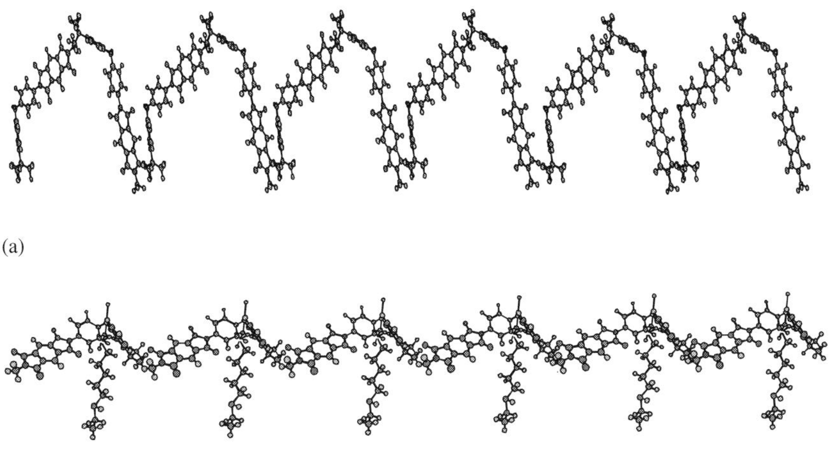

Fig. 2.26 Illustrations of two types of mechanism for pretilt angle generation (a) Asymmetric distribution of the main chains. (b) Side chains push up the main chain.

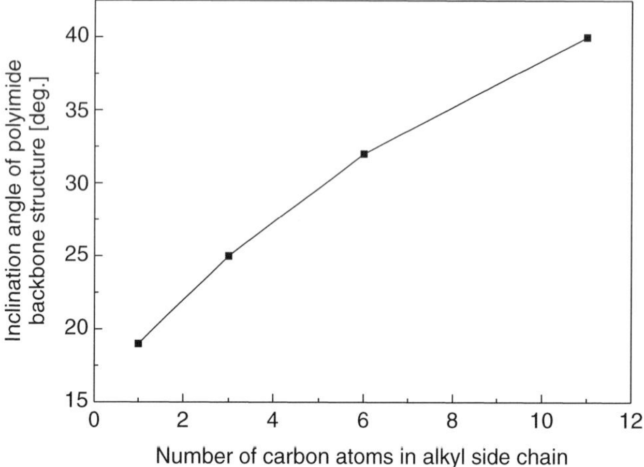

Fig. 2.27 Best-fit parameters obtained from fitting the data for the incident angle dependence of IR absorption of rubbed *An*-PI films. Reproduction by permission from [59].

mechanism of the pretilt angle generation of the LC might be the tilt angle of the polymer main chains, and the asymmetric distribution of the side chains might be a side effect [41, 59]. Shirota *et al.* believed that the asymmetric distribution of the side chains was the main mechanism causing the pretilt angle of the LC based on the results of the SHG measurements [52].

The relationship between the rubbing strength and the pretilt angle of the LC depends on the alignment materials. In some cases, the pretilt angle decreases with increase in rubbing strength. In other cases, it increases with increase in rubbing strength and decreases after showing a peak [62]. Nishikawa *et al.* calculated the molecular conformation, and explained these phenomena by using the model shown in Fig. 2.28 [27]. Paek *et al.* considered that the relationship between the pretilt angle of the LC and the rubbing strength is caused by increasing the interaction between the LC and the alignment material due to the increase in the surface polarity caused by rubbing, or the decrease in the average tilt angle of the polymer caused by changes of the surface wave structure [62].

Methods to increase the pretilt angle of the LC are the introduction of long side chains and the use of fluorinated moieties in the polymer main chain of the alignment layer. This was confirmed by molecular conformational calculations as shown in Fig. 2.29 where the introduction of the fluorinated moiety increases the roughness of the surface, and an increasing average tilt angle of the main chains seems to cause a larger pretilt angle of the LC [33]. Some studies have mentioned a relationship between the surface tension of the alignment layer and the pretilt angle of the LC. However, the changes in the surface tension might be a side

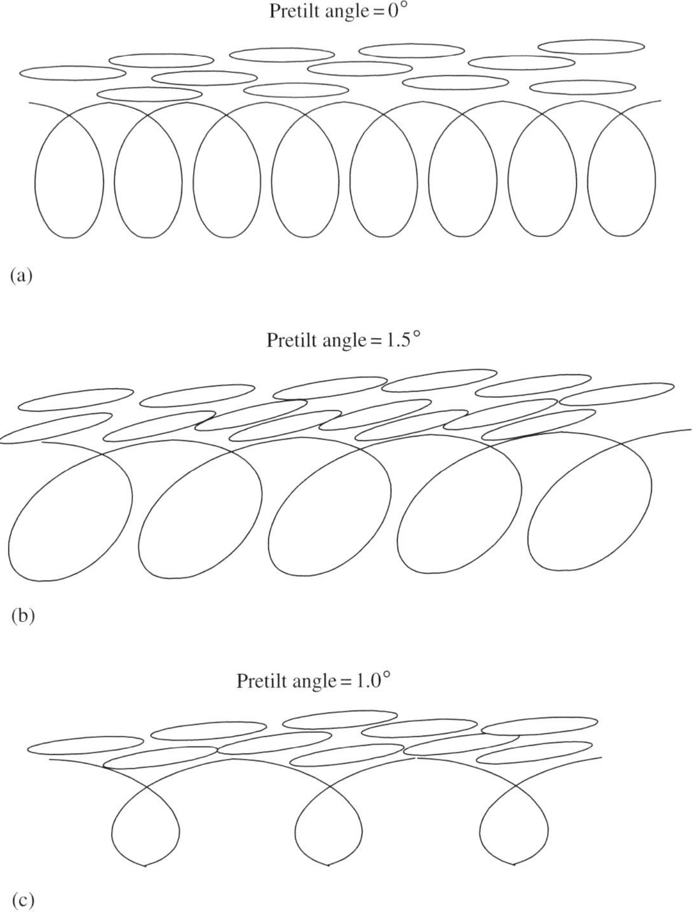

Fig. 2.28 Alignment models of nematic LC on rubbed polyimide surface. Reproduction by permission from [33].

effect. The relationship between the surface tension of the alignment layer and the pretilt angle of the LC does not always hold. The relationship holds only in the case of certain materials and certain conditions. The surface tension depends on the surface roughness. Rubbing increases the surface polarity because of merging of the side chains into the bulk and also increases the surface roughness in the case of strong rubbing. In this way, the surface tension increases as a result of rubbing, but this can be regarded as an unimportant side effect. The dominant mechanism for the pretilt angle of the LC seems to be an asymmetrical distribution of the polymer main chains in the alignment layer.

(a)

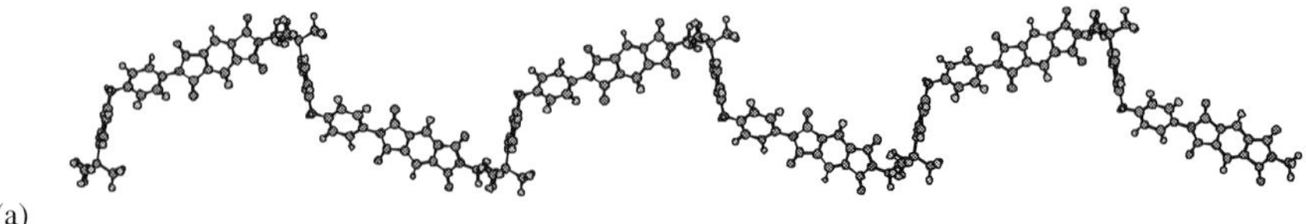

(b)

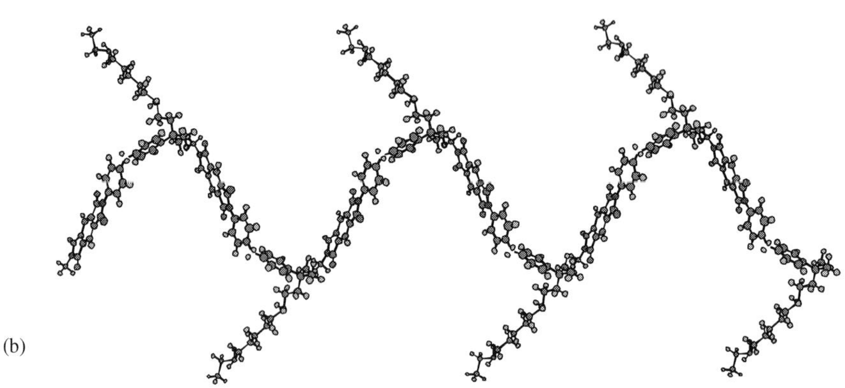

Fig. 2.29 Calculated PI conformations by molecular mechanics method. Reproduction by permission from [33].

2.3 Applications

Almost all of the present LCD manufacturers adopted the rubbing process to control the LC alignment. If the alignment of the LC is not uniform or the alignment strength is weak, alignment-related defects appear, such as disclination lines, reverse twist, and reverse tilt. These defects have large effects on the image quality of the LCDs because of light reflection from the defect interfaces or because of the electro-optical characteristic differences. Therefore, controlling the alignment defects is a very important issue for LCD manufacturing. Here, we will discuss the issues related to the LC alignment of actual LCD panels.

2.3.1 Alignment defects of actual devices

As the resolution and the size of the LCDs becomes higher and larger, the number of causes of alignment defects increases, depending upon for example the rubbing strength evenness, the lateral electric field, and the interfaces of the spacers. Some defects appear soon after the LCD panels are manufactured, and others appear after the electric field is applied to the LC. These differences are caused by the origin of each defect. Knowing the characteristics and origin of each defect is important in order to manufacture LCDs without these defects. Here, we will introduce the origin and structure of the typical disclinations and alignment defects of the twisted nematic (TN) configuration of LCDs.

2.3.1.1 Disclinations

Néel walls, twist- and tilt-reverses are the three major disclinations in actual LCDs. A Néel wall disclination may be caused by weak azimuthal anchoring. A twist-reverse is a defect in which the twist direction is opposite to the chirality of the added chiral agent. A tilt-reverse is a defect in which the inclination direction induced by applying a voltage is the opposite of the intended direction. These tilt-reverse defects may be caused by the lateral electric field generated by the fringe field of the pixel electrodes.

1. Néel wall

A Néel wall appears after injecting the LC into the display panel because of the effects of the LC flow during injection. The alignment situation for the Néel wall is shown in Fig. 2.30. Usually the panels are heated to around 100 °C to cancel the flow alignment effect during the injection process. If the anchoring energy is strong enough, all of the Néel wall disclinations vanish, but when the anchoring energy is weak or when impurities are attached to the surface of the alignment membrane, some disclinations will remain.

2. Tilt-reverse

Tilt-reverse occurs when the pretilt angle is not large enough to prevent inclination opposite to the intended direction caused by the lateral electric field coming from the edges of the pixel electrodes. The alignment in a tilt-reverse defect is shown in Fig. 2.31. Recently, the pixel size has become smaller as the resolution increases.

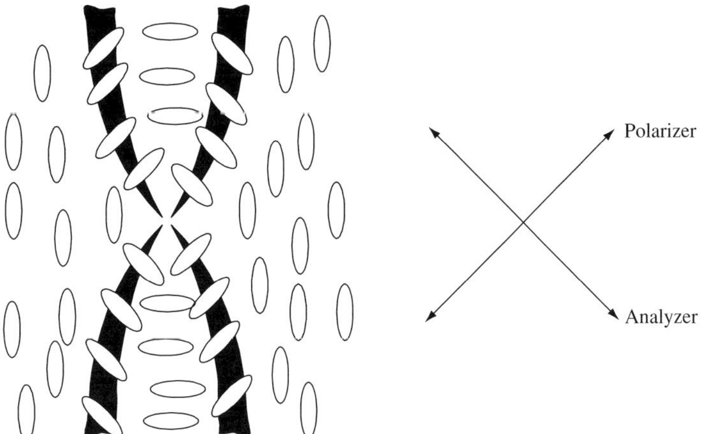

Fig. 2.30 Alignment in a Néel wall defect.

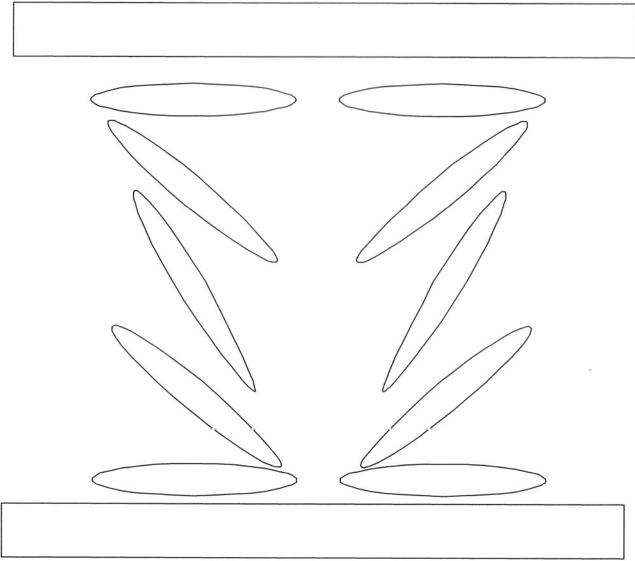

Fig. 2.31 Alignment in a tilt-reverse defect.

Therefore, the effect of this fringe field increases, and the pretilt angle should be large. A large pretilt angle is usually generated by using a polyimide having long alkyl side chains as the alignment material.

3. Twist-reverse

To fix the twist direction of the twisted nematic (TN) display mode, several methods are used. One is to add a small amount of a chiral agent. A typical chiral pitch for the 5 μm gap TN mode is 100 μm. The other method involves the rubbing direction. The rubbing direction determines the inclination direction of the LC, and it is selected so as to avoid a splay alignment as shown in Fig. 2.32. If the pretilt

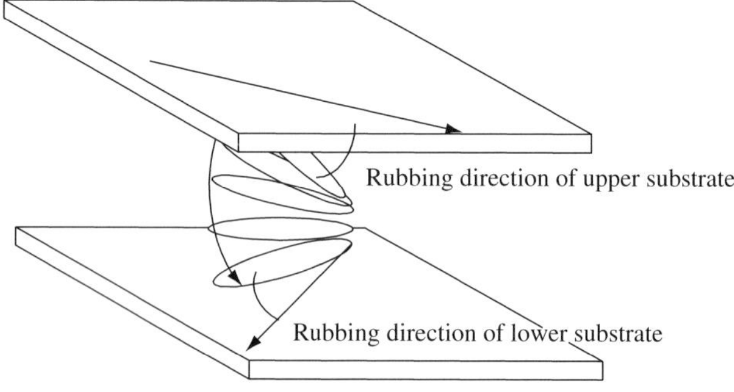

Fig. 2.32 Rubbing direction of TN mode.

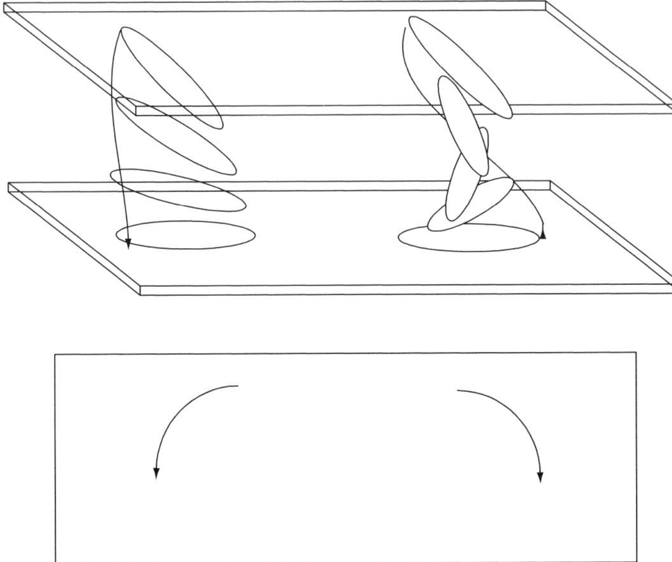

Fig. 2.33 Alignment for a twist-reverse defect.

angle is not large enough or if the effect of the fringe electric field is strong, the twist direction will be reversed. The alignment for the twist-reverse defect is shown in Fig. 2.33. Increasing the amount of chiral agent to decrease the chiral pitch is effective in eliminating the twist-reverse, but increasing the chiral agent concentration increases the threshold voltage of the display. Therefore, a large pretilt angle is preferred in order to suppress twist-reverse defects.

2.3.1.2 Alignment defects

1. Impurity attached to surface

LCDs consist of many types of materials. The LCs, the TFTs which are made of metals, the glass substrates, the alignment polyimides, the epoxy glue for sealing, the spacers made of plastic, and the colour filters made of acrylic resins. Many of these materials contaminate the LC with impurities that cause alignment defects. Impurities in the LCD panels may be solid or volatile. Solid impurities disturb the LC alignment by their physical presence, while volatile impurities affect the alignment membrane surface. Suspected solid impurities can be confirmed by microscopic observation when a defect is found, but volatile impurities cannot be seen. Such impurities are attached to the substrates before the alignment layer coating step, blocking the polyimide, and this affects the alignment process. These defects cannot be detected without applying a voltage.

2. Pretilt angle fluctuation
Pretilt angle fluctuations may be caused by variations in the dispersion speed of the polyimide solvent, or caused by unevenness in the rubbing strength. Recently developed polyimides for the alignment layer consist of several components to control the contact specifications, affecting such properties as the alignment and electrical characteristics. These components may become separated because of changes in the volatility of the solvent. For example, one component of the polyimide that influences the alignment may be exposed at the surface. The dispersion rate of the solvent is affected by the temperature and by the air flow while in the pre-baking process. This very active process strongly affects the molecular distribution of the alignment material, which causes pretilt angle fluctuations.

3. Impurity from sealing glue
The defects caused by impurities sometimes appear as white spots when a voltage is applied to the panel. The epoxy glue used for sealing and the alignment layer materials are the main sources of the impurities in LCD panels. The impurities from these sources soak into the LC gradually. Water from the outside also soaks into the LC through the epoxy glue or through the interface between the epoxy seal and the alignment layer. Therefore, defects caused by these impurities appear several days or months after assembly. The impurities have two major effects on the panels. One is the creation of secondary electric layers that decrease the effective voltage in the LCs, and the other is a weakening of the alignment strength by bonding to the surface of the alignment material. Fluorinated LCs are often used for the LCD panels because of their high resistivity, and their dispersion constant is very small, making it hard for the impurities to disperse into the bulk of the LC. The impurities migrate in response to the DC component of the driving voltage, which moves them towards the surface where they are absorbed. The alignment strength of the surface that absorbs the impurities is thereby weakened, and this causes alignment defects.

4. Injection hole
Since the LC injection speed near an injection hole is very high, alignment defects frequently appear in this area. The high speed flow of the LC affects the orientation of the alignment material, and causes alignment defects. After injecting the LC, the injection hole is covered by a photocurable resin. Therefore, the injection hole has a higher risk of being not only the locus of flaws, but also the source of impurities that may be present in the photocurable resin.

2.3.1.3 Defects around spacer beads

Spacer beads are sprayed on the substrate to make the cell gap uniform, and they can cause alignment defects in the LC because of their surface alignment strength and movement. Recently, the density of spacer beads has tended to increase in order to produce lightweight panels with thinner glass substrates, and also for more precise cell gap control for several electrically controlled birefringence (ECB) display modes, such as the vertical alignment (VA) mode and the in-plane switching (IPS) mode. Therefore, light leakage from the spacer beads has become a serious problem. Their alignment strength is comparable with that of the alignment layer, and this can cause disclinations in the LC. Figure 2.34 shows a photomicrograph of the spacers in a TN-LCD with voltage applied. The alignment model around the spacers is shown in Fig. 2.35. In case of the 90° twist configuration, the LC molecules in the middle of the cell align in between the top and bottom rubbing directions. However, the surfaces of the spacer beads tend to align the LC molecules perpendicular to the bead surfaces. Therefore, the disclination lines appear parallel to the upper and lower rubbing directions. When the electric field is applied to the LC cell, it aligns the LC molecules including those at the surfaces of the beads parallel to it. By the effect of the pretilt angle of the LC molecules on the glass substrates, the disclination lines then extend asymmetrically to the rubbing direction. The alignment orientation and its strength depends on the material of the spacer beads. Their movement also makes scratches on the alignment layer. The number of these defects is influenced by the pressure during assembly. At high pressures, the defects around the spacers increase, and at a low pressures, the defects caused by scratches increase. Several studies of the alignment around the spacers have been reported [60, 64]. To eliminate light leakage from the spacer beads, a technique using small pillars made of photoresist to control the cell gap has been developed [53].

Fig. 2.34 Photomicrograph of the spacers in a TN-LCD (5 V applied).

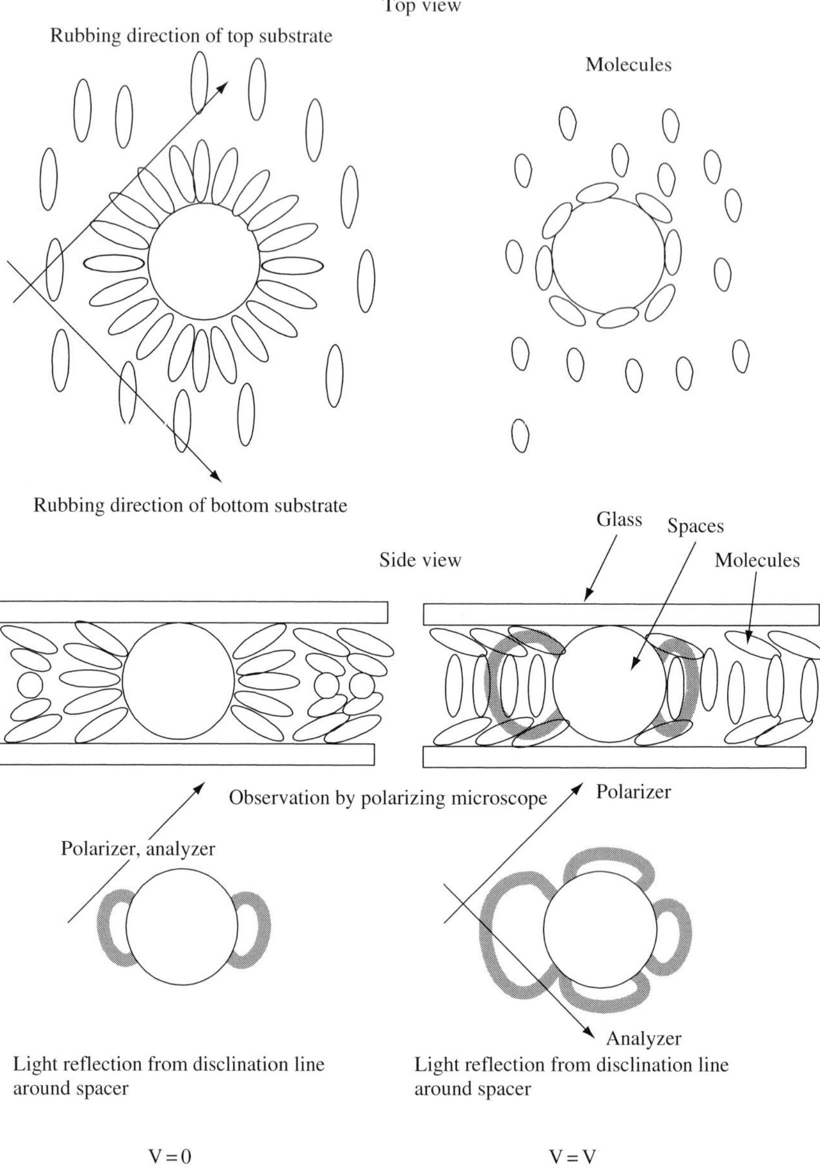

Fig. 2.35 Alignment model around the spacer beads. The left side is the alignment model when there are no electric fields. The right side is that when the electric field is applied to the LC cell. When the electric field is applied to the LC cell, the disclination lines extend asymmetrically to the rubbing direction on the glass substrates because of the pretilt angle of the LC molecules.

2.3.2 Characteristics of the rubbing process

It has been explained quite often in papers and books related to LCDs that rubbing has many problems, such as the generation of static electricity and dust. However, it is still the dominant alignment process for manufacturing LCDs. Understanding the characteristics of the process is very important to control the alignment process. Here, we will introduce the characteristics of the rubbing process, and the materials used in the alignment process for LCDs.

2.3.2.1 Rubbing machine and process sequence

Figure 2.1 shows a photograph of a typical rubbing machine. The metal cylinder covered by cloth rotates 100–800 times a minute, and moves up and down relative to the substrate stage. The rubbing pressure is adjusted by changing the gap between the cylinder and the substrate. The glass substrates are fixed to the stage by air suction. The rubbing direction on the substrate can be controlled by rotating the angle of the substrate on the stage. The stage moves back and forward at a constant velocity. The cylinder moves down to the substrate while rotating, applies pressure to the substrate, and the stage carrying the substrate moves forward with a constant velocity. After the entire area is buffed, the cylinder moves away, and the stage returns to the initial position.

The size of the cylinder and the stage have become larger as the substrate area has become larger. The cloth on the cylinder is changed after processing 2000–4000 substrates, because the cloth becomes too soft. The controllable parameters of the rubbing machine are the gap between the substrate and the cylinder, the cylinder rotation speed, the stage movement velocity, and the number of times each substrate goes through the rubbing process. The most significant parameter is the gap between the cylinder and the substrate. The gap controls the pressure of the buffing cloth on the substrate, and it affects the anchoring strength, pretilt angle, and surface morphology of the substrate.

2.3.2.2 Problems and current solutions

The rubbing process is employed for the manufacturing of LCDs with compensation being made for its problems. Typical problems and the current methods to eliminate or control them are discussed in this section.

1. Static electricity and dust

Rubbing a high resistance material, such as polyimide, generates static electricity. Therefore, during the rubbing process the static electricity is neutralized by injecting ionized N_2 gas to protect the TFTs. In addition to this, the TFT substrate has short rings to connect the gate lines and the signal lines of each TFT to ground so as to avoid applying the high voltages generated by the rubbing process to the TFTs.

The rubbing machine is covered with a hood, and an exhaust fan is used to prevent the diffusion of the dust from the rubbing cloth and alignment material. After the rubbing process, a cleaning process is used to wash away particles of the rubbing cloth and alignment material from the surface of the substrate.

2. Uniformity

The rubbing process controls not only the alignment direction, but also the pretilt angle of the LC. The pretilt angle is greatly affected by the pressure of the rubbing cloth against the substrate. Usually, this pressure is controlled by the mechanical gap between the cylinder and the substrate. The rubbing cloth is changed after several thousand substrates have been processed. After changing the cloth, a test run is required to check the setting status of the cloth and the gap. If the cloth is not positioned evenly, the LC panel brightness will not be uniform because of dispersion of the pretilt angles, and the electro-optical characteristics are greatly affected by the pretilt angles of the LC. Monitoring the rubbing status is very difficult, because evaluation of the effects of rubbing is possible only after assembling the substrates and observing the images on the finished LCD panels. Therefore, when alignment defects caused by the rubbing process are found on the manufacturing line, a lot of LCD panels have already been affected by any failure in the rubbing process. Developing a method to monitor the rubbing status of the alignment layer *in situ* has been a high priority goal. Recently, the optical retardation measured by reflective ellipsometry has been used for monitoring the rubbing condition in situ. The printing and baking of the alignment layer also affect the pretilt angle of the LC. Therefore, the LC manufacturer chooses an alignment material that shows little change in the LC alignment condition due to undulations.

How strictly do we need to control the LC alignment to keep the electro-optical characteristics of the panel uniform? Figure 2.36 shows the relation between applied voltage and transmittance when the pretilt angle is changed from 2° to 4°. It shows only a 2% change in the grey level, and this level is acceptable. Figure 2.37 shows the relation between rubbing strength and the pretilt angle for two types of alignment materials. The type A material shows only a 1° change, and can be used for production.

How large is the effect of anchoring strength on the LC panel characteristics? Figure 2.38 shows the relation between the anchoring energy and the actual twist angle when the LC of the cell is configured to a 90° twist. The azimuthal anchoring energy of the rubbing has a range from $1 \times 10^{-4}\,\mathrm{J\,m^{-2}}$ to $5 \times 10^{-4}\,\mathrm{J\,m^{-2}}$. By transferring the anchoring energy to the twist angle, the relationship between the applied voltage and the transmittance is obtained and shown in Fig. 2.39. The transmittance change in the grey level caused by changing the twist angle is less than 2%, which is acceptable. As shown here, a uniform alignment is obtained by choosing an alignment material whose alignment characteristics show small dependences on the processing conditions.

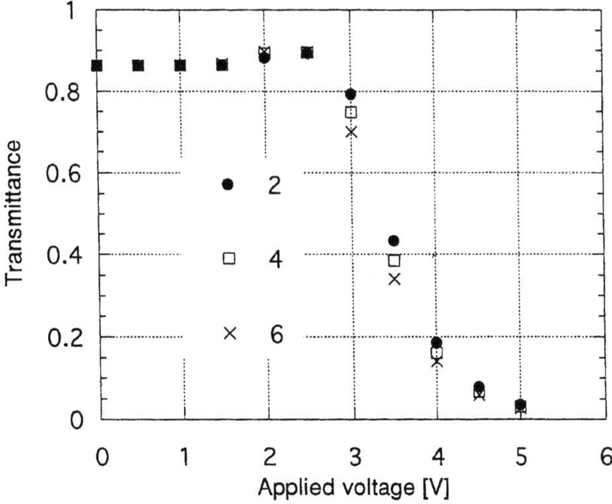

Fig. 2.36 Tilt angle effect on applied voltage–transmittance (V–T) curve: calculated V–T curves for 2°, 4°, and 6° pretilt angles.

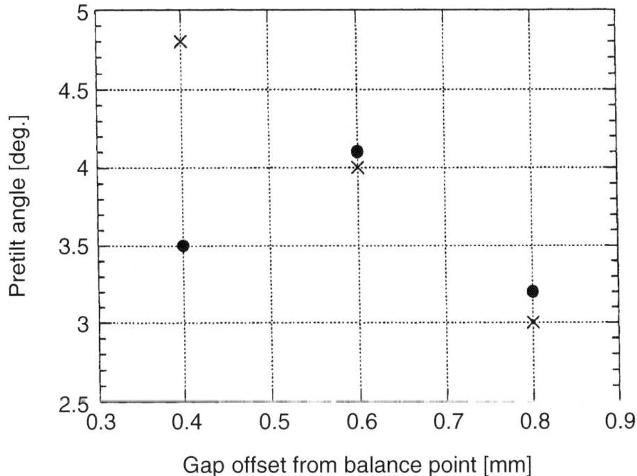

Fig. 2.37 Rubbing strength and pretilt angle relations of two types of polyimide (●–A, ×–B polyimide): rubbing strength represents the offset height from the standard position.

3. Limited viewing angle caused by unidirectional alignment

The electro-optical characteristics of a single domain TN LC display have a strong viewing angle dependency, because the director of the LC rotates in the same plane over the entire area of each pixel. To improve this, a multi-domain

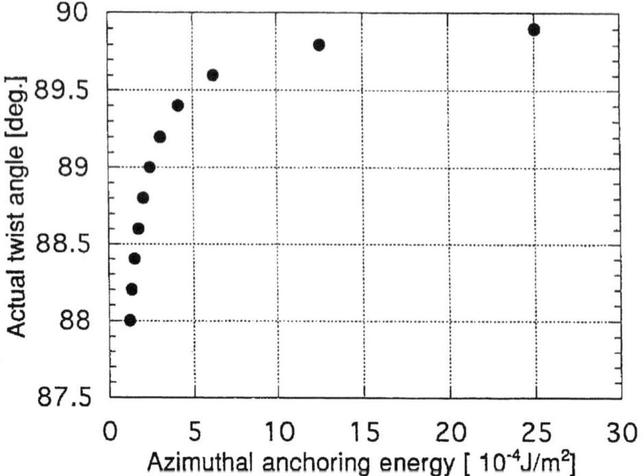

Fig. 2.38 Relation between azimuthal anchoring energy and actual twist angle of a 90° TN cell.

technology was developed, in which each pixel is divided into two or four areas, and the LC in each area rotates in a different direction when the electric field is applied. There are several techniques to divide the pixel, for example mask rubbing [15], use of plural alignment material printing [18], pretilt angle modification by exposure to UV light [36], applying voltage during the LC injection [47], and the axially symmetric aligned microcell (ASM) mode [40]. In addition to these multi-domain techniques, a compensation film using discotic LC can be applied

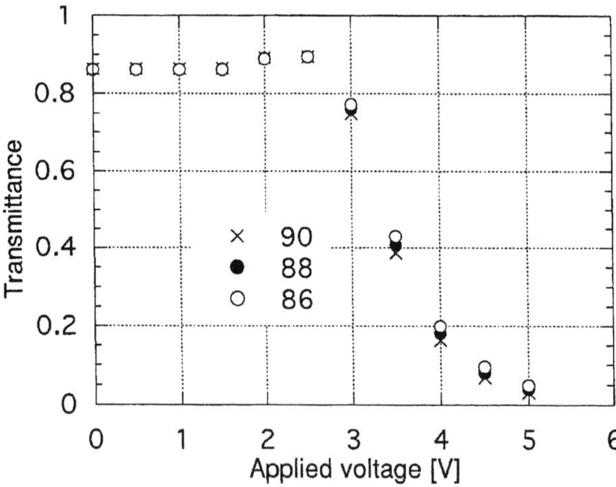

Fig. 2.39 Actual twist angle effect on V–T curve: calculated V–T curves of 86°, 88°, and 90° actual twist angles.

to a single domain type TN display to enhance the viewing angle [45]. Recently, the in-plane switching (IPS) mode was developed, which in principle has a large viewing angle [28]. In the vertical alignment display (VA) mode, triangular structures or electrodes having a slit in the centre are used to obtain a multi-domain effect [63, 42]. In these ways, the limitation of the viewing angle is addressed, even though rubbing is still used for the alignment process.

2.3.3 Alignment material

The dominant material for display devices using rubbing alignment processes is a polyimide, not only because of the alignment characteristics, but also for stability, together with its electrical properties, such as resistively, which are suitable for these devices. Especially for thin film transistor (TFT) devices, the voltage retention ratio is one of the important issues. Table 2.2 summarizes the characteristics of the present commercially available alignment materials for both parallel and perpendicular alignment. Two types of alignment material, polyimides and organosiloxanes, are on the market for perpendicular alignment. The polyimide used for this kind of alignment has long side chains.

2.3.3.1 Process

Two types of polyimide are used for actual devices; solvent soluble pre-imidized polyimide and polyamic acid type. Though the pre-imidized polyimide is already imidized, a baking process is required for both type of polyimide. The alignment-related process consists of a coating of polyimide, a baking step, rubbing, and cleaning of the substrates. The sensitivity of the alignment material to the manufacturing process conditions is one of the important issues when selecting it for manufacturing use. Stability of the pretilt angle and alignment in the face of normal variations in the processing conditions is required. Major process parameters are the rubbing process strength, the baking conditions for the alignment material, and the annealing conditions of the panels. Nishikawa *et al.* reported that a soluble pre-imidized polyimide shows better stability against process instabilities [33]. Here, we explain and discuss each process sequencially.

1. Coating

Spin coating and printing are the two major methods for coating polyimide on the glass substrate. Printing is the dominant process for LCD panels, because printing can coat just the desired pattern on the substrate. Around the edges of the substrates of the LCD panels there are places for mounting driver chips and electrodes for connections which are not coated with the alignment material. The accuracy of the printing pattern is within several tens of microns.

2. Baking

The baking process consists of two stages, pre-baking and hard-baking. The pre-baking is to heat the substrate after coating it with polyimide to evaporate

Table 2.2 Characteristics of alignment materials on the market. VHR, R-DC and a represent Voltage Holding Ratio, Residual DC voltage and refractive index, respectively

Manufacture	Type	Pretilt angle	VHR	R-DC	n	Application
JSR	AL1454	0.7	99	200	1.6	TFT-parallel
	JALS-146-R39	1	99	100	1.59	TFT-parallel
	AL5056	2	99	100	1.59	TFT-parallel
	AL3046	3.5	99	130	1.59	TFT-parallel
	AL1J508	4.7	98	120	1.58	TFT-parallel
	AL8254	3	99	200	1.54	TFT-parallel
	AL1F408	4.5	98	80	1.58	TFT-parallel
	JALS-1068-R2	4.3	99	0	1.68	TFT-parallel
	JALS-1077-R2	5	99	0	1.6	TFT-parallel
	JALS-682-R6	88	98.9	250		Vertical
	JALS-2021-R2	89	98.9	10		Vertical
	JALS-2022-R2	82	99.4	20		Vertical
	JALS-204	89	97	550		Vertical
	JALS-9800-R1	4–5				STN
	JALS-9005-R1	5–6				STN
	JALS-1024-R1	4–5				STN
	AL3408	3–4				STN
Nissan Chem.	SE-410	2			1.71	TN
	SE-130	2			1.63	TN
	SE-2170	2			1.64	TN
	SE-150	4–5			1.63	STN
	SE-3310	4–5			1.66	STN
	SE-610	7–8			1.61	STN
	SE3140	5–6			1.67	STN
	SE3510	7–8			1.66	STN
	SE7992	4–5	high		1.62	TFT-parallel
	SE7492	6–7	high		1.62	TFT-parallel
	SE5291	5–6	high		1.61	TFT-parallel
	SE7511L	90			1.60	Vertical
	SE1211	90			1.54	Vertical
Hitachi Chem. Dupont	LQ-T120–03	3.5	98.9	55		TFT-parallel
	LQ-T120–04	6.8	97.2	57		TFT-parallel
	LQ-1800	8.0				STN
	LQ-C100	3.1				STN
	LX-1400	2.6				TN
	LQ-2200	0.8				TN

the solvent at about 90 °C for one minute. The hard-baking heats the substrate to 200 °C for 30 minutes after the pre-baking. Hard-baking evaporates the solvent completely, and imidizes the polyamic acid to make the thin film hard. In LCD panel manufacturing, these processes are done by using a hot plate.

3. Rubbing
Details of the rubbing process are as introduced above.

4. Cleaning
Since bits of the rubbing cloth and pieces of alignment material are embedded on the substrate by the rubbing process, a cleaning step is required after rubbing. The substrates are cleaned by using a normal process with purified water and ultrasonic vibration, and then dried.

2.3.3.2 Electrical characteristics

In addition to the alignment properties, the electrical characteristics of the alignment material that affect the image quality are important for display devices. The alignment materials for display products are selected by evaluating such electrical characteristics as the voltage holding ratio, the residual DC, the image sticking time, and the common level shift. These electrical properties are seen in the flicker and image sticking characteristics of the LCD panels. Table 2.3 summarizes the electrical characteristics of the alignment materials.

1. Voltage holding ratio
The applied voltage on a pixel is refreshed at 60 Hz for normal TFT-driven LCD panels. Therefore, each pixel should hold an electric charge for 1/60 sec = 16.7 msec, like a capacitor. If the charge of the pixel changes during this

Table 2.3 Electrical characteristics of alignment materials and their related display characteristics

Characteristics	Related display characteristics	Description
Voltage holding ratio	Flicker	Voltage retention ratio for 16 msec after charging the electrode for 20 μsec.
Residual DC voltage	Image sticking	Residual voltage after DC voltage is applied and removed.
Common voltage shift	Flicker	Shifted voltage after AC voltage is applied several times.
Image sticking time	Image sticking	Residual image time after a black box is displayed.

period due to current leakage in the LC, the transmittance of the pixel will change. Evaluation of this charge holding time is done by monitoring the pixel voltage for 16.7 msec after charging the electrode. The definition of the voltage holding ratio (VHR) is the ratio of the final voltage to the initial charging voltage, or the ratio of the integral of the changing voltage of the pixel over time divided by the initial voltage multiplied by 16.7 msec. The actual voltage holding ratio of the material used for display products is over 98%.

2. Residual DC
Though the voltage holding ratio evaluates the charge leakage over a short period, the residual DC value measures it over several minutes. After applying a DC voltage to the pixel for several minutes, the pixel electrodes are shorted briefly, and then the electrodes of the test pixel are opened, and the voltage is measured. This voltage is defined as the residual DC voltage, and is related to the image sticking characteristics of the display panel.

3. Common level shift
When a pixel is driven by a TFT, a small DC voltage is applied to the pixel, caused by the asymmetric characteristics of the circuit. This DC voltage causes a migration of ions within the materials making up the LC panel and dissolved in the LCs themselves. Between the migrating ions and the electrodes an internal electric field is generated. This internal electric field changes the effective voltage applied to the LC, because each data frame reverses the polarity of the pixel drive voltage. This phenomenon causes flicker. To counteract the internal voltage, a DC voltage is added to the driving voltage of the pixel so that the flicker is suppressed. This added DC voltage that cancels the internal voltage is defined as the common level shift. In some cases, the common level shift strongly correlates with the image sticking time.

4. Image sticking time
To evaluate the image retention characteristics directly, the image retention time must be measured by using an actual display panel. A black box on a white background is displayed on the panel for a defined time, for example, one hour. After showing this fixed pattern, the entire screen is changed to a uniform grey level image, and the time until the residual image of the box disappears is measured. This test is useful to evaluate the image retention time under actual operating conditions.

2.3.3.3 Rubbing cloth

Table 2.4 summarizes specifications for rubbing cloths used for display manufacturing. The groove structure caused by rubbing resembles the structure of the textile of the rubbing cloth, and this has secondary effects on the alignment of the LC.

Table 2.4 Specifications for rubbing cloths in the market

Product	YA-20-R	YA-18-R	YA-19-R	YA-25-C
Material	Rayon	Rayon	Rayon	Cotton
Pile diameter	120D	100D	100D	265D
Filament diameter	3D	2.5D	2.5D	2.5D
Filament number/cm^2	24000	32000	28000	28000
Pile number/(3.78 cm)^2	8600	11500	1000	4000
Total thickness (mm)	1.8	1.65	1.8	2.5
Width (cm)	92	92	110	92110

References

[1] C. Mauguin: Bull. Soc. Fr. Min. 34 (1911) 71.

[2] D.W. Berreman: Phys. Rev. Lett. 28 (1972) 1683.

[3] J. A. Castellano: Mol. Cryst. Liq. Cryst. 94 (1983) 33.

[4] Y. R. Shen: Annu. Rev. Mater. Sci. 16 (1986) 69.

[5] J. M. Geary, J.W. Goodby, A. R. Kmetz, and J. S. Patel: J. Appl. Phys. 62 (1987) 4100.

[6] W. Chen, M. B. Feller, and Y. R. Shen: Phys. Rev. Lett. 63 (1989) 2665.

[7] Y. R. Shen: Nature 337 (1989) 519.

[8] Y. R. Shen: Annu. Rev. Phys. Chem. 40 (1989) 327.

[9] Y. R. Shen: Nature (London) 337 (1989) 519.

[10] Y. R. Shen: Liq. Cryst. 5 (1989) 635.

[11] T. Uchida, M. Hirano, and H. Sasaki: Liq. Cryst. 5 (1989) 1127.

[12] M. Suzuki, T. Maruno, F. Yamamoto, and K. Nagai: J. Vac. Sci. Technol. A8 (1990) 631.

[13] M. B. Feller, W. Chan, and Y. R. Shen: Phys. Rev. A43 (1991) 6778.

[14] H. Nejoh: Surf. Sci. 256 (1991) 94.

[15] K. H. Yang: IDRC 91 Digest (1991) 68.

[16] M. Barmentlo, R. W. J. Hollering, and N. A. J. M. van Aerle: Phys. Rev. A46 (1992) R4490.

[17] H. Kado, K. Yokoyama, and T. Tohda: Rev. Sci. Instrum. 63 (1992) 3330.

[18] Y. Koike, T. Kamada, K. Okamoto, M. Ohashi, I. Tomita, and M. Okabe: SID 92 Digest (1992) 798.

[19] M. Nishikawa, T. Miyamoto, S. Kawamura, Y. Tsuda, and N. Bessho: Proc. Japan Display 1992 (1992) 819.

[20] D.-S. Seo, K. Muroi, and S. Kobayashi: Mol. Cryst. Liq. Cryst. 213 (1992) 223.

[21] D.-S. Seo, S. Kobayashi, and M. Nishikawa: Appl. Phys. Lett. 61 (1992) 2392.

[22] N. A. J. M. van Aerle, M. Barmentlo, and R. W. J. Hollering: J. Appl. Phys. 74 (1993) 3111.

[23] E. S. Lee, Y. Saito, and T. Uchida: Jpn. J. Appl. Phys. 32 (1993) L1822.

[24] E. S. Lee, T. Uchida, M. Kano *et al.*: SID '93 Digest (1993) 957.

[25] H. Mada and T. Snoda: Jpn. J. Appl. Phys. 32 (1993) L1245.

[26] D.-S. Seo, T. Oh-ide, H. Matsuda, T. Isogami, K. Muroi, Y. Yabe, and S. Kobayashi: Mol. Cryst. Liq. Cryst. 231 (1993) 95.

[27] M. Nishikawa, Y. Yokoyama, N. Bessho, D.-S. Seo, Y. Iimura, and S. Kobayashi: Jpn. J. Appl. Phys. 33 (1994) L810.

[28] M. Oh-e, M. Ohta, K. Kondo, and S. Oh-hara: 15th ILCC (1994) K-P15.

[29] K. Sakamoto, R. Arafune, N. Ito, S. Ushioda, Y. Suzuki, and S. Morokawa: Jpn. J. Appl. Phys. 33 (1994) L1323.

[30] Y.-M. Zhu, L. Wang, Z.-H. Lu, Y. Wei, X. X. Chen, and J. H. Tang: Appl. Phys. Lett. 65 (1994) 49.

[31] X. Zhuang, L. Marrucci, and Y. R. Shen: Phys. Rev. Lett. 73 (1994) 1513.

[32] K.-Y. Han and T. Uchida: J. SID 3 (1995) 15.

[33] M. Nishikawa, Y. Matsuki, N. Bessho, Y. Iimura, and S. Kobayashi: J. Photopolymer Sci. and Tech. 8 (1995) 233.

[34] J. Y. Huang, J. S. Lie, Y.-S. Juang, and S. H. Chen: Jpn. J. Appl. Phys. 34 (1995) 3163.

[35] Y. B. Kim, H. Olin, S. Y. Park, J. W. Choi, L. Komitov, M. Matuszezyk, and S. T. Lagerwall: Appl. Phys. Lett. 66 (1995) 2218.

[36] A. Lien, R. A. John, M. Angelopoulos, K. W. Lee, H. Takano, K. Tajima, A. Takenaka, K. Nakagawa, Y. Momoi, and Y. Saitoh: Asia Display '95 (1995) 593.

[37] Y. Ouchi, I. Mori, M. Sei, E. Ito, T. Araki, H. Ishii, K. Seki, and K. Kondo: Physica B 308/209 (1995) 407.

[38] K. Shirota, K. Ishikawa, H. Takezoe, A. Fukuda, and T. Shiibashi: Jpn. J. Appl. Phys. 34 (1995) L316.

[39] M. S. Toney, T. P. Russel, J. A. Logan, H. Kikuchi, J. M. Sands, and S. K. Kumar: Nature 374 (1995) 709.

[40] N. Yamada, S. Kohzaki, F. Fukuda, and K. Awane: SID '95 Digest (1995) 575.

[41] R. Arafune, K. Sakamoto, D. Yamakawa, and S. Ushioda: Surface Science 368 (1996) 208.

[42] N. Koma, R. Nishikawa, and K. Tarumi: SID '96 Digest (1996) 39.

[43] K.-W. Lee, S.-H. Paek, A. Lien, C. J. During, and H. Fukuro: Macro-molecules 29 (1996) 8894.

[44] K.-W. Lee, S.-H. Paek, A. Lien, C. J. During, and H. Fukuro: SID '96 Digest 27 (1996) 273.

[45] H. Mori, Y. Itoh, Y. Nishimura, T. Nakamura, and Y. Shinagawa: IDW/AM-LCD '96 (1996) 189.

[46] I. Mori, T. Araki, H. Ishii, Y. Ouchi, K. Seki, and K. Kondo: J. Electron Spectrosc. Relat. Phenom. 78 (1996) 371.

[47] H. Murai, M. Suzuki, T. Suzuki, T. Konno, and S. Kaneko: AM-LCD 96 (1996) 185.

[48] A. J. Pidduck, G. P. Bryan-Brown, S. D. Haslam, R. Bannister, and I. Kitely: J. Vac. Sci. Technol. A14 (1996) 1723.

[49] A. J. Pidduck, G. P. Bryan-Brown, S. D. Haslam, and R. Bannister: Liq. Cryst. 21 (1996) 759.

[50] K. Sakamoto, R. Arafune, S. Ushioda, Y. Suzuki, and S. Morokawa: Appl. Surf. Sci. 100/101 (1996) 124.

[51] K. Sakamoto, R. Arafune, N. Ito, and S. Ushioda: J. Appl. Phys. 80 (1996) 431.

[52] K. Shirota, M. Yaginuma, T. Sakai, K. Ishikawa, H. Takezoe, and A. Fukuda: Appl. Phys. Lett. 69 (1996) 164.

[53] H. Yamashita, Y. Saitoh, S. Matsumoto, and M. Kodate: SID '96 Digest (1996) 600.

[54] R. Arafune, K. Sakamoto, and S. Ushioda: Appl. Phys. Lett. 71 (1997) 2755.

[55] K.-W. Lee, A. Lien, J. Stahis, and S.-H. Paek: Jpn. J. Appl. Phys. 36 (1997) 3591.

[56] T. Sakai, J.-G. Yoo, Y. Kinoshita, K. Ishikawa, H. Takezoe, and A. Fukuda: Appl. Phys. Lett. 71 (1997) 2274.

[57] D.-S. Seo, L.-Y. Hwang, and B.-H. Lee: Proc. 5th Int. Conf. on Properties and Applications of Dielectric Materials (1997) 946.

[58] K. Wako, K.-Y. Han, and T. Uchida: Mol. Cryst. Liq. Cryst. 304 (1997) 235.

[59] R. Arafune, K. Sakamoto and S. Ushioda, S. Tanioka, and S. Murata: Phys. Rev. E58 (1998) 5914.

[60] M. Hasegawa and H. Yamanaka: Proc. IDW '98 (1998) 45.

[61] M. P. Mahajan and C. Rosenblatt: J. Appl. Phys. 83 (1998) 7649.

[62] S.-H. Paek, C. J. Duming, K.-W. Lee, and A. Lien: J. Appl. Phys. 83 (1998) 1270.

[63] A. Takeda, S. Kataoka, T. Sasaki, H. Chida, H. Tsuda, K. Ohmuro, and Y. Koike: SID '98 Digest (1998) 1077.

[64] Y. Utsumi, S. Komura, Y. Iwakabe, S. Matsuyama, and K. Kondo: Proc. IDW '99 (1999) 289.

Chapter 3

Non-rubbing Methods

3.1 Introduction

Masaki Hasegawa

The rubbing process [1] has many problems, such as unevenness, lack of controllability, failure to achieve multiple domain alignment, and difficulties of status monitoring, all related to its mechanical contact operating principle. To overcome these problems caused by the buffing mechanism, several liquid crystal (LC) alignment methods have been developed. Here, we summarize the previously developed LC alignment methods in Tables 3.1, 3.2, and 3.3. These methods can be divided into two categories. One uses surface alignment caused by the anisotropy of the surface. The other method aligns the LCs based on an electric or magnetic field coming from outside the cell, and the surface of the substrate becomes joined to the aligned LC. After the electric or magnetic field is removed, the aligned LC on the surface aligns the bulk of the LC. Nematic LCs can be easily aligned by a small anisotropy of the surface. However, when the order parameter of the surface alignment or the anchoring energy is small, this alignment tendency is comparable to other causes of anisotropy, e.g. the flow effect of the injection, and many disclination faults can appear. In this specific case, the influence of flow can be eliminated by using an isotropic phase injection of the LC.

Recently, studies of stabilized LC alignment obtained by using polymer networks have been reported. A photocurable monomer is added to the LC, and it is polymerized by UV exposure after the LC has been aligned either by the surface method or by using electric or magnetic forces. An application for an FLC was reported based on this approach, and good alignment of the FLC was reported [31].

Another approach, to eliminate all contact processes includes a printing process for the alignment layer, and the alignment process, using a deposited carbon with Ar^+ ion beam exposure, is applied to the aligned LC [36, 47]. A diamond-like carbon is deposited on the glass substrate, and exposed to the Ar^+ ion beam. The Ar^+ ion beam anisotropically decomposes bonds in the diamond-like carbon layer.

Table 3.1 Summary of alignment methods I

Method	Description	Characteristics	Ref.
Oblique evaporation	Obliquely evaporate SiO onto a substrate. Alignment direction depends on incident angle.	High tilt angles near 80° can be generated.	[2]
Micro-groove	Make photolithographic micro-grooves on a substrate.		[3]
Stretched polymer film	Attach a stretched polymer film to the substrate.		[4]
Dip-coat	Dip a substrate into a polymer solution, and pull it up slowly.	Alignment strength is even weaker than flow alignment.	
Transcription	Transcribe a rubbing surface structure onto the resin surface.	Multiple domains can be obtained easily by using a prefabricated multiple domain master.	[29, 34]
Surface stripped film	Strip a polymer film from the substrate.	Tilt angle depends on strip direction and speed.	
LB film	Stack LB films on a substrate.	Tilt angle depends on pull up speed of the LB film.	[16]

Table 3.2 Summary of alignment methods II

Alignment method	Description	Characteristics	Ref.
Photo isomerization	Isomerization	Reversible	[7, 8]
Photo dimerization	Dimerization	Stable	[12, 20, 28,33, 37, 41, 44]
Photo decomposition	Decomposition	Wide selection of materials	[15, 18, 19, 22, 43]
Photocrosslinking	Crosslinking	Stable	[32]
Combination of photocrosslinking and photodecomposition	Combination of crosslinking and decomposition.		[21]
Magnetic field	Inject LC into a cell in the isotropic phase, apply a magnetic field, and cool down to the nematic phase.	High tilt angles can be established.	[5, 9]
Flow alignment	Inject LC into a cell with a homogeneous alignment layer in the nematic phase.	Weak anchoring	

Table 3.3 Summary of alignment methods III

Alignment method	Description	Characteristics	Ref.
Temperature gradient	Inject LC into a cell with a homogeneous alignment layer in the isotropic phase, and cool following a temperature gradient. Polymer edge is used in the cells for nucleation of the alignment.	Used for FLC	[6]
Shear	Put FLC polymer between film, and apply shear stress.	Used for FLC polymer alignment.	[13]
Amorphous	Inject LC with chiral agent in the isotropic phase into a cell with a homogeneous alignment layer, and cool down.	TN configurations having multiple domains can be established.	[29, 34]
Ion beam	Substrate covered with polymer exposed to low energy argon ion beam.		[23, 36]

Both the coating of the alignment material and the alignment processes can be freed from any mechanical contacts by using this technique.

3.2 Photoalignment

Masaki Hasegawa

3.2.1 Introduction

Photoalignment has been a focus of research since Gibbons *et al.* reported their photoalignment results in 1991 [8]. They described an azobenzene-mixed polyimide exposed to linearly polarized UV light that aligned the LC. In 1992, Schadt reported photoalignment by dimerization of poly (vinyl cinnamate) exposed to linearly polarized UV light [12]. Actually, prior to their reports, Ichimura had reported an isomerization of azobenzene which showed reversible photoalignment in 1988 [7]. In these projects, special photochemical reactions, such as dimerization or isomerization, were used, and only a few limited types of photoalignment materials were involved. In 1994, it was found that polyimide exposed to linearly polarized deep UV light showed LC alignment, and that a wider variety of materials could be used for photoalignment [15]. At that time, viewing angle improvement was a big issue for LCD manufacturers, and the multiple domain technique was developed to enhance the viewing angle. Photoalignment received a lot of attention because it can be used to establish multiple domains easily.

However, multi-domain technology cannot address the colour shift problem and requires additional processes and investments. Therefore, the photoalignment approach did not become widespread as a manufacturing process. Instead, an IPS display mode was developed, which can enhance the viewing angle without creating a colour shift problem [17, 21]. In addition to this, an optical compensation film using a discotic LC was developed [27]. The compensation film is expensive, but it does not require any additional processes or investments in manufacturing. Therefore, the need for multi-domain technology was reduced. With a weaker demand for the multi-domain approach, there was a lower acceptance of the photoalignment approach, since its main virtue was for establishing the multi-domain configurations more easily.

However, recent trends towards larger display panel sizes and higher resolutions, and the advent of more display modes have increasingly exposed the limitations of the rubbing process. The desire for a new and improved alignment technology steadily became stronger, and today many types of photoalignment have been developed, and the choices of materials have increased. All ranges of pretilt angles can be established by photoalignment. Not only linearly polarized light, but also a non-polarized light can be used as the light source for photoalignment [10, 35]. Figure 3.1

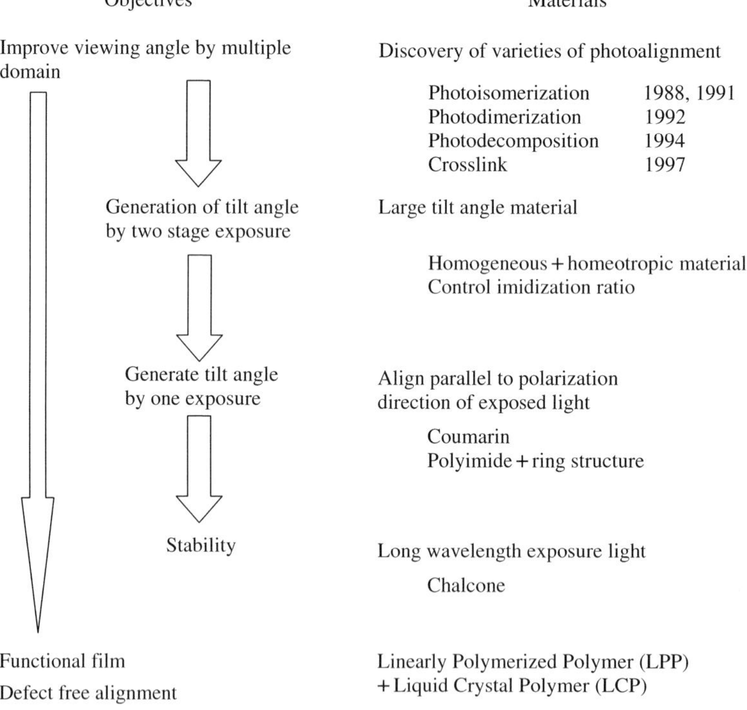

Fig. 3.1 History of the development of photoalignment.

summarizes the main progress and trends in the photoalignment method. The light sources for photoalignment have also progressed. The requirements for the light sources for photoalignment are almost the same as those for lithography except for the polarization state. Recently, high energy density light sources have been developed that can provide linearly polarized light for a wide exposure area.

3.2.2 Overview

The most common technique for photoalignment involves the generation of an anisotropic distribution of the molecules of the alignment materials by using the dependence of the polarization direction of the absorption of light by the molecules. When the electric vector of the photons coincides with the transition moment of the molecule, the molecule absorbs light strongly, and a photochemical reaction occurs. The creation of anisotropy by using polarized light is called the Weigert effect, and has been known since 1920. However, the photoalignment mechanisms in some materials are not yet understood because the photochemical reactions are different in each photoalignment material. In some cases, the LC alignment direction depends on the exposure energy of the light on the photoaligning material. Two similar LC alignment mechanisms coexist, one caused by residual molecules not affected by the light, and the other relying on photochemically produced structural changes [24, 41]. Here, we will overview photoalignment as categorized by the mechanism, conformational change, and photochemical reaction. The photochemical reaction type can be categorized as photodimerization, photodecomposition, photolinking, and a combination of decomposition and linking.

3.2.2.1 Photoisomerization

Photoalignment using photoisomerization phenomena controls the LC alignment by a conformational change between *cisoid*- and *transoid*-type azobenzene functional units. The photoisomerization caused by changing the wavelength or polarization direction of the exposed light can switch the LC alignment reversibly between homeotropic and homogeneous, or change the direction of the homogeneous alignment [10, 11]. The principle of LC alignment by photoisomerization is shown in Fig. 3.2. The important phenomena here are the axis-selective anisotropic absorption and the reorientation of the molecules. The molecules whose transition moments are parallel to the electric field of the exposed light change their conformation because of their anisotropic absorption. An azobenzene molecule whose long axis is parallel to the polarization axis of the exposed light reversibly changes its conformation between *cisoid* and *transoid*, and the ratio of the two forms is different. For azobenzene, linearly polarized light with a wavelength of 365 nm, whose polarization axis is parallel to the molecular long axis, isomerizes *transoid*-type molecules to the *cisoid*-type which has a small absorbance for the polarized light of that orientation. Therefore, the number of *transoid* molecules whose axes are perpendicular to the polarization axis increases in relative terms,

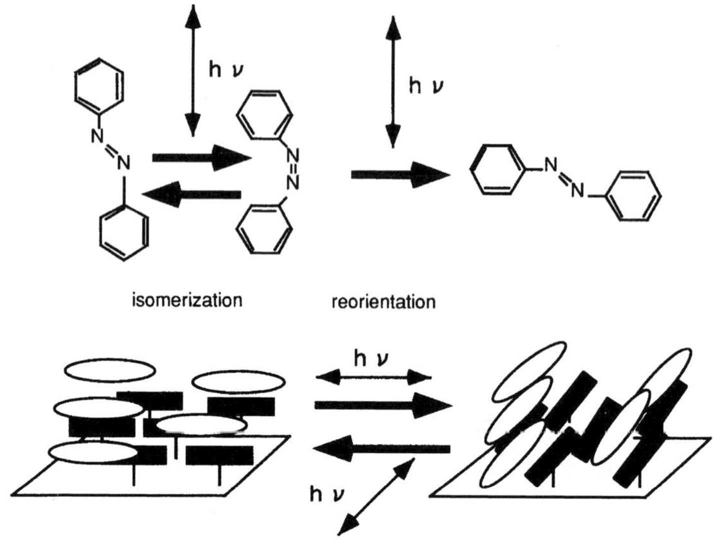

Fig. 3.2 Principle of photoisomerization in-plane alignment.

and the LC becomes aligned along the molecular axis, which is perpendicular to the polarization axis. The axially anisotropic photoisomerization alone is sufficient to control the LC alignment. However, for polymethacrylate a molecular reorientation was observed [38]. The molecular reorientation involves a molecular axis that actually moves during its conformation change. If the molecular axis rotates perpendicular to the polarization axis during the photoisomerization, the absorbance of light decreases and it tends to remain stably in that orientation. This reorientation phenomena for other materials has not been well studied.

A reversible change between homeotropic and homogeneous alignment can be achieved by aligning the azobenzene units perpendicular to the substrate [7]. Figure 3.3 shows switching between the homeotropic and homogeneous alignment of a LC using photoisomerization. The photoisomerization materials used for the alignment layer of display panels is determined by the stability of the electrical and alignment characteristics, because photoalignment using photoisomerization is limited to materials having special functional units, such as azobenzene. To solve this problem, some techniques to mix an azobenzene dye into a polyimide and to select the photosensitive wavelength have been described. Another technique to establish a stable structure using azobenzene is to synthesize a polyimide whose main chain contains azobenzene units [8]. The photoisomerization material, called the "command surface" [7], has the characteristic capability of reversibly controlling the LC alignment through control of the wavelength or polarization axis of the light to which it is exposed. Using this peculiar characteristic, it can be applied to photonic materials [14].

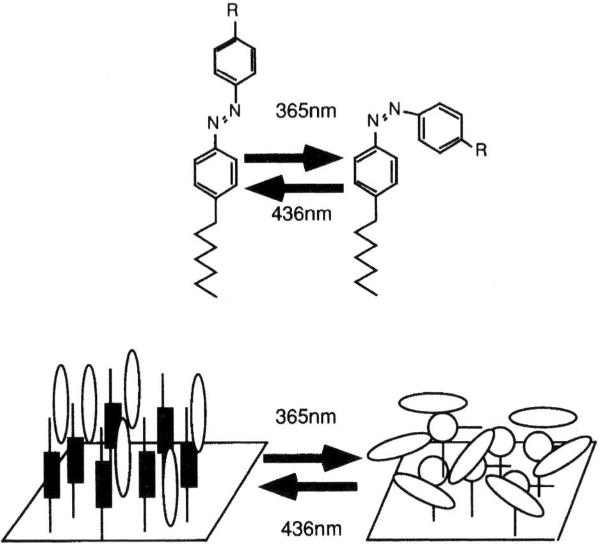

Fig. 3.3 Principle of switching between homeotropic and planar alignment by photoisomerization.

3.2.2.2 Photochemical reaction

1. Photodimerization

Poly (vinyl cinnamate) (PVCi) shown in Fig. 3.4 is a well-known material that aligns LCs by photodimerization [12, 20]. Double bonds parallel to the polarization axis of the linearly polarized (LP) light absorb the energy and become scissored. The two opened bonds generated by the scissoring are reconnected and dimerized. The alignment direction of the LC is perpendicular to the polarization direction of the light in case of PVCi. The alignment mechanism of the PVCi is not completely understood, because it has not only been dimerized, but also

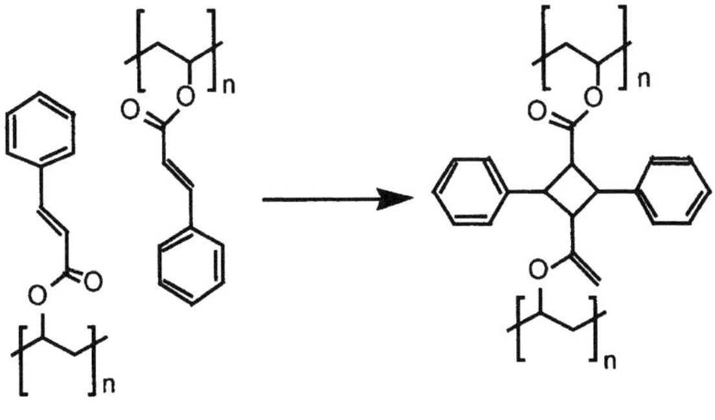

Fig. 3.4 Dimerization of poly (vinyl cinnamate) – PVCi.

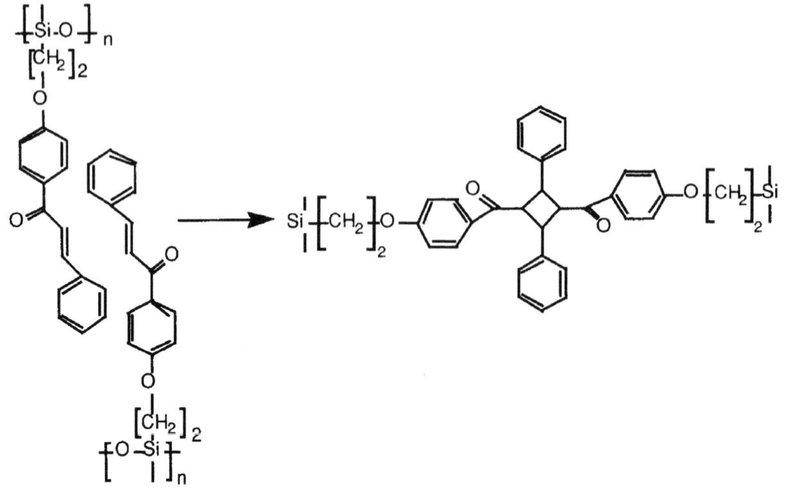

Fig. 3.5 Dimerization of a polysiloxane-based photopolymer.

isomerized by the exposure to light. Stabilization of the electrical characteristics by introducing a cinnamoyl unit as shown in Fig. 3.5 was used to create a polysiloxane as recently reported [33, 44]. However, the cinnamoyl moiety also gives photoisomerization, and its alignment stability has also been a subject of concern. It was reported that coumarin molecules as shown in Fig. 3.6, which do not give photoisomerization, would align LCs parallel to the polarization axis of the light [28], and a test display panel which used a coumarin-like material was described [37]. The chalcone structure shown in Fig. 3.7 was developed to stabilize the alignment characteristics. Chalcone has a sensitivity to relatively long wavelength light (365 nm) when used as a dimerization material. Long wavelength light poses a smaller risk than shorter wavelength light of photodecomposition of the photoalignment material [41].

2. Photodecomposition

Photoalignment by using a photodecomposition mechanism has been applied to a wide variety of materials compared with the other two methods, photoisomeriza-

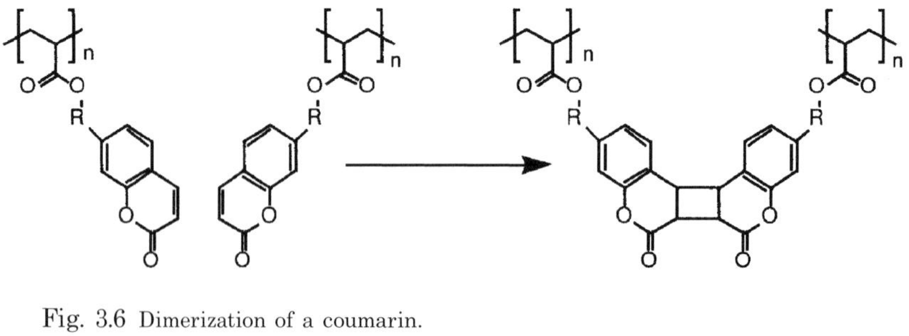

Fig. 3.6 Dimerization of a coumarin.

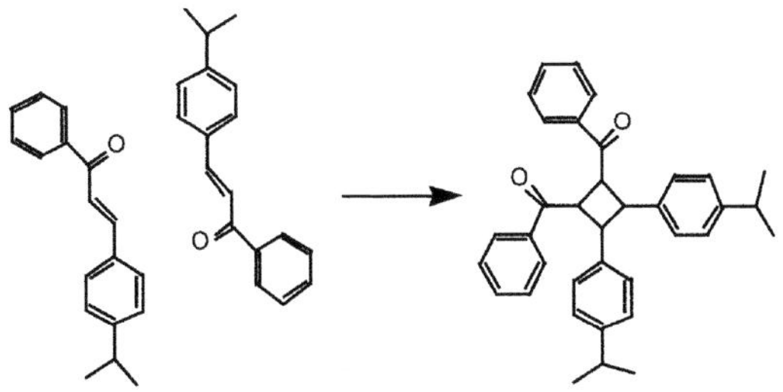

Fig. 3.7 Dimerization of a chalcone.

tion and photodimerization. The polyimide shown in Fig. 3.8 has been studied as a material for photodecomposition-based alignment, because it is stable electrically, thermally, and chemically [15, 18, 22, 43]. Polyimide also has a lot of associated evaluation data from its use as an LC alignment material. When the polyimide is exposed to LPUV, an anisotropic photodecomposition occurs because of the dependence of the polarization direction of the light absorption of the molecules. The anisotropic absorption of the UV light changes the anisotropic molecular length distribution of the polyimide, which controls the alignment of the LC.

3. Crosslink
In this method, radicals generated by the UV exposure connect to other parts of molecules and generate a new structure. If this type of material shows a preferred photoalignment, then this new structure can be used to align the LC. This type of photoalignment was found in the benzophenone structured polyimide shown in Fig. 3.9 [32].

4. Hybrid (decomposition and crosslinking)
In some cases involving photodecomposition-type materials, broken bonds create new structures that cause an alignment of the LC. The polyimide containing the cyclobutane ring structure shown in Fig. 3.10 was described as this type of material [24]. In this case, alignment competition occurs between the residual structure and the newly created structure. The LC alignment caused by this photoalignment competition changes the LC alignment direction based on the exposure energy or wavelength of the irradiating light [24].

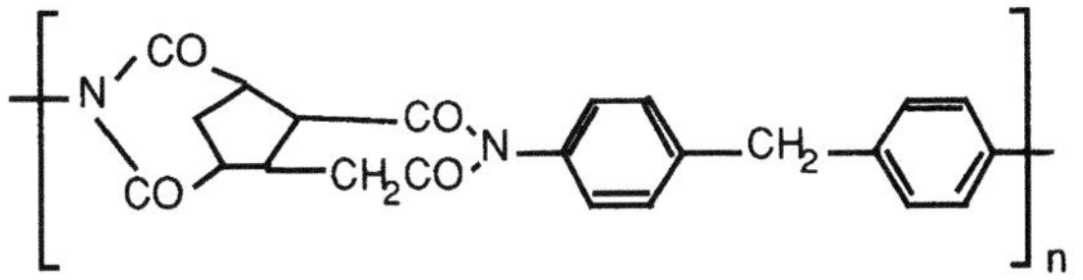

Fig. 3.8 Molecular structure of soluble polyimide AL1254 (JSR).

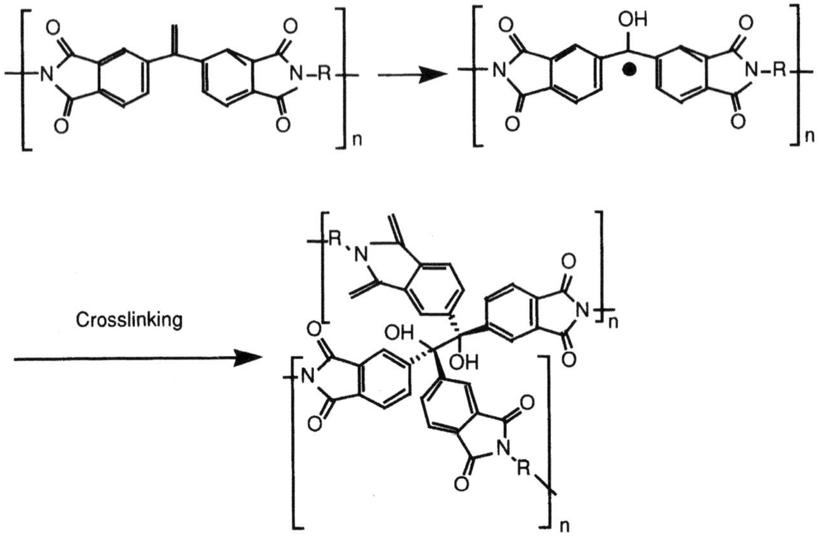

Fig. 3.9 Crosslinking of a benzophenone-type polyimide.

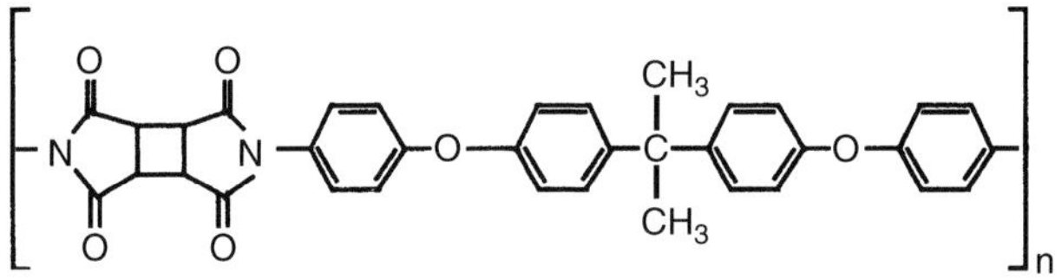

Fig. 3.10 Molecular structure of CBDA/BAPP: Polyimide containing a cyclobutane ring.

3.2.2.3 Progress in photoalignment materials

Here we summarize the progress in developing photoalignment materials. In the early stages of development of the photoalignment approach, when it first became a focus of research, many different photoalignment mechanisms were reported, such as photoisomerization, photodimerization, and photodecomposition. The direction of the LC alignment was perpendicular to the polarization axis of the exposed light in all of these early cases. Therefore, it was necessary to develop techniques to control the pretilt angle. For example, the substrate could be exposed to linearly polarized light twice, first with the light perpendicular to the substrate, and the second time with the light shining obliquely and with the polarization axis rotated by 90° [19]. A large pretilt angle could be created by using a homeotropic alignment material, or by changing the imidization ratio of the polyimide for homogeneous alignment [22]. Materials for photo-isomerization, -dimerization, and -decomposition which showed parallel alignment along the polarization axis were also developed, making it possible to create a pretilt angle with only one exposure to polarized light [28, 43]. By these methods, photoalignment materials were developed which show a wide range of pretilt angles for the LCs, extending from parallel to homeotropic alignment. Recently, the focus of research has moved to alignment stability, and materials were developed which are sensitive to longer wavelengths of light, in order to prevent photodecomposition by short wavelength UV [41, 42]. The photoalignment materials progressed while targeting new objectives such as a larger variety of materials, the generation of larger pretilt angles, alignment parallel to the polarization axis, and improved stability.

3.2.3 Photoalignment using polyimide

3.2.3.1 Experimental results

Here we will describe some experimental results for photodecomposition-based alignment to explain photoalignment using polyimide. First we spin-coated a polyimide, SE7792 (Nissan Chemical), which has a long side chain, on a glass substrate and baked it for an hour at 180°. For the primary photoalignment experiment, we used a main chain-type polyimide, but such a material cannot be used to generate a pretilt angle using the two-stage exposure method. The substrate was exposed to linearly polarized deep UV light (LPDUV). The LPDUV, with a wavelength of 257 nm, is generated by a non-linear optical crystal, BBO, as a second harmonic wave when excited by 514.5 nm light generated by an Ar^+ ion laser. The light generated by this method is originally linearly polarized, and the extinction ratio is more than 1,000, with a power density of 20 mW cm^{-2}. When a high pressure mercury lamp is used as a light source, a polarizer and a bandpass filter are required. A typical polarizer for UV is a Glan-Thompson prism, or stacked fused quartz plates having a Brewster angle suitable for the objective wavelength. First, the substrate is exposed to LPDUV light perpendicular to the substrate to determine the azimuthal alignment direction. There is no pretilt angle for the LCs in the cell assembled from this substrate. To add a pretilt angle to the LCs, the substrate is next exposed to the UV light with its polarization axis rotated by 90°, and with the substrate tilted (inclined) to change the incident angle. The resulting inclined P wave creates distribution of polyimide molecules anisotropic in polar angle. An inclined single exposure with non-polarized collimated deep UV can also create a pretilted alignment for the LC [30]. However, in our experiment, the alignment of the LC from that approach was not as good as that from using the two-stage exposure. Two substrates exposed to LPDUV in two stages were assembled and injected with the LC at room temperature. The LC we used was 5 CB, and the cell thickness was 5 μm. The alignment direction was determined by the reflection angle of the polarized light from the wedge cell. The pretilt angle of the LC was measured by the crystal rotation method. Figure 3.11 shows the relationships among the pretilt angles of the LCs, the exposure energies, and incident angles of the second exposure. The tilt

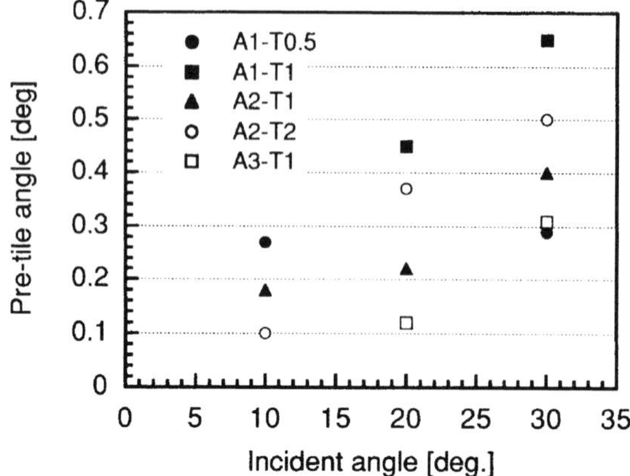

Fig. 3.11 Relationship between incident angle of the UV light and the pretilt angle of the LC for different exposure energies: A2-T1 represents 2 and 1 Jcm^{-2} dosages for first and second exposures, respectively.

direction of the LC was toward the incident direction of the second exposure. In this experiment, the maximum tilt angle was small, no larger than 0.65°; however, there were no reverse tilt disclination lines when voltage was applied to the assembled cells.

3.2.3.2 Pretilt angle control

The required pretilt angle for the LC depends on the display modes and pixel sizes in the actual display panels. In some cases, more than 10° of pretilt angle is required. To satisfy such requirements, the generation of large pretilt angles was studied. One technology is to change the ratio of the imidization. It has been reported that a wide range of pretilt angles from 0° to 90° [22] can be created by changing the baking temperature and the imidization ratio. Another technique is to use a homeotropic alignment material. Two-stage exposure of homeotropic alignment material has been used to establish pretilt angles from 0° to 90°.

Materials that show parallel LC alignment along the polarization axis have been developed to eliminate the need for the two-stage exposure. Polyimide with a cyclobutane ring in the main chain and polyimide with the ring structure shown in Fig. 3.12 have been reported to be such materials [43]. By using these materials, a pretilted LC alignment with strong anchoring can be established with one LPUV exposure.

3.2.3.3 Alignment mechanism

Polyimide exposed to LPUV aligns LCs as described above. But why does the LPUV exposed polyimide align the LCs? Here we will consider the alignment mechanisms for polyimide exposed to LPUV.

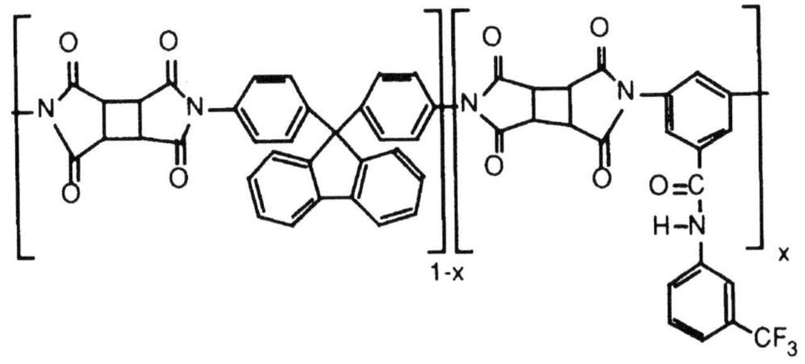

Fig. 3.12 Polyimide with fluorine atoms that shows high tilt angle photoalignment.

A photodecomposition-type material does not include optically sensitive structures, and the determination of the photochemical reaction for such materials is rather more difficult than for others. Therefore, determination of the photoalignment mechanism for a photodecomposition-type material is difficult to compare with that for the other two mechanisms.

We have studied the mechanism of the photodecomposition-type alignment by using IR measurements and characterization techniques for LC alignment. The polyimide used was a CBDA-ODA with the molecular structure shown in Fig. 3.13. We baked a CBDA-ODA coated glass substrate for an hour at 250 °C and exposed it to LPUV. The cell was assembled from this substrate, and was injected with a guest-host LC, MLC-6069 (Merck), and then with 5 CB. The alignment characteristics were evaluated by measuring the anchoring energy, and from the order parameter of the doped dye in the LC from the guest-host LC. The chemical reactions were analyzed by measuring the IR and UV-visible absorption spectra of the polyimide when exposed to LPUV. Figure 3.14 shows how the UV-visible absorption spectra change before and after exposure to LPUV. The absorption of the short wavelengths decreased while that of longer wavelengths increased. This absorption increase is caused by the creation of new structures because of the photochemical reaction. The wavelength of the LPUV was 257 nm, and the absorption peaks are caused by π–π transitions involving the benzene rings, and by the carbonyl structure. Figure 3.15 shows the difference in the IR absorption spectra obtained, by subtracting the spectrum before exposure from the spectrum after

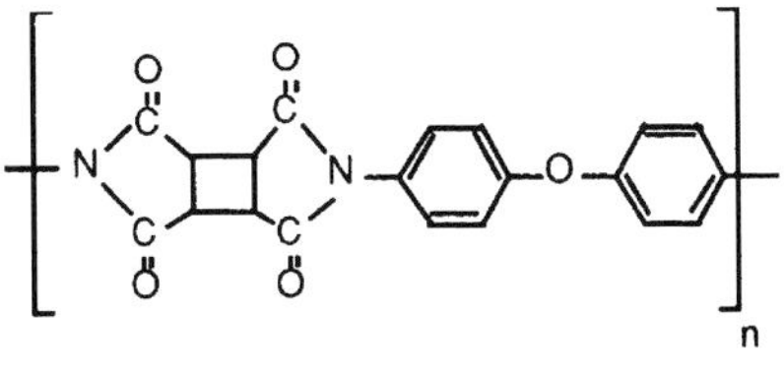

Fig. 3.13 Molecular structure of CBDA/ODA.

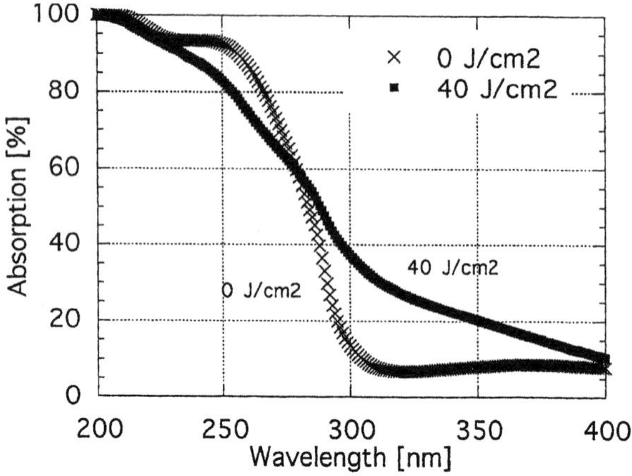

Fig. 3.14 UV absorption spectra of CBDA/ODA before and after UV exposure.

exposure. The solid line represents the initial absorption spectrum. Values above zero represent a reduction in absorption caused by the UV exposure, and negative regions show the production of new structures during the exposure. From Fig. 3.15, we can see a reduction of absorption due to benzene rings and C=O, C—N—C, and C—O—C structures. Figure 3.16 shows the dichroism of the IR absorption. The solid line represents the initial absorption spectrum. The dotted line represents the differences between the polarization IR spectrum perpendicular to the polarization of the exposed LPUV and that parallel to the polarization of the exposed LPUV. Benzene rings, C=O, C—N—C, and C—O—C moieties in the main chain perpendicular to the polarization axis of the exposed light are less affected than when the main chain

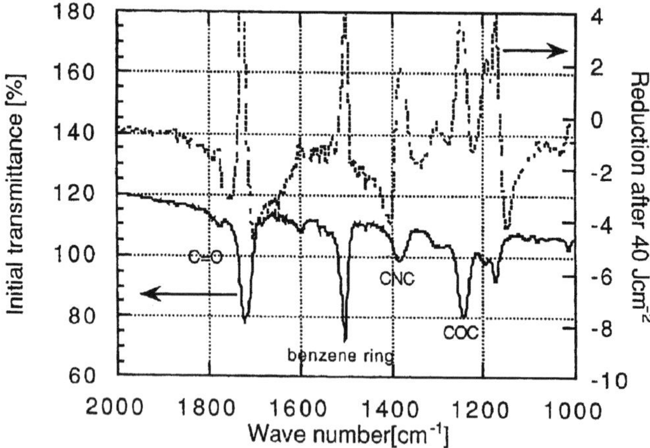

Fig. 3.15 IR spectra of CBDA/ODA: Initial transmittance and reduction after 40 Jcm^{-2} of UV exposure.

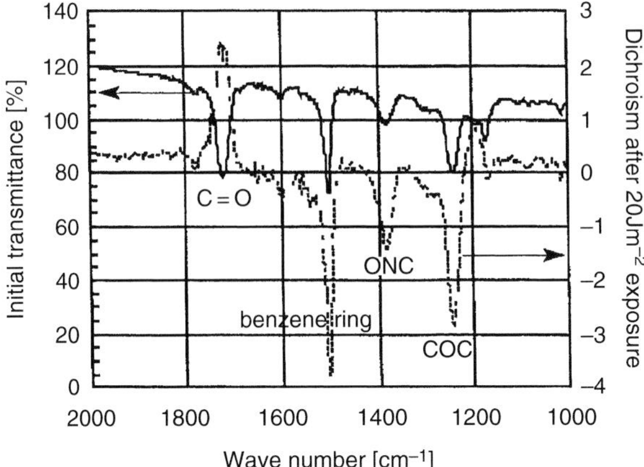

Fig. 3.16 IR spectra of CBDA/ODA: Initial transmittance and dichroism after 20 J/cm^2 LPDUV exposure.

is parallel to the polarization. These spectral measurements confirms that there are anisotropic reductions in benzene ring and C = O content caused by the polarization axis dependency of the UV absorption.

Why do these chemical changes align the LCs? Figure 3.17 shows the relationship between the order parameter of the dye added as dopant to the LC and the exposure energy. In this figure, since all values are positive, it shows alignments perpendicular to the polarization axis of the exposure light. A CBDA-ODA polyimide exposed to 257 nm UV light shows perpendicular alignment, resulting in negative values. Figure 3.18 shows the relationship between the azimuthal anchoring energy and the exposure energy. This figure also shows the same peak as that

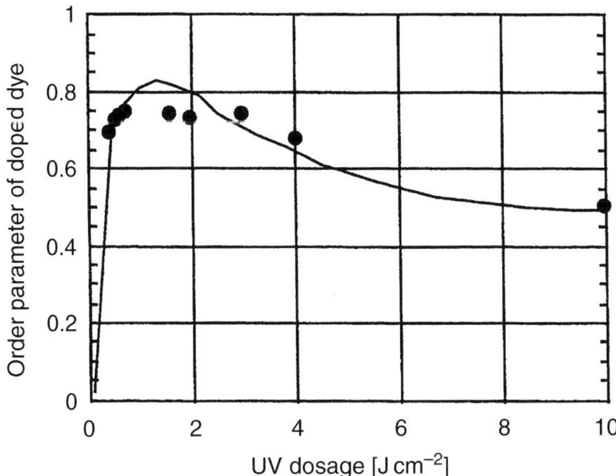

Fig. 3.17 Relation between UV dosage and order parameter of the dye dopant.

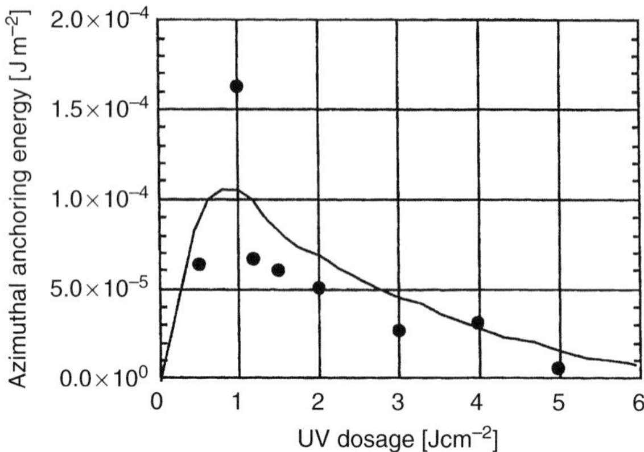

Fig. 3.18 Relation between UV dosage and azimuthal anchoring energy.

in Fig. 3.17. Surface examination of the LPUV exposed polyimide using AFM did not reveal any morphological changes. From these results, we concluded that the LC alignment from photodecomposition is caused by the anisotropic decomposition of benzene, C=O, C—N—C, and C—O—C moieties. Other groups of researchers have reported that the cause of the alignment in photodecomposition was a surface anisotropic morphological change that is caused by anisotropic decomposition, and this surface morphological change causes the alignment [39, 40]. However, in our experiment, the alignment of LCs is established by even a very small dosage of UV, and the order of the LC increased rapidly with small dosages. Even if surface morphological changes occur, we believe they might not be the main cause of the alignment. Our research leads us to believe that anisotropic chemical changes are probably the major cause of the alignment. A surface analysis technique is needed to determine the actual chemical reactions occurring near the surface exposed to LPUV. Recently, detailed IR absorption and NEXAFS measurements on polyimides exposed to LPDUV polyimide have been reported [45].

3.2.4 Influence of UV light on display device characteristics

Degradation of the polymer by deep UV exposure is a concern when photoalignment is used for the alignment process of LCD manufacturing. The generation of many radicals was confirmed by ESR measurements after UV exposure [26]. Do these radicals affect the display characteristics of the LCD panels? Here, we will describe our study of the effects of UV exposure on the characteristics of the LCD [25].

Image sticking, or other residual image phenomena, is one of the big problems for LCDs. We studied the effects of UV exposure on this phenomenon. Figure 3.19 shows the relationship between the offset voltage inside the cell and the exposure dosage of UV light. The offset voltage, which causes the image sticking, increases

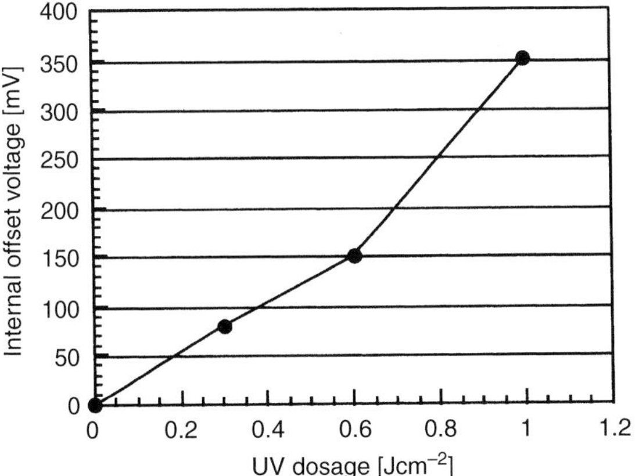

Fig. 3.19 Relationship between UV dosage and offset voltage.

linearly with dosage. The radicals generated by the UV exposure absorb ions or impurities in the LC, and generate an electrical double layer. This double layer generates an internal voltage inside the cell. Figure 3.20 shows the relationship between image sticking time and the resistivity of the LC cell using six kinds of polyimide. Here, the image sticking time represents the residual time for the image of a box after showing a black box on the display panel for 24 hours. The

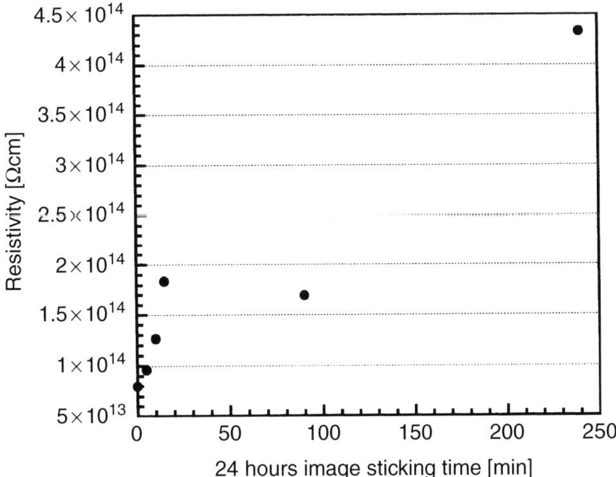

Fig. 3.20 Relationship between image sticking time and resistivity of the cell: image sticking time represents the residual time for a pattern after showing a fixed black box on the panel for 24 hours.

resistivity represents the relational constant between the voltage and current of the display panel at low frequency. The lower the resistivity, the shorter the image sticking time for all polyimides. When the UV light dosage increases, the resistivity tends to become low, and the image sticking times become shorter. Even at the lowest resistivity, the voltage holding ratio is more than 98%. Such a high voltage holding ratio does not cause flicker. The generation of polar functional groups, such as OH, caused by the UV exposure was confirmed by using IR and water contact angle measurements [26]. We found the best LC alignment at a $1\,\mathrm{J\,cm^{-2}}$ dosage, and such a small dosage does not affect any of the display characteristics. From these experiments, we confirmed that a small exposure to UV light for photoalignment does not affect the display characteristics.

3.2.5 Light source

Table 3.4 summarizes light sources for photoalignment. The critical wavelengths for photoalignment depend on the principle involved and the material. Only small

Table 3.4 Characteristics of UV light sources. BBO represents β-BaB_2O_4 crystal, and is a non-linear optical crystal

UV light source	Wavelength	Shape	Energy ($\mathrm{mWcm^{-2}}$: shorter than 300 nm)	Lifetime (hours)	Other required devices
High pressure Mercury Lamp	254, 365, 405, 436	Point	10	1000	Filter, polarizer
Low pressure Mercury Lamp	185, 254	Line	5	1000	Polarizer, scanning
Hg-Xe lamp	254, 313, 365	Point	5	1000	Filter, polarizer
Excimer lamp	172, 222, 308	Line	10	1000	Polarizer, scanning
Metal-halide	250–450	Line		1500	Filter, polarizer, scanning
UV laser	257 (Ar^+x2), 355 (YAGx3)	Small diameter collimated light	30–100 mW/beam	Depends on non-linear optical crystal, for example BBO	Two axes scanning

polarizers originally existed for UV, so the UV exposure equipment could only create a low energy density. Such low energy density light sources required long process times. Recently, large polarizers based on using the Brewster angle have been developed, and it is now possible to expose a large area to LPUV. Stacked fused quartz plates are placed in the light path, configured to use the Brewster angle for the required wavelength of the desired light, and the polarization direction of the incident light can thus be controlled.

3.2.6 Comparison of photoalignment and rubbing

Alignment mechanisms and materials for photoalignment have been described above. How can we apply the photoalignment technology to the practical manufacturing process for LCDs? Here, we discuss the use of photoalignment as an actual manufacturing process in comparison to the rubbing process.

3.2.6.1 Cost

In this section we estimate the cost of new machines and maintenance when a photoalignment process replaces the present rubbing process. We assume the panel size to be 300 mm × 400 mm, and the light source is a high pressure mercury lamp.

1. Rubbing
   (a) Instrument: 50 million yen
   (b) Processing area: 400 mm wide
   (c) Processing time: 20 min for coat and bake, 20 sec for rubbing
   (d) Life: Change cloth every several thousand substrates
   (e) Others: Exhaust fan and post-process cleaning required
   (f) Monitoring is difficult in situ
   (g) Number of control factors is large.

2. Photoalignment
   (a) Instrument: 100 million yen
   (b) Processing area: 400 mm × 400 mm, 30 mW cm^{-2}
   (c) Processing time: 20 min for polyimide, 40 sec for exposure
   (d) Life: Change lamp every 1,000 hours (40 days in 24 hour-operation)
   (e) Others: Exhaust fan and post-process cleaning required
   (f) Monitoring exposure energy is easy
   (g) Lamp is 0.5 million yen.

The cost of rubbing and photoalignment is therefore comparable. When the rubbing process is controlled well and no defects are found in the products, it works very well. However, once defects caused by the rubbing process are found, the influence of failures on the process is very large. This is because the rubbing process is a contact alignment system, and failures in the rubbing process can be spread widely. In comparison with rubbing, photoalignment is easy to control in a consistent way, because only the exposure energy needs to be controlled. In addition to this, the influence of a defect on one substrate upon other substrates is small, because photoalignment is a non-contact alignment process. The baking time for the alignment material is the longest processing step in the alignment process. Therefore, there is little meaning in a comparison of the actual processing time between rubbing and photoalignment. Reduction of the processing time could potentially be achieved by using a photoalignment material other than polyimide, for example some substrate which does not require a long baking time.

3.2.6.2 Controllable parameters

This section summarizes some of the controllable characteristics of each alignment process.

1. Rubbing
   (a) Unidirectional alignment
   (b) Stable pretilt angles can be modified only by changing the alignment materials
   (c) Anchoring energy cannot be controlled. Only strong anchoring.

2. Photoalignment
   (a) Alignment direction can be controlled by changing the polarization axis or incident angle of the light
   (b) Pretilt angle can be easily controlled from 0° to 90° without changing the alignment material
   (c) Anchoring energy can be controlled by changing dosage.

From the economic point of view, photoalignment costs more facility investments and maintenance. Therefore, the replacement of the present unidirectional alignment process using rubbing by a photoalignment process is not immediately feasible. However, the trends towards high resolution and large size panels require much more uniformity in the alignment process. It is questionable whether rubbing can be applied to higher resolution and larger display panel sizes. It will be better to use photoalignment where its characteristics are superior, and when the present rubbing techniques cannot be used. For example, small optical elements using LCs, alignment processes for structured surfaces, and flexible substrates are good applications for photoalignment. Photoalignment has great potential for creating new display modes and new optical devices.

3.2.7 Current status of photoalignment

At this moment, there are no products mass-produced using photoalignment. LG.Philips LCD Co. Ltd. showed a prototype of a wide viewing angle multi-domain LCD using photoalignment in an exhibition in 1996. However, this product is apparently not yet being produced in large numbers. Recently, photoalignment materials for production use and the related exposure equipment have appeared on the market. Table 3.5 summarizes the available photoalignment materials.

One material is "OptoAlign" developed by Elsicon in the USA. This was developed by Gibbons, who first reported on photoalignment using azobenzene mixed in polyimide [8]. The base material of "OptoAlign" is polyimide, and the processes for this material are the same as for any typical polyimide. This material can be used for the TN, STN, multi-domain, homeotropic alignment, and the alignment for in-plane switching (IPS) modes. The wavelength of the exposure light is 360 nm. Elsicon has also developed an exposure light source for their material. It can expose a $250 \times 350\,mm^2$ wide substrate, using a high pressure mercury lamp as the light source.

Another photoalignment approach is Linear Photo-Polymerization (LPP), developed by Rolic in Switzerland. This is a name for a complete technology using a photopolymerizable polymer. This was developed by Schadt, who first described photoalignment using photodimerization of polyvinylcinnamate [12]. The material for LPP can improve productivity, reduce baking time, and produce multi-domain products. However, Rolic sells this technology not only for the alignment of LCs, but also as a manufacturing technology for specialized function optical films using combinations with a liquid crystal polymer (LCP). They suggest that by using LPP and an LCP, optical films with new functions can be developed. For example, using this technology and these materials, optical retardation films that are optimized for a colour filter, for an interference filter, and for optical storage media can be developed [46].

Ushio Ltd. in Japan developed a light source for photoalignment exposure. Its basic configuration is the same as for photolithography. However, photoalignment requires a large exposure area and a polarization element. Ushio's equipment uses

Table 3.5 Photoalignment materials on the market

Manufacture	Name	Main material	Alignment	Wavelength of exposure (nm)	Coating process
Elsicon (USA)	OptoAlign	Polyimide	Parallel, vertical	360	Same as polyimide
Rolic (Switzerland)	LPP	Dimerization material	Parallel, vertical	320	Baking time is shorter than for polyimide

stacked fused quartz plates, which control the polarization direction by using the Brewster angle. This equipment offered an exposure area of $650 \times 800\,\mathrm{mm}^2$ wide in 1998. The wavelength can be adjusted by changing the interference filter. The extinction ratio is more than 10:1, and is constant for a wide range of wavelengths, from 230 nm to 400 nm. The extinction ratio can be adjusted by changing the number of stacked plates. The light source is an 8 kW high pressure mercury lamp, and the energy density is $10\,\mathrm{mW\,cm^{-2}}$.

3.3 Oblique Evaporation Method

Masanori Sakamoto

Practically, liquid crystal alignment is required on an active matrix like an amorphous Si-TFT array or on microstructures like colour filters. Therefore a surface contact treatment like the rubbing method tends to be dangerous because of substrate surface damage like electrostatic damage, chemical contamination and mechanical damage of the patterns on the substrate surface. Thus a non-contact alignment treatment has advantages compared with the conventional rubbing method. Oblique evaporation of SiO is one of the few methods in this category [1] in which device quality alignment is realized other than by rubbing.

3.3.1 Alignment mechanism

In the oblique evaporation process, a micro columnar structure is realized on the substrate surface, due to the self shadowing effect as shown in Fig. 3.3.1. When a nematic liquid crystal contacts such a surface, elastic deformation of the liquid crystal along the surface induces an interaction energy between the surface and the nematic material. This is thought to be the driving force for alignment of the nematic director.

The surface structure of the oblique evaporated film changes with the evaporation angle (the angle between evaporation beam and substrate normal). The structure of stepwise grooves and the array of columns change their angle and density.

The surface structure of the oblique evaporated film is illustrated schematically as in Fig. 3.3.2. There are many grooves and columns, the features of which vary with evaporation angle and change in the nematic alignment direction and pretilt angle.

The relation between the evaporation angle and the alignment direction is summarized as shown in the Fig. 3.3.3. A homogeneous alignment is realized by an evaporation angle less the 20°, but the in-plane direction of the alignment is not determined uniquely. At an evaporation angle of ca. 50°, homogeneous alignment with a 0° pretilt angle is realized due to the grooves perpendicular to the beam plane – the plane made by the substrate normal and the evaporating beam [2].

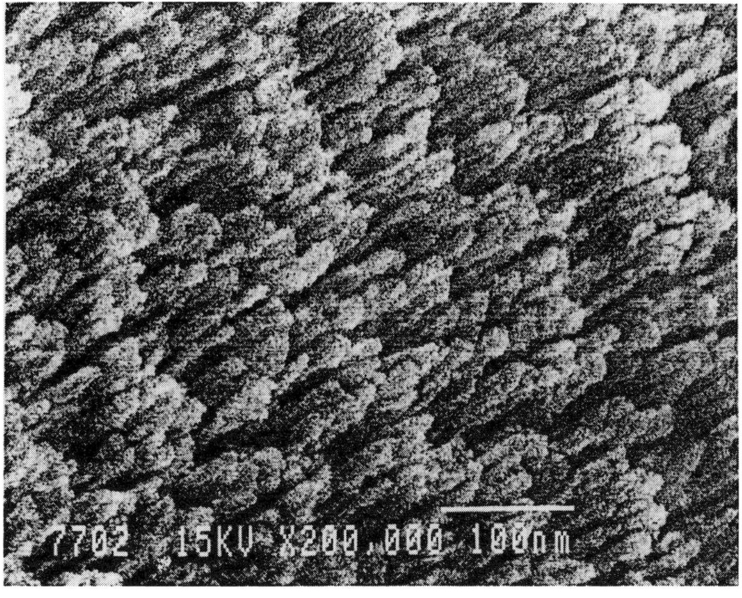

1000 nm

Fig. 3.3.1 SEM image for the columnar structure of an SiO oblique evaporated film surface.

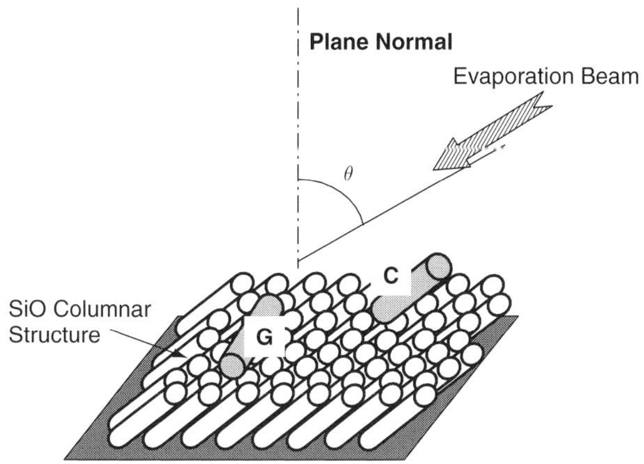

Fig. 3.3.2 Liquid crystal molecules lie in the position G (along the grooves) or in position C (along the columns) according to the evaporation angle θ as indicated.

Evaporation Angle (deg)	50	80	90
Liquid crystal alignment	Position G	Position C	
Tilt angle (deg)	0	ca. 5 to 50	

Fig. 3.3.3 Evaporation angle dependence of the liquid crystal alignment. Position G and C are illustrated in Fig. 3.3.2.

Over 80°, the self shadowing effect develops a columnar structures having the column axis aligned parallel to the beam plane with a certain angle between the substrate surface, and the tilted homogeneous nematic alignment appears. An alignment technology which realizes a tilt angle higher than 20° has not been established other than by the oblique evaporation technology and therefore it is useful to obtain high tilt angles at least on a laboratory scale, though there are several troubles for mass-production application.

3.3.2 Evaporation apparatus

The SiO, usually a dark brown powder, is used for the evaporation source because it is a very easily evaporated oxide using conventional resistive heating. In the case of SiO_2, high temperature heating by the electron beam, for example, is indispensable, and is not suitable for TFT array substrates due to the damage by recoiling electrons.

As the evaporation angle must be controlled accurately, the angle difference at both edges of the substrate must be less than several degrees. Thus the distance between the substrate and the linearly aligned evaporation boats must be rather large. Due to this geometrical requirement, evaporation chambers of sufficient size are necessary.

Moreover, in-plane uniformity of the film thickness has to be carefully controlled for device manufacturing which means taking extreme care over the evaporation control of the boats, together with the charging amount and the temperature. Thus there are a lot of problems to be solved for handling large sized substrates of almost 1 m square and the continuous process required for the mass production.

3.3.3 Material scientific view point

Using materials other than SiO attains nematic alignment by oblique evaporation, but a well grown columnar structure is essential. Generally the oblique evaporation film has a large surface area and is chemically active in gas adsorption for example. Detailed information about the chemical interaction between the surface and the liquid crystalline materials on it is not fully clarified at present, and therefore there might be many unknown factors involved in achieving the reliability of alignment required in device application.

3.4 Liquid Crystal Alignment on Microgroove Surfaces

Kohki Takatoh

3.4.1 Liquid crystal alignment on microgroove surfaces

It is well known that the main reason for liquid crystalline alignment on rubbed polymer surfaces is the anisotropic characteristics of the stretched polymer surfaces, although it is also due to microgroove structures [1].

The microgroove structure itself can cause liquid crystalline materials to align in the direction parallel to the groove structure as shown in Fig. 3.4.1. Several processes to produce surfaces with microgroove structures for liquid crystalline alignment have been proposed, for example: reactive ion-etching on glass surfaces with chromium masks [2, 3]; the photolithographic technique for photoreactive polymers using holographic exposure and reactive etching on the SiO_2 surface [4]; the pressing on of the microgroove surface fabricated by holographic light exposure; the development and metal evaporation of photoreactive material onto the epoxy resin layer [5, 6]; the exposure of UV light onto photocurable polymer film through masks with a grating pattern [7].

In all cases, it has been reported that microgrooves with a pitch smaller than 1 μm can realize liquid crystalline alignment giving a practical contrast. Figure 3.4.2 shows the temperature dependence of the liquid crystal order parameter on microgroove surfaces with different pitches.

Microgroove surfaces with a pitch smaller than 1 μm can realize the alignment better than those from the practical alignment method using SiO oblique evaporation. Not all reported methods can be used in LCD production lines. Two methods, the photolithographic method and the press method using a replica with a microgroove surface, should have practical potential. The photolithographic methods are especially convenient for LCD production, because they have already been used in TFT production lines.

3.4.2 Control of the pretilt direction by a "hybrid cell"

For LCD production, the direction of the liquid crystalline molecular inclination relative to the alignment surface, as well as the liquid crystalline direction on

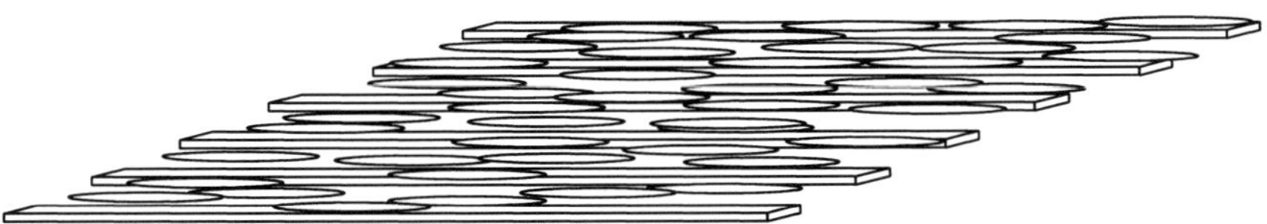

Fig. 3.4.1 Liquid crystalline alignment on microgroove structures.

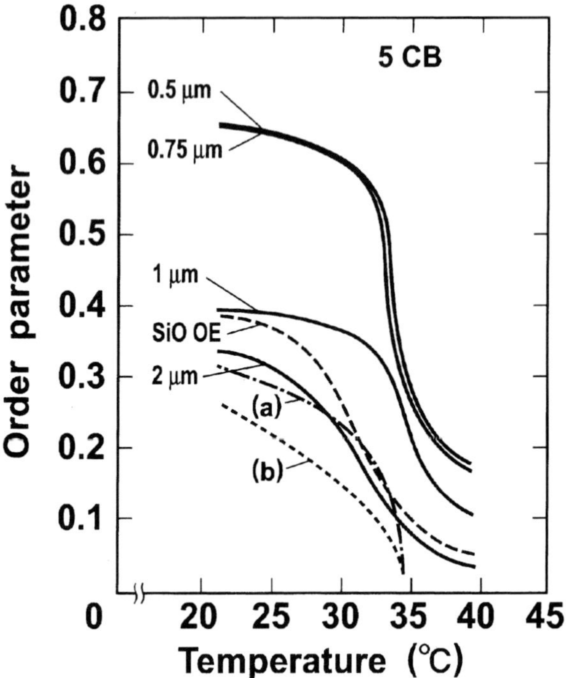

Fig. 3.4.2 The temperature dependence of the magnitude of the liquid crystal order parameter on grooved surfaces with various pitches [4]. Solid lines show the cases using grating surfaces with pitches of 0.5, 0.75, 1, and 2 μm. The broken line shows the case for the surface made by SiO oblique evaporation. The lines, (a) and (b), show the values in the bulk for the LC cells using grating surfaces with pitches of 0.5 and 2 μm, respectively.

the alignment surface must be controlled. In the case of the rubbing process, the relationship between the rubbing direction and the direction of inclination are shown in Fig. 3.4.3(b) [1].

On the other hand, in the case of alignment on microgroove surfaces, no factor exists to determine the directions of the molecular alignment or the pretilt angle direction. As a result, disclinations called "tilt reverse" are observed on applying a grey-level electric field to TN cells with a microgroove surface alignment layer. The existence of tilt reverse causes the degradation of the LCD contrast. In Fig. 3.4.3 (a), the configuration for the liquid crystalline molecules around a tilt reverse defect is shown. This disclination appears on the boundary between two areas in which the pretilt angle directions are opposite to each other.

To prevent tilt reverse disclination occurrence on microgroove surfaces, "hybrid type" cells with a rubbed alignment layer and a microgroove alignment layer have been proposed [7]. In Fig. 3.4.4(a), the liquid crystal configurations on

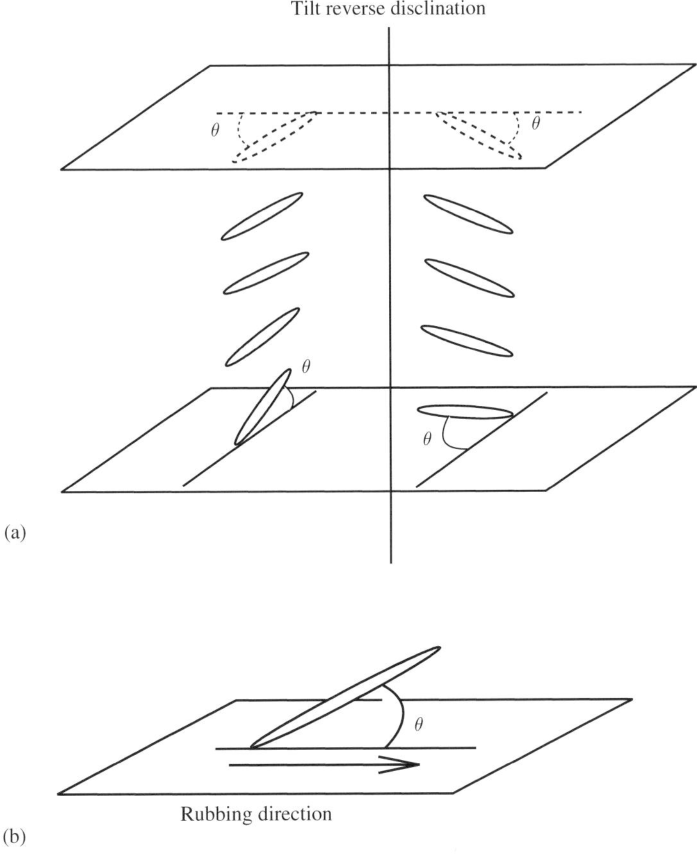

Fig. 3.4.3 Liquid crystalline alignment on alignment layer surfaces. (a) Two possible directions of molecular inclination on an alignment layer surface and a tilt reverse disclination. (b) The direction of molecular inclination on a rubbed alignment layer.

the left and right sides are twisted in the opposite direction. In this liquid crystalline material, either no or only a little chiral compound is included, so the twist direction is determined by the combination of the tilt directions on the upper and lower alignment layers.

Usually, in the case of TN and STN LCDs, small amounts of optically active compounds are added to prevent the coexistence of two kinds of twisted states. In Fig. 3.4.4(b), the left and right sides show two possible configurations in hybrid cells with liquid crystalline material containing an optically active compound. In both configurations, the liquid crystalline materials are twisted in the same sense, as determined by the optically active compound.

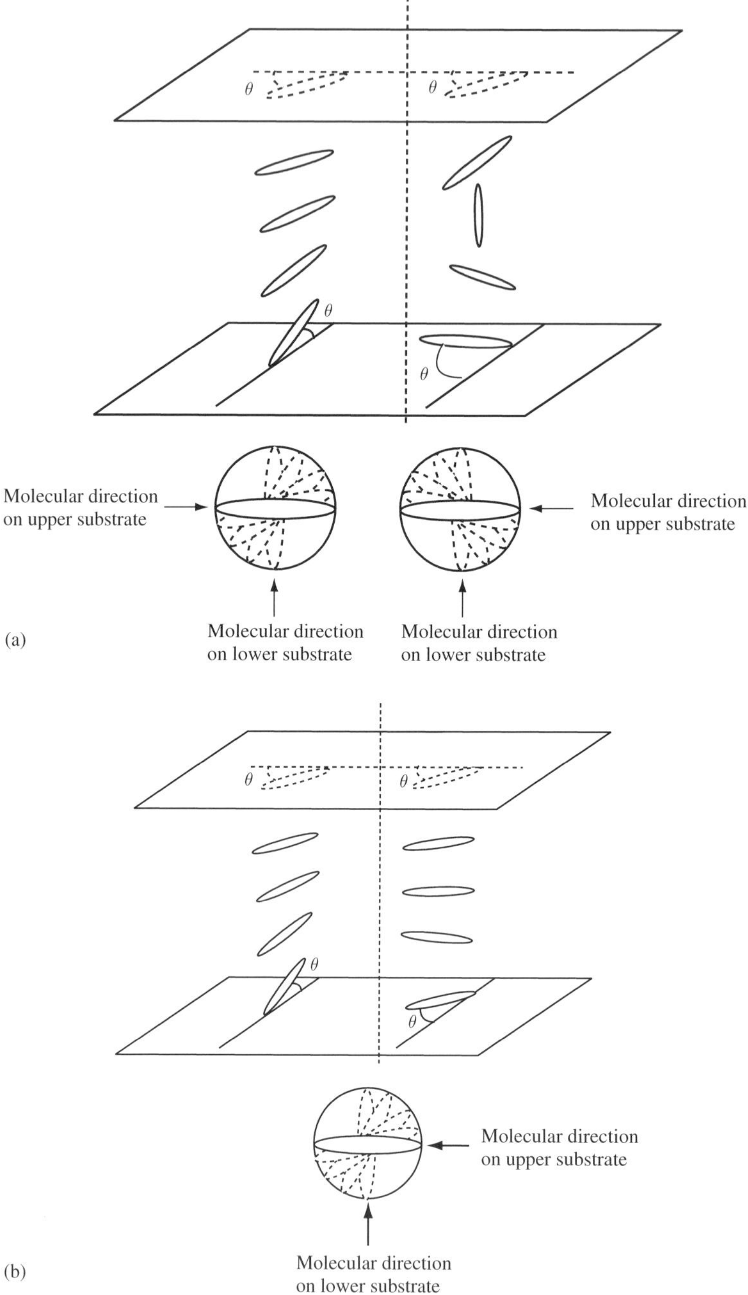

Fig. 3.4.4 (a) Hybrid cell structure and liquid crystalline configurations without a chiral compounds. (b) Liquid crystalline configuration in a hybrid cell with added chiral compound.

In Fig. 3.4.4(b), the configuration on the right is more unstable than that on the left because of the additional splay deformation. As a result, only the left configuration is formed. Through this mechanism, the pretilt angle direction is determined in one way even on the surface with microgrooves. "Hybrid cells" have the disadvantage that different procedures are necessary to produce two kinds of alignment layers, resulting in increased production costs. However, "hybrid cells" with a rubbed alignment layer on a colour filter and a microgroove alignment layer on an the TFT devices, have a practical significance in that the occurrence of static electricity causing deterioration in TFT devices can be eliminated. This "hybrid cell" has also been used for LCDs using an alignment layer produced by polarized UV light radiation [8].

3.4.3 Microgroove surface control of pretilt angle direction

Microgroove alignment layers which can control the pretilt angle direction (Fig. 3.4.5) have also been proposed [9].

To obtain this structure, the grating is fabricated by holographic interference and ion beam etching. Next, the microgroove structure is fabricated on it by pressing the replica with the microgrooved surface. Figure 3.4.6 shows the relationship between the inclination angle (θ_i) of the alignment layer and the pretilt angle (θ_p) of the liquid crystalline material. The larger the inclination angle, the larger the pretilt angle. The magnitudes of both angles have almost the same value.

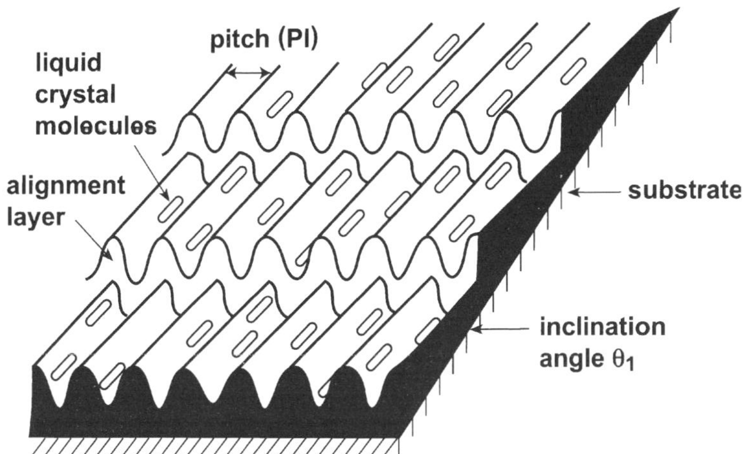

Fig. 3.4.5 The structure of a microgroove alignment layer to control pretilt angle direction.

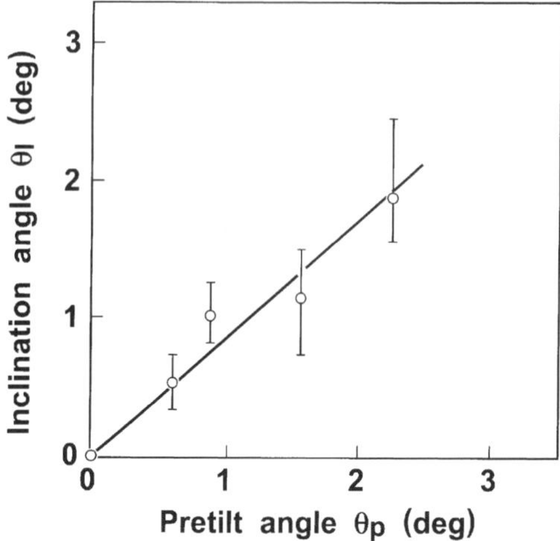

Fig. 3.4.6 Relationship between the inclination angle (θ_i) of the alignment layer and the pretilt angle (θ_p) of the liquid crystal.

3.5 LB Membranes for the Alignment Layer

Kohki Takatoh

3.5.1 LB membranes for the alignment layer

It is well known that surface active agents with hydrophobic groups such as long alkyl chains with hydroxy or carboxy groups form monomolecular films on the surface of water. By removing the formed film onto a substrate, the monomolecular film can exist on the substrate. Monomolecular layers can also be stacked to form multilayers with several molecular layers. The monomolecular film is named a Langmuir-Blodgett film after the names of the inventors, or usually LB film for short.

To move the monomolecular layers from water surface to substrates, two methods have been investigated. During the process of one method, the substrate is moved up or down perpendicular to the water surface to move the monomolecular film from the water surface to the substrate (method A), as shown in Fig. 3.5.1. In another method, the substrate surface is held parallel to the water surface and touches onto the monomolecular film parallel to the water surface so shifting the film to the substrate (method B).

Substrate is moved up and down perpendicular to the water surface to move the monomolecular film from the water surface to the substrate (method A).

Also, there are many categories for method A. In the method shown in Fig. 3.5.1, the LB film is transferred onto the substrate only when the substrate is lifted upwards. During this process, pressure is added to the edge of the monomolecular film on the

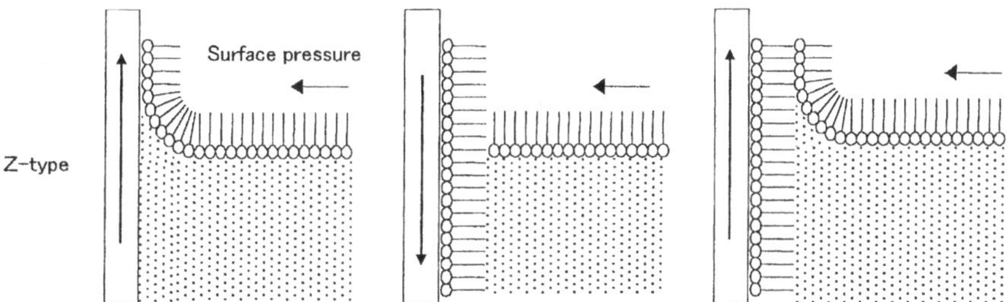

Fig. 3.5.1 Formation of a LB film.

water surface. LB films formed by this method are called Z type. On the other hand, LB films formed only when a substrate is moved downwards are called "X-type". LB films can also be formed by transferring the monomolecular film to a substrate when the substrate is moved both upwards and downwards. This type of LB film is called "Y-type".

It is well known that liquid crystalline materials align in the extension direction on an extended polymer film surface. The polymer LB film formed by moving the substrates perpendicularly is considered to be a uniaxially extended polymer film, and therefore, liquid crystal alignment on the surface of such a polymer LB film can be expected.

3.5.2 Polyimide LB alignment film

For practical alignment layers for LCDs, only polyimides have been used because of their efficiency in the rubbing treatment and their thermal and chemical resistance. Polyimides have also been investigated for LB film alignment layers. The polyimide molecular structures and syntheses investigated for such LB alignment films are shown in Fig. 3.5.2 [10, 11]. To form polyimide LB films, the polyamic acids are changed into ammonium salts or esters with alkyl chains of polyamic acids, as the precursor of the polyimide. Next, LB film formation is carried out by the usual method for surface active agents. The obtained precursors of LB films are converted into polyimides by heat or by treatment under acidic conditions. The acid treatment is carried out by immersing the substrates into a solution of acetic anhydride, pyridine and benzene (1:1:3).

3.5.3 Structure of polyimide LB films and liquid crystalline alignment on the film

By investigations using IR and UV spectroscopy, it has been confirmed that the polyimide molecular configurations in LB films are uniaxially aligned in the direction parallel to the substrate displacement during the LB film formation. Although the degree of alignment for polyimide LB films does not depend on the

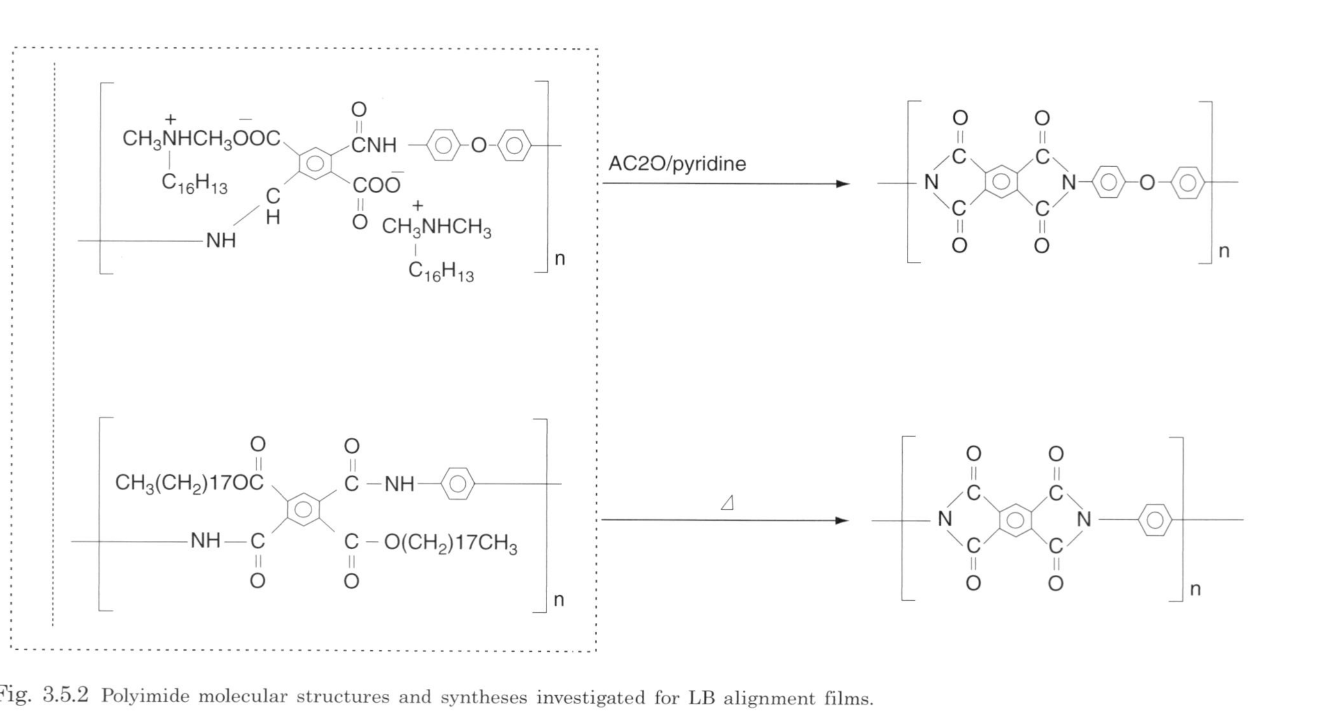

Fig. 3.5.2 Polyimide molecular structures and syntheses investigated for LB alignment films.

substrate displacement speed, it increases with the surface pressure of monomolecular film by which the monomolecular film is compressed on the water surface. For the mechanism of polyimide LB film alignment, two possibilities can be considered. One is that the polyimide materials align during the shift from the water surface onto the substrate. Another possibility is that the monomolecular films on the water surface already have the orientational order. However, an LB film formed by shifting the monomolecular film onto the substrate parallel to the water surface (method B) does not show any orientational order. Therefore, it is concluded that the orientational order of polyimide LB films can only be realized during the process in which the monomolecular film on the water surface is moved onto the substrate by moving the substrate perpendicularly, as in method A. By evaluation using the dichroic ratio of guest-host liquid crystalline materials, it was found that in the LB film, the liquid crystalline materials align in the direction parallel to the shift of the substrate. The liquid crystalline alignment property depends also on the number of LB layer stackings. The best quality alignment was observed in the case of a 5-layer film. In this case, the dichroic ratio of the liquid crystal is equal to that on the alignment layer produced by the rubbing process. Generally, the degree of alignment on an LB film is considered to be comparable to that on alignment films produced by weak rubbing [12, 13].

Table 3.5.1 shows the relationship between the chemical structures of polyimide materials, the degree of alignment for polyimide LB films evaluated by UV and IR spectroscopy, and the contrast estimated from TN-LCDs.

Concerning the contrast of TN-LCDs, in one state (light state), the directions of the polarization axes are perpendicular to each other. In the other state (dark state), the two polarization directions are parallel. In each case, at least one alignment direction is parallel to one polarizer axis. The higher the degree of liquid crystal alignment on the alignment layer, the greater the contrast becomes. The polyimide materials of rigid and highly linear structure show the highest degree of alignment.

It is reported that the memory properties of SSFLC are improved by the usage of polyimide LB film, because an extremely thin layer is realized, and the resultant electrical resistance is low [12, 14].

3.6 PTFE Drawn Films for Alignment Layers

Kohki Takatoh

Wittmann and Smith [15] reported that thin PTFE layer deposition is achieved by rubbing a bar of PTFE on a smooth surface and that this induces a well oriented chain axis of PTFE macromolecules. (See Fig. 3.6.1)

The chain axes align parallel to the substrate surface and along the sliding direction. The PTFE alignment was confirmed by electron diffraction studies, showing that the orientation and crystalline perfection of the layer were high. It

Table 3.5.1 The relationship between the chemical structures of polyimide materials, the degree of alignment for polyimide LB films evaluated by UV and IR spectroscopy and the contrast evaluated from TN-LCDs [13]

Molecular structures of PI LB films	UV dichroic ratio	IR dichroic ratio	TN liquid crystal cell	
			Observation by a polarizing microscope	Degree of alignment
PI-1	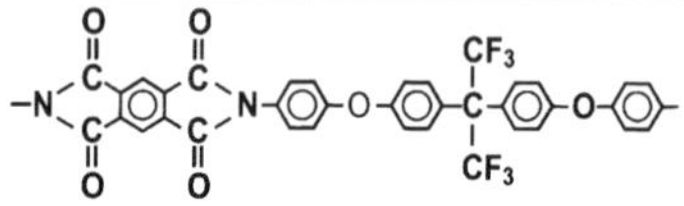 1.00	0.97	randomly aligned	0.5
PI-2	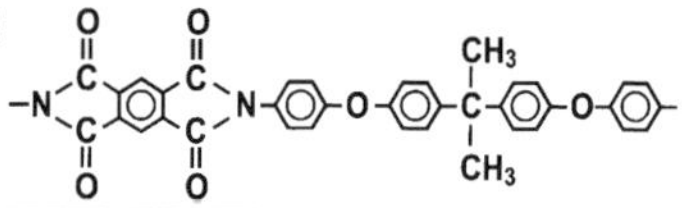 1.08	1.04	randomly aligned	0.6
PI-3	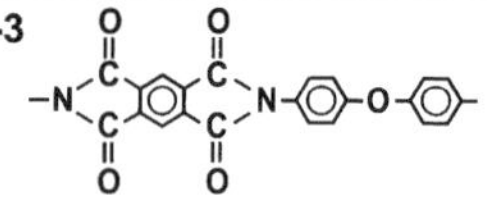 1.11	0.93	randomly aligned	0.8
PI-4	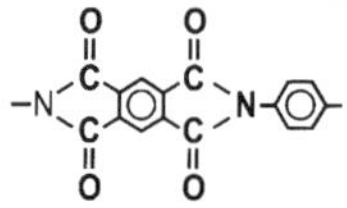 1.45	0.68	well aligned	60
PI-5	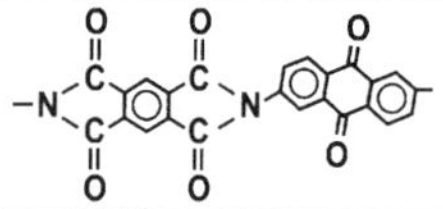 2.36	0.41	well aligned	20

can be compared in fact to that found in extended chain whiskers of PTFE directly formed during polymer synthesis or in high-performance, gel-processed, ultra-high molecular mass polyethylene. On the resultant surface, a 5 Å periodically grooved order perpendicular to the molecular chain could be observed. By deposition of organic and inorganic materials from the vapour, the molten state or the solution, highly ordered crystals are formed on the PTFE layers. Orientated polymers are also obtained by the polymerization of monomers on the PTFE oriented layers, as exemplified by poly-*p*-xylene. This improved orientational capacity and versatility originates in a specific surface topography combined with a favourable crystal structure for each crystalline material. Liquid crystal alignment was also realized on the PTFE oriented layers, and practical application has been investigated. The method requires only a simple, clean process without using a cloth. In the usual rubbing process, multiple stages, such as the coating of polyimide or polyamide solution, the baking of formed polyimide or polyamide layers, and a rubbing process are necessary. Another advantage of the PTFE method is that it can realize very thin films which can lower the threshold and saturation voltage in spite of the low dielectric constant, because the capacitance

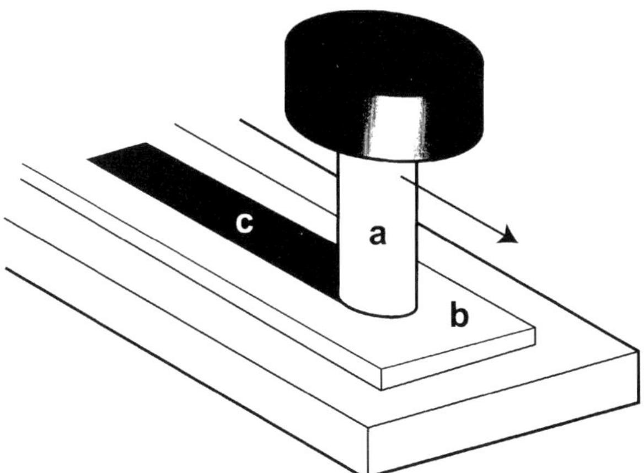

Fig. 3.6.1 Thin PTFE layer deposition on a smooth surface: PTFE bar (a), substrate (b), aligned macromolecules of PTFE (c).

of the layer is inversely proportional to the layer thickness. The layer thickness is reported to be 20–1000 Å. It was reported that Lester *et al.* [16] constructed a practical machine for this process. Practical PTFE film can be produced at velocities of 1 mm; therefore, it is estimated that an A4 panel can be treated in 30 sec. The temperature of the substrate affects the resultant alignment condition, especially the tilt angle. In the case of deposition at 125 °C, the surface obtained was smooth and no pretilt angle could be observed. On the other hand, deposition at 75 °C caused the alignment layer to have a pretilt angle of ca. 1°. The difference between the two alignment layers can be explained as follows: at high temperatures, PTFE has a very low coefficient of friction and a thin layer of highly oriented chains is deposited. At lower temperatures, PTFE is deposited in irregular packets of hundreds of angstroms. The observed pretilt angle on the alignment layers formed at lower temperature is thought to be due to the rough surface structures.

3.7 Liquid Crystalline Alignment on Chemically Treated Surfaces

Kohki Takatoh

Surface active agents are often used in industry to form homogeneous polymer layers or to increase adhesion for paints and adhesives. In the case of liquid crystalline materials, surface active agents like silicone compounds are used to obtain homeotropic alignment.

3.7.1 Addition of surface active agents into liquid crystalline materials

In the case of polymer materials, surface active agents are used both for surface treatment of the substrates and as additives to the polymer materials. When using surface active agents for liquid crystal materials, two methods are again reported to control the liquid crystal alignment-treatment of the surface of substrates with surface active agents and addition of surface active agents to the liquid crystalline materials. In the latter case, a solution of surface active agents in an appropriate solvent is added to the liquid crystalline materials and then the solvent is removed under reduced pressure. In Table 3.7.1, the relationship between the surface active agent concentration and the liquid crystal alignment are shown when the surface active agent, cetylammonium bromide is added to a nematic liquid crystalline material [1]. Low concentrations of the surface active agent result in homogeneous alignment, and high concentrations result in homeotropic alignment.

In Table 3.7.1, an ionic surface active agent is used. However, when ionic materials are mixed into liquid crystalline materials, several LCD driving problems are known to occur. Saturated voltage increases because of the decrease in the electrical resistance of the liquid crystalline layer. During driving of the LCDs, the absorption of ionic materials onto the substrate surface occurs depending on the magnitude of the signal voltage. The absorption increase or decrease of the threshold voltage compared to that of the initial state causes the defect known as the, "sticking phenomenon". In this phenomenon, the signal information remains after the signal on the liquid crystal disappears. In the injection process of liquid crystalline materials, ionic compounds are also absorbed onto the surface near the injection point, and this causes an inhomogeneous threshold voltage. Although the application of ionic surfactants in the dynamic scattering (DS) mode was examined, potential use is limited for the above reasons.

3.7.2 Treatment of the substrate surface by active surface agents

To obtain homeotropic alignment by surface active agent treatment of the substrate surface, silicon compounds are usually used as the surface active agent. Silicon compounds, such as alkoxysilanes or chlorosilanes, react with the substrate surface following Scheme 3.7.1 and polymerize to form polysiloxane structures near the surface.

Table 3.7.1 Alignment and conductivity (σ) of the nematic mixture TN103, with added CTAB (cetyltrimethylammonium bromide)

%CTAB	0	0.01	0.1	0.25	0.5	1
σ	1.8×10^{-10}	2×10^{-10}	10^{-9}	1.7×10^{-9}	2.9×10^{-9}	7.4×10^{-8}
LC alignment	‖	‖	‖	—	⊥	⫫

J. Cognard, Mol. Cryst. Liq. Cryst. Suppl. Ser. (1), 1(1982)

$$R-Si(OCH_3)_3 \xrightarrow[\text{-3CH}_3\text{OH, hydrolysis}]{+3H_2O} R-Si(OH)_3$$

adsorption

dehydration
condensation polymerization

scheme 3.7.1

In practice, the surface treatments are carried out by (1) dipping the substrates into a 1–5% aqueous or organic solution of the silane compounds, (2) exposing the substrate surface to the vapour of pure silane compounds or of a toluene solution (3) spreading silane compounds on the substrate surface.

In the case of polymer materials, such as paints or adhesives, the wetting properties can be anticipated by using the critical surface tension γ_c. Let the surface tension of the polymer material be γ_l.

If $\gamma_l > \gamma_c$, it is difficult for the surface to become wetted.

If $\gamma_l < \gamma_c$, the surface can be easily wetted.

Thus, the critical surface tension, γ_c, can be used to measure the ease of wetting of the solid surface.

When homologous series of liquids like alcohols or alkyl halides give a contact angle θ on some solid surface and surface tension is γ, $\cos\theta$ can be expressed as

$$\cos\theta = 1 - b(\gamma - \gamma_c)$$

where b is a constant, usually 0.03–0.04. The surface tension of the liquid for which $\cos\theta$ is equal to 1, or θ is equal to 0, is the critical surface tension γ_c. Liquid crystalline materials also give a similar relationship, and, the critical surface tension, γ_c, is used as the criterion for homeotropic or homogeneous alignment.

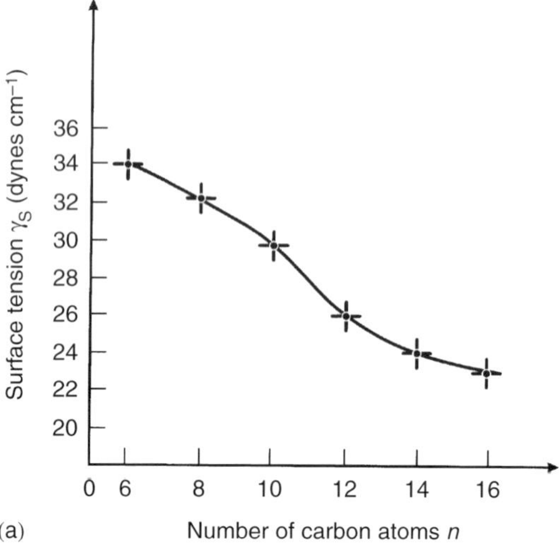

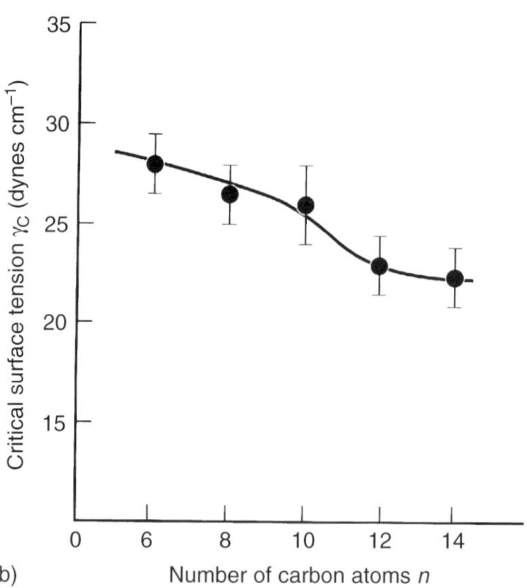

Fig. 3.7.1 Variation of (a) surface tension γ_s and (b) critical surface tension γ_c of surfactant treated glass surfaces with increasing carbon content of the surfactant alkyl chain. (a) Surface tension γ_s for substrates with n-alkyl monoamine layers [6]. (b) Critical surface tension for substrates with n-alkyltrimethylammonium bromide layers [7].

Where γ_{lc} is the surface tension of a liquid crystalline material,

if $\gamma_{lc} > \gamma_c$, the liquid crystalline material shows homeotropic alignment
if $\gamma_{lc} < \gamma_c$, the liquid crystalline material shows homogeneous alignment.

In another expression, by using the solid surface tension, γ_s,

if $\gamma_{lc} > \gamma_s$, liquid crystalline materials show homeotropic alignment
if $\gamma_{lc} < \gamma_s$, liquid crystalline materials shows homogeneous alignment.

These experimental rules are called the "Friedel-Creagh-Kmetz (FCK)" rule. In most cases, this rule can be confirmed, with some exceptions.

The strong relationship between the alkyl chain length of surface active agents and the surface tension and critical surface tension of the treated surface can be observed. In Fig. 3.7.1a,b, the relationship between the chain length of the surface active agent and the magnitude of (a) γ_s and (b) γ_c of the treated surface is shown. Generally speaking, a surface treated by a surface active agent with short alkyl chain shows a large surface tension value and vice versa. This suggests that to obtain homeotropic alignment, surface active agents with long alkyl chains should be applied. Usually, *N*, *N*-dimethyl-*N*-octyl-*N*-3-trimethoxysilylpropyl-ammoniumchloride (DMOAP) is generally selected to obtain homeotropic alignment.

Polynuclear complex salts of chromium, aluminum, iron and cobalt are reported to have been used to form the alignment layers for liquid crystalline materials [2–3]. In Fig. 3.7.2 are the structures of layers on glass surfaces treated by carboxylatochromium complexes of (a) an alkanoic acid and (b) an alkandioic acid. Glass surfaces treated by n-alkanoic acids with more than 3 carbon atoms induce homeotropic liquid crystalline alignment as indicated in Fig. 3.7.2(a). On the other hand, Fig. 3.7.2(b) shows the structure of glass surfaces treated by carboxy-latochromium complexes of an alkandioic acid. These surfaces show homogeneous

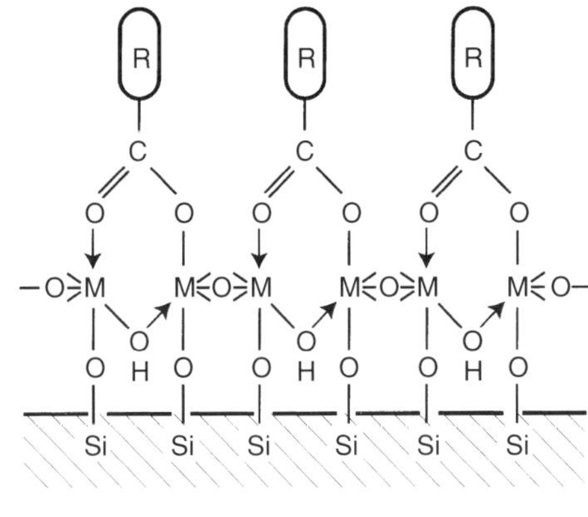

Fig. 3.7.2 (Continued)

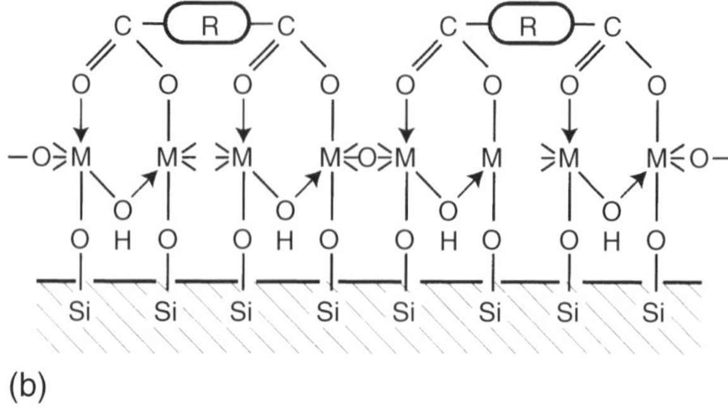

Fig. 3.7.2 Schematic representation of polymerized layers of carboxylatochromium complexes chemisorbed on glass surfaces: (a) monocarboxylatochromium complex; (b) dicarboxylatochromium complex.

liquid crystalline alignment. In the above cases, the alkyl group conformation of the carboxylic acid determines the alignment of the liquid crystalline materials.

An application using homeotropic alignment by surface active agents and involving treatment of the alignment layer surface produced by SiO oblique

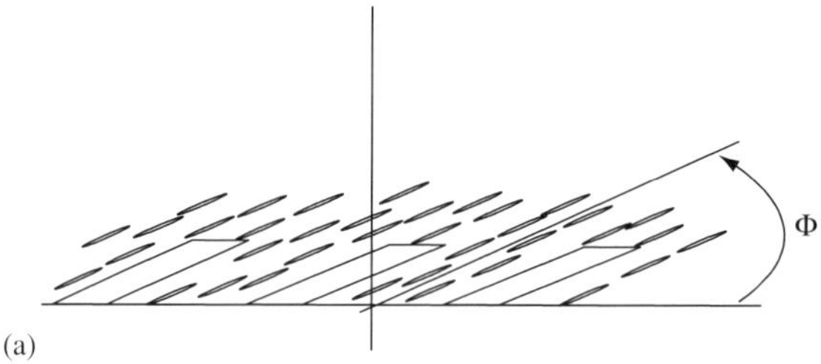

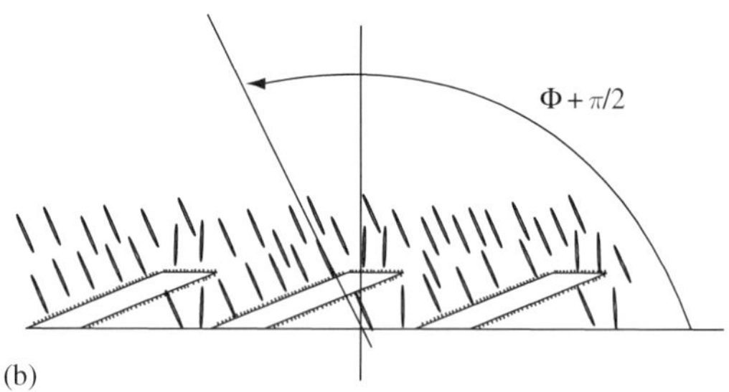

Fig. 3.7.3 Liquid crystal molecular orientation on obliquely evaporated SiO surfaces (a) without treatment by surface active agent, (b) with treatment by surface active agent.

evaporation has been reported. When the pretilt angle on the alignment layer produced by SiO oblique evaporation is Φ, the pretilt angle, $\Phi + \pi/2$, can be obtained by this method (Fig. 3.7.3). Generally, it is difficult to obtain a high pretilt angle by the rubbing process on polyimide polymer layers, and this method has some potential for application in practical processes to realize a specific range of high pretilt angles [4, 5].

The method of obtaining homeotropic alignment by processing the substrate surface with a surface active agent is well used as a basic laboratory procedure. Trials continue an adding special functions into the alignment layers by using functional groups instead of alkyl groups in active surface agents. By introducing the azo group whose structure can be reversibly changed by light, homogeneous alignment or homeotropic alignment can be changed by irradiation with two different wavelengths of light.

References

3.1–3.2 Introduction

[1] C. Mauguin: Bull. Soc. Fr. Min. 34 (1911) 71.

[2] J. Janning: Appl. Phys. Lett. 21 (1972) 173.

[3] D. Flanders, D. Shaver, and H. Smith: Appl. Phys. Lett. 32 (1978) 15.

[4] H. Aoyama, Y. Yamazaki, N. Matsuura, H. Mada, and S. Kobayashi: Mol. Cryst. Liq. Cryst. 72 (1981) 127.

[5] N. Kishida and S. Kikui: Appl. Phys. Lett. 40 (1982) 541.

[6] K. Ishikawa, K. Hashimoto, H. Takezoe, A. Fukuda, and E. Kuze: Jpn. J. Appl. Phys. 23 (1984) L211.

[7] K. Ichimura, Y. Suzuki, T. Seki, A. Hosoki, and K. Aoki: Langmuir 4 (1988) 1214.

[8] W. Gibbons, P. Shannon, S. Sun, and B. Swetlin: Nature 351 (1991) 49.

[9] K. Hiroshima, T. Maeda, and T. Furihata: Proc. 12th IDRC (Japan Display '92) (1992) 831.

[10] Y. Kawanishi, T. Takimiya, and K. Ichimura: Polym. Mater. Sci. Eng. 66 (1992) 263.

[11] Y. Kawanishi, T. Tamaki, M. Sakuragi, T. Seki, Y. Suzuki, and K. Ichimura: Langmuir 8 (1992) 2601.

[12] M. Schadt, K. Schmitt, V. Hozinkov, and V. Chigrinov: Jpn. J. Appl. Phys. 31 (1992) 2155.

[13] S. Hachiya, K. Tomoike, K. Yuasa, S. Togawa, T. Sekiya, K. Takahashi, and K. Kawasaki: J. Soc. Inf. Disp. 1 (1993) 295.

[14] Y. Tabe and H. Yokoyama: J. Phys. Soc. Jpn. 63 (1993) 2472.

[15] M. Hasegawa and Y. Taira: Proc. 14th IDRC (1994) 213.

[16] K. Kawahara, Y. Nakajima, T. Udagawa, H. Fuji, and H. Morimoto: Proc. 14th IDRC (1994) 180.

[17] M. Oh-e, M. Ohta, K. Kondo, and S. Oh-hara: 15th ILCC (1994) K-P17.

[18] M. Hasegawa and Y. Taira: J. Photopolymer Science and Technology 8 (1995) 241.

[19] T. Hashimoto, T. Sugiyama, K. Katoh, T. Saitoh, H. Suzuki, Y. Iimura, and S. Kobayashi: SID '95 Digest (1995) 877.

[20] Y. Iimura, T. Saitoh, S. Kobayashi, and T. Hashimoto: J. Photopolymer Sci. and Tech. 8 (1995) 257.

[21] M. Oh-e, M. Ohta, S. Aratani, and K. Kondo: Proc. 15th IDRC (Asia Display '95) (1995) 577.

[22] J. West, Z. Wang, Y. Ji, and J. Kelly: SID '95 Digest (1995) 703.

[23] N. Yamada, S. Kohzaki, F. Fukuda, and K. Awane: SID '95 Digest (1995) 575.

[24] E. Endo, T. Shinozaki, H. Fukuro, Y. Iimura, and S. Kobayashi: Proc. AM-LCD '96 (1996) 341.

[25] M. Hasegawa, H. Takano, A. Takenaka, Y. Momoi, K. Nakayama, and A. Lien: SID '96 Digest (1996) 666.

[26] K. Lee, A. Lien, J. Stathis, and S. Paek: SID '96 Digest (1996) 638.

[27] H. Mori, Y. Itoh, Y. Nishiura, T. Nakamura, and Y. Shinagawa: AM-LCD '96 (1996) 189.

[28] M. Schadt: Nature 381 (1996) 212.

[29] Y. Toko, B. Y. Zhang, T. Sugiyama, K. Katoh, and T. Akahane: 16th ILCC Abstracts (1996) B1P53.

[30] T. Yamamoto, M. Hasegawa, and H. Hatoh: SID '96 Digest (1996) 642.

[31] H. Furue, Y. Iimura, H. Hasebe, H. Takatsu, and S. Kobayashi: IDW '97 (1997) 73.

[32] Y.-K. Jang, H.-S. Yu, J. K. Song, B. H. Chae, and K.-Y. Han: SID '97 Digest (1997) 703.

[33] M. S. Nam, J. W. Wu, Y. J. Choi, K. H. Yoon, J. H. Jung, J. Y. Kim, K. J. Kim, J. H. Kim, and S. B. Kwon: SID '97 Digest (1997) 933.

[34] Y. Toko, B. Y. Zhang, T. Sugiyama, K. Katoh, and T. Akahane: Mol. Cryst. Liq. Cryst. 304 (1997) 1985.

[35] T. Yoshida, T. Tanaka, J. Ogura, H. Wakai, and H. Aoki: SID '97 Digest (1997) 841.

[36] P. Chaudhari, J. A. Lacey, S.-C. A. Lien, and J. Speidell: Jpn. J. Appl. Phys. 37 (1998) L55.

[37] E. Hoffmann, H. Klausmann, E. Ginter, P. M. Knoll, H. Seiberle, and M. Schadt: SID '98 Digest (1998) 734.

[38] K. Ichimura, S. Morino, and H. Akiyama: Appl. Phys. Lett. 73 (1998) 921.

[39] J-H. Kim and S. Kumar: Phys. Rev. E 57 (1998) 5644.

[40] J-H. Kim, S.-D. Lee, and S. Kumar: ILCC '98 (1998) 110.

[41] Y. Makita, T. Natsui, S. Kimura, S. Nakata, M. Kimura, Y. Matsuki, and Y. Takeuchi: SID '98 Digest (1998) 750.

[42] Y. Makita *et al.*: J. Photopolymer Sci. Tech. 11 (1998) 187.

[43] M. Nishikawa, B. Taheri, and J. L. West: SID '98 Digest (1998) 131.

[44] D. Shenoy, K. Gruenenberg, J. Naciri, M.-S. Chen, and R. Shashidhar: SID '98 Digest (1998) 730.

[45] N. Matsuie, H. Oji, E. Ito, H. Ishii, Y. Ouchi, K. Seki, M. Hasegawa, and M. Zharnikov: 18th ILCC Abstract (2000) 182 (24D-65-P).

[46] M. Schadt: Euro Display '99 (1999) 27.

[47] Y. Nakagawa *et al.*: SID '01 Digest (2001) 1346.

3.3 Oblique Evaporation Method

[1] J. L. Janning: Appl. Phys. Lett. 21, 173 (1972).

[2] L. A. Goodman, J. T. Mcginn, C. H. Anderson, and F. Digernimo: IEEE, ED-24, 795 (1977).

3.4–3.6 Liquid Crystal Alignment on Microgroove Surfaces–PTFE Drawn Films for Alignment Layers

[1] J. M. Geary, J. W. Goodby, A. R. Kmetz, and J. S. Patel: J. Appl. Phys. 62 (10) 4100 (1987).

[2] D. C. Flanders, D. C. Shaver, and H. I. Smith: Appl. Phys. Lett. 32 (10) 579 (1978).

[3] H. V. Kaenel, J. D. Lister, J. Melngailis, and H. I. Smith: Phys. Rev. A24 (5) 2713 (1981).

[4] A. Sugimura, N. Yamamoto, and T. Kawamura: Jpn. J. Appl. Phys. 20 (7) 1343 (1981).

[5] K.Toda, N.Watanabe,T.Takemoto, andT. Nakamura: SharpTech. Rep. 39,68 (1988).

[6] E. S. Lee, T. Uchida, M. Kano, M. Abe, and K. Sugawara: Proceedings of Japan Display' 92, 880 (1992).

[7] Y. Kawata, K. Takatoh, M. Hasegawa, and M. Sakamoto: Liq. Cryst. 16 (6) 1027 (1994).

[8] M. Schadt, K. Schmitt, V. Kozinkov, and V. Chigrinov: Jpn. J. Appl. Phys. 31, 1(7) 2155 (1992).

[9] K. Sugawara, Y. Ishitaka, M. Abe, K. Seino, M. Kano, and T. Nakamura: Proceedings of Japan Display '92, 815 (1992).

[10] M. Suzuki, M. Kakimoto, T. Konishi, Y. Imai, M. Iwamoto, and T. Hino: Chem. Lett. 395 (1986).

[11] M. Kakimoto, M. Suzuki, T. Konishi, Y. Imai, M. Iwamoto, and T. Hino: Chem. Lett. 823 (1986).

[12] H. Ikeno, A. Oh-saki, N. Ozaki, M. Nitta, K. Nakaya, and S. Kobayashi: SID 88 Digest 45 (1988).

[13] M. Murata, H. Awaji, M. Isurugi, M. Uekita, and Y. Tawada: Jpn. J. Appl. Phys. 31, L189 (1992).

[14] H. Ikeno, A. Oh-saki, M. Nitta, N. Ozaki, Y. Yokoyama, K. Nakaya, and S. Kobayashi: Jpn. J. Appl. Phys. 27 (4) L475 (1988).

[15] J. C. Wittmann and P. Smith: Nature, 352, 415 (1991).

[16] G. Lester, J. Hanmer, and H. Coles: Mol. Cryst. Liq. Cryst. 262, 149 (1995).

3.7 Liquid Crystalline Alignment on Chemically Treated Surfaces

[1] J. Cognard: Mol. Cryst. Liq. Cryst. Suppl. Ser. (1) 1(1982).

[2] S. Matsumoto, M. Kawamoto, and N. Kaneko: Appl. Phys. Lett. 27 (5) 268 (1975).

[3] S. Matsumoto: Chemistry and Chemical Industry 32 (10) 753 (1979) (in Japanese).

[4] W. R. Heffner, D.W. Berreman, M. Sammon, and S. Meiboon: Appl. Phys. Lett. 36, 144 (1980).

[5] T. Uchida, M. Ohgawara, and M. Wada: J. J. Appl. Phys, 19, 2127, (1980).

[6] G. Porte and J. de Physique 37, 1245 (1976).

[7] S. Naemura and J. de Physique, Colloque C3, 40, C3-514 (1979).

Chapter 4

Applications of Nematic Liquid Crystals

4.1 Summary of Molecular Alignment and Device Applications

Ray Hasegawa

4.1.1 Molecular alignment in nematic phases

In LCD devices, liquid crystal materials are usually sandwiched between two glass substrates carrying alignment film with a gap of 1–10 μm. By the influence of the alignment film on the substrates, liquid crystal molecular orientations are determined. Typical orientations are shown in Fig. 4.1.1.

These orientations are classified into two groups. The directors of the liquid crystal molecules in the Homogeneous, Tilted and Homeotropic cases are aligned in one fixed direction, while, the director of the liquid crystal molecules in the Splay, Twist, Bend, Hybrid and Super-twisted nematic cases are not fixed in one direction. In the latter orientations, the liquid crystals are under stress.

The liquid crystal molecules align parallel to the substrates in the Homogeneous and perpendicular to the substrates in the Homeotropic alignments. Tilted alignment is an intermediate state between Homogeneous and Homeotropic, and the molecules are tilted at a certain pretilt angle. Actually, the Tilted alignment with low pretilt angle (less than about 10°) is often called Homogeneous. In the Hybrid alignment, the liquid crystals on one substrate have Homogeneous alignment and those on the other substrate Homeotropic alignment.

In order to obtain Homogeneous alignment (including the Tilted alignment with less than about 10° pretilt angle), rubbed polyimide films are usually used in mass production. Non-rubbing methods such as photoalignment and those using microgrooves have been developed (see Section 3).

Homeotopic alignment is obtained by substrates coated with hydrophobic films such as silane compounds. Recently, polyimide films with hydrophobic side chains

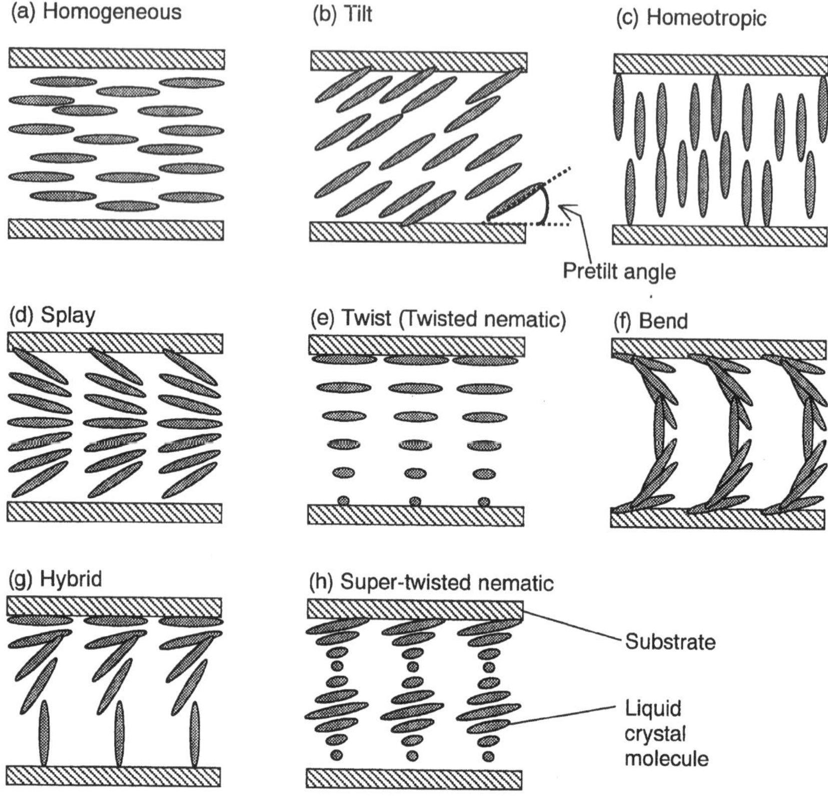

Fig. 4.1.1 Typical orientations of nematic liquid crystals.

have been developed and used in the multi-vertical aligned nematic (MVA) mode (see Section 4.5). The reason why hydrophobic surfaces give Homeotropic alignment is explained by the FCK model [1] (see Section 3.7). The SiO evaporation method (see Section 3) can give any of the Homogeneous, Homeotropic and Tilted alignments by changing the evaporation angle.

The liquid crystal molecules near the substrate surface affect the liquid crystal orientation in the bulk (in the middle of a liquid crystal cell) by molecular–molecular interactions, and the orientation in the entire liquid crystal cell is so determined. In most cases, the interaction between the substrate surface (alignment layer) and the liquid crystal molecules near the surface is very strong. By the application of electric or magnetic fields, the alignment of the liquid crystals near the substrate is hardly changed. Such a relationship between the substrate surface and the liquid crystals is called "strong anchoring". In the case of the strong anchoring, after removing the electric or magnetic field, the liquid crystal orientation recovers to that of the initial orientation. In contrast, when the liquid crystal orientation after removing the field is different from the initial orientation, the relationship is called "weak anchoring".

Table 4.1.1 Typical display modes of nematic liquid crystals

Display modes		Principle	Numbers of polarizers	Dielectric anisotropy	Molecular orientation (Initial → Applied electric field)
Twisted nematic (TN) Multi domain TN		Optical rotation	2	$\Delta\varepsilon > 0$	Twist (= 90 °) → Perpendicular to substrate surface
Super-twisted nematic (STN)		Retardation	2	$\Delta\varepsilon > 0$	Twist (> 90 °) → Perpendicular to substrate surface
In-plane switching (IPS)		Retardation	2	Usually $\Delta\varepsilon > 0$	Homogeneous → Homogeneous
Electrically controlled birefringence (ECB) [4, 5]	Homogeneous type	Retardation	2	$\Delta\varepsilon > 0$	Homogeneous → Perpendicular to substrate surface
	VA, MVA		2	$\Delta\varepsilon < 0$	Homeotropic → Parallel to substrate surface
	HAN [6]		2	$\Delta\varepsilon > 0$	Hybrid → Perpendicular to substrate surface
PI cell, Optically compensated bend (OCB)		Retardation	2	$\Delta\varepsilon > 0$	Bend → Bend
Polymer dispersed liquid crystal (PDLC)		Light scattering	0	$\Delta\varepsilon > 0$ or $\Delta\varepsilon < 0$	Random → Perpendicular/Parallel to substrate surface
Guest-hosts (GH) [7]		Dichroism	1 or 0	$\Delta\varepsilon > 0$	Homogeneous/Twist → Perpendicular to substrate surface
				$\Delta\varepsilon < 0$	Homeotropic/Twist → Parallel to substrate surface

VA: Vertical aligned nematic
MVA: Multi vertical aligned nematic
HAN: Hybrid aligned nematic

4.1.2 Device applications

Nowadays, liquid crystal displays (LCDs) based on nematic liquid crystals are widely used for watches, calculators, cellular phones and pagers, display monitors in personal computers, TV screens and so on. Although the liquid crystal state was discovered in 1888, it was 80 years before the first liquid crystal display was developed. In 1968, Heilmeier *et al.* [2] first applied the dynamic scattering (DS) effect to LCDs. After this application, many electro-optic effects of liquid crystals were discovered and many applications to display devices were developed. The twisted nematic (TN), the most widely used LCD mode, was invented by Schadt and Helfrich in 1970 [3]. TN displays provide a high contrast ratio, analog grey scale and low driving voltage. Especially, the twisted nematic LCD combined with an array of thin film transistors (TFTs), the TN mode TFT-LCD, produces a high quality full colour display, and is used for high information content displays such as computer monitors and TV.

Typical display modes using nematic liquid crystals are shown in Table 4.1.1. Super-twisted nematic (STN) modes give sharp electro-optical properties by increasing the twist angle of the liquid crystal molecules from the 90° used in the TN mode to approximately 240°, and achieve high multiplexable driving without TFT elements. The in-plane switching (IPS) mode shows very wide viewing angles because the liquid crystal molecules rotate in-plane under the applied electric field. The multi-vertical aligned nematic (MVA) mode is a type of electrically controlled birefringence (ECB) mode and its initial alignment is Homeotropic. Since the liquids crystals are inclined in 4 directions by the application of the electric field, the viewing angle of the MVA mode is wide. The optically compensated bend (OCB) mode has a bend alignment. The response speed of the OCB mode is widely thought to be the fastest among nematic LCDs.

The important display modes from the viewpoint of device applications and alignment techniques are described in more detail in the following sections.

4.2 Twisted Nematic (TN) [1, 2]

Ray Hasegawa

The TN is the most widely used LCD mode for applications ranging from watches to computer monitors. The advantages of the TN mode are high contrast ratio, analog grey scale and low driving voltage. Additionally, the TN mode has a wide gap margin, because the principle of the TN mode utilizes not retardation but optical rotation. Although viewing angle and response speed are shortcomings of the TN mode, new technologies to resolve them have been developed. Multi-domain methods [3–9] and an optically anisotropic film with negative birefringence using a discotic liquid crystal [10] have improved the viewing angle of

TN-LCDs. Fast response speed was attained by a narrow gap (about 2 μm) TN-LCD with a novel driving scheme for impulse-type displays [11].

4.2.1 Basic operation

The basic operation of a TN-LCD in normally the white mode is illustrated in Fig. 4.2.1. In the inside of each glass substrate, a thin layer of indium-tin oxide (ITO) is present, which acts as a transparent electrode. Then it is coated with an alignment layer, which promotes the alignment of the liquid crystals parallel to the glass substrate plane. Rubbed polyimide film is mostly used as the alignment layer. The rubbing directions of the alignment layers of the upper and lower glass substrates are perpendicular and the liquid crystal molecules perform a 90° twist through the thickness of the liquid crystal cell. In the case of the normally white mode, each polarizing film (polarizer) is placed on the outside of the substrates, so that the transmissive axis of each polarizer is parallel to the rubbing direction of each alignment layer (i.e. the relationship between the upper and lower polarizers is crossed).

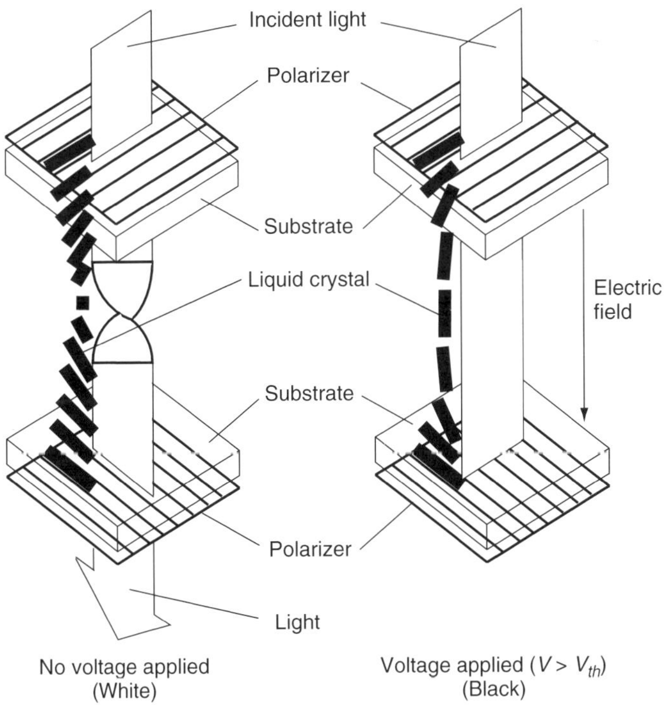

Fig. 4.2.1 Basic operation of a TN-LCD in the normally white mode.

The light incident on the TN display is polarized by the first polarizer and the light passes through the liquid crystal layer with its polarization direction rotated by 90°. Consequently, the polarization direction of the light becomes parallel to the transmissive axis of the second polarizer and the light is transmitted.

In the case that the transmissive axes of the upper and lower polarizers are placed parallel, the OFF state of the display is black and this configuration is called normally black mode. Since the normally black mode gives some light leakage in the black state, the normally white mode is adopted in a TFT-LCD.

4.2.2 OFF state

The optical properties of a TN cell were calculated using Jones matrices by Gooch and Tarry [12]. The normalized transmittance T_{TN} in the OFF state in the normally white mode is given by

$$T_{\mathrm{TN}} = \frac{1}{2}\left(1 - \frac{\sin^2(\pi/2)\sqrt{1+u^2}}{1+u^2}\right) \tag{4.2.1}$$

where

$$u = 2\Delta nd/\lambda \tag{4.2.2}$$

d is the thickness of the liquid crystal layer, Δn is the anisotropy of the refractive index (birefringence) and λ is the wavelength of the incident light. The transmittance T_{TN} calculated from equation (4.2.1) is plotted in Fig. 4.2.2. This indicates that

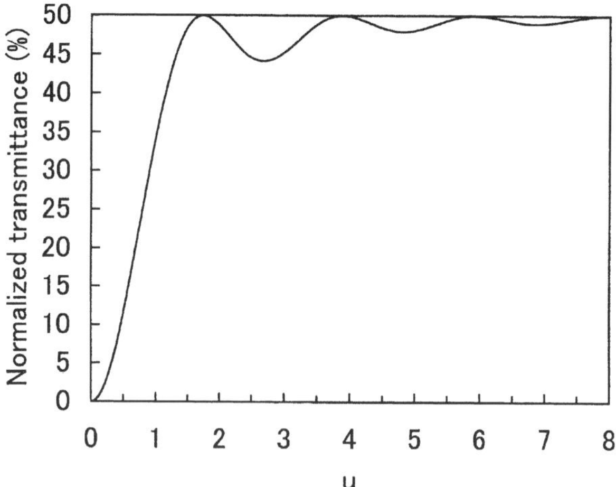

Fig. 4.2.2 Transmittance of monochromatic light through a TN cell in the normally white mode. The transmittance value normalized to that of the incident light was 100%. A TN cell requires two polarizers and this reduces their OFF-state transmittance by half. Thus, the maximum value of the transmittance is 50%.

a series of transmission maxima is obtained when $u = \sqrt{3}, \sqrt{15}, \sqrt{35}$. Since a smaller u gives a faster response speed because of the smaller cell gap and wider viewing angle, the cell condition, $u = \sqrt{3}$, is usually selected in the case of a TFT-LCD.

4.2.3 ON state

The liquid crystal in the TN mode has a positive dielectric anisotropy ($\Delta\varepsilon > 0$). The director of the liquid crystal tends to reorient parallel to the electric field by the Fréedericksz effect. In the ON state, the majority of the liquid crystal molecules except for those near the alignment layers have their molecular long axes reoriented parallel to the electric field and rotation of the polarization direction does not occur. The light passing through the liquid crystal layer is blocked by the second polarizer and the pixel appears dark. Since the liquid crystal molecules very near the alignment layers are strongly anchored to the alignment layers, these molecules hardly reorient towards the electric field. Consequently, on the removal of the electric filed, the liquid crystal molecules are restored to a 90° twist and the light is transmitted again.

The threshold voltage V_{th} of the TN cell is given by equation (4.2.3) [13, 14].

$$V_{th} = \pi\sqrt{\frac{\mathrm{K}_{11} + \frac{1}{4}(\mathrm{K}_{33} - 2\mathrm{K}_{22})}{\varepsilon_0 \cdot \Delta\varepsilon}} \tag{4.2.3}$$

$\Delta\varepsilon$ is the dielectric anisotropy of the liquid crystal. K_{11}, K_{22} and K_{33} are the elastic constants of bend, twist and splay, respectively. The molecular direction in the liquid crystal layer at various applied voltages was calculated by Berreman [15].

4.2.4 Dynamic response

The response time τ_{ON} from the OFF state to the ON state and the response time τ_{OFF} from the ON state to the OFF state are given by equations (4.2.4) and (4.2.5), respectively [16].

$$\tau_{\mathrm{ON}} = \frac{\gamma \cdot d^2}{\varepsilon_0 \cdot \Delta\varepsilon\left(V^2 - V_{th}^2\right)} \tag{4.2.4}$$

$$\tau_{\mathrm{OFF}} = \frac{\gamma \cdot d^2}{\pi^2 \cdot \mathrm{K}} \tag{4.2.5}$$

$$\mathrm{K} = \mathrm{K}_{11} + \frac{\mathrm{K}_{33} - 2\mathrm{K}_{22}}{4} \tag{4.2.6}$$

where γ is viscosity of liquid crystal and d is the TN cell gap.

As indicated by equations (4.2.4) and (4.2.5), reducing γ and d is useful for obtaining fast response speeds. When a smaller gap TN-LCD is fabricated to make its response speed faster, it is necessary to use a higher Δn liquid crystal material in order to match the cell condition to $u = \sqrt{3}$.

4.2.5 Reverse domains

To avoid the occurrence of domains with reverse tilt and domains with reverse twist in a TN-LCD, the use of an alignment layer with a pretilt angle more than about 5° and the addition of cholesteric dopants to the nematic liquid crystal mixture are preferable (see Section 2.1.1).

4.3 Super Twisted Nematic (STN)

Nobuyuki Itoh

The simple multiplexed LCD has the advantage of a simple electrode configuration as illustrated in Fig. 4.3.1. Two glass plates with striped transparent electrodes

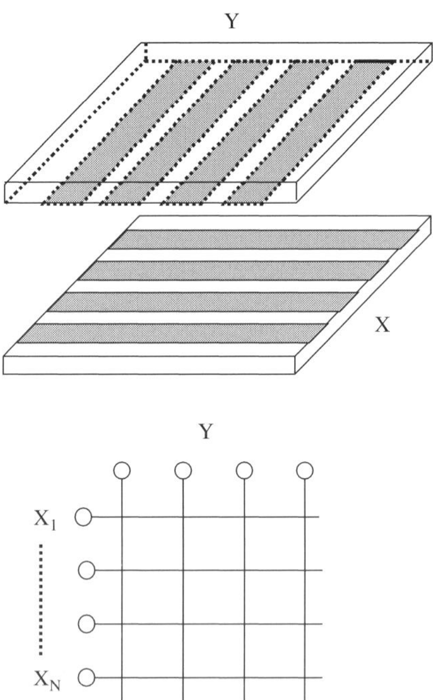

Fig. 4.3.1 Electrode configuration of a simple multiplexed LCD.

made of, for example, indium/tin oxide (ITO), are placed together with the stripes running at right angles to each other so that they make an X-Y dot matrix. The main drawback of this configuration is that it produces cross talk; that is, sufficient contrast cannot be achieved in a high-multiplex operation by using the conventional TN-LCD because of the unwanted half-selected turn on state of non-selected pixels. Moreover, the number of addressing lines that can be provided by this configuration is limited by the so called "iron law of multiplexing". Its operating tolerance can be expressed as a selection ratio R [1],

$$R = \mathrm{V_{ON}}/\mathrm{V_{OFF}} = \sqrt{\frac{\sqrt{N}+1}{\sqrt{N}-1}}, \tag{4.3.1}$$

which is the ratio of the highest root-mean-square (rms) voltage across a pixel to the lowest rms voltage (Fig. 4.3.2). This figure shows the typical electro-optic (EO) response of the TN-LCD, where N is the number of addressing lines and its inversion is the duty ratio. It is clear from Fig. 4.3.2 and Eq. (4.3.1) that a sharp EO response is necessary in order to achieve the high-multiplexing operation required for a graphics display. The TN-LCD cannot provide such a sharp response and the number of addressing lines is still limited to below 100 in spite of improvements to materials and the optimization of the device structure and addressing scheme.

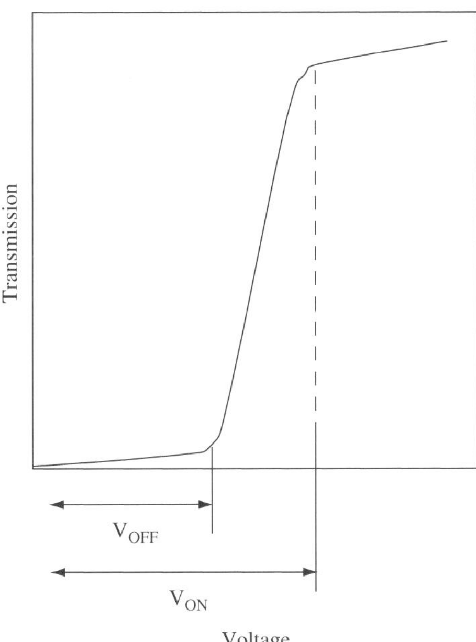

Fig. 4.3.2 Typical electro-optic (EO) response of a liquid crystal.

It was found that the sharpness of the EO response can be increased by increasing the twist angle; accordingly, the super-twisted nematic (STN)-LCD was developed [2]. The molecular orientation of the STN-LCD is schematically shown in Fig. 4.3.3. The LCD has a twist angle from 180° to 270° and the pretilt angle is 5° to 20°. Such a wide range of high pretilt angles is necessary in order to produce a defect-free homogeneous orientation so that an unwanted striped texture is avoided [3]. The steep voltage-transmission response of the STN-LCD in comparison with that of a TN-LCD is shown in Fig. 4.3.4. In the fabrication of the STN-LCD, first, a high pretilt angle is formed by SiO oblique evaporation. Then, the aligning films are formed using suitable aligning materials and rubbing technology. This fabrication process can produce STN-LCDs for high-multiplexing displays with over 400 addressing lines that are suitable for laptop PC monitors and word processors.

A coloured background caused by optical interference is the most serious disadvantage of the STN-LCD because it degrades the colour reproduction of an LCD display. To solve this problem and produce an achromatic black-and-white background, the double-layered STN-LCD (D-STN-LCD), which comprises a liquid crystal optical compensator mounted on the STN-LCD, has been developed [4]. Figure 4.3.5 shows the principle of the D-STN-LCD. The compensation layer, which is placed ready fabricated on the driven cell, compensates the optical dispersion caused by the birefringence effect and decolourizes the STN-LCD as long as the following three conditions are met:

(i) the total consecutive twist angles of the layers must be the same but of opposite sense,
(ii) the directors on the matching layer surfaces must meet at right angles,
(iii) the retardation products of the layers must be the same.

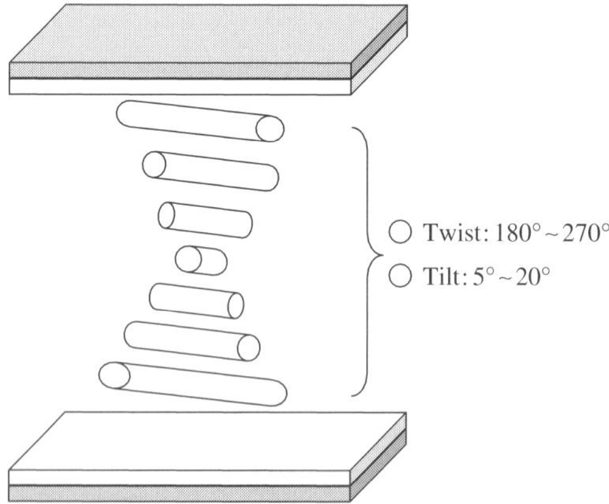

Fig. 4.3.3 Schematic molecular orientation of the STN-LCD.

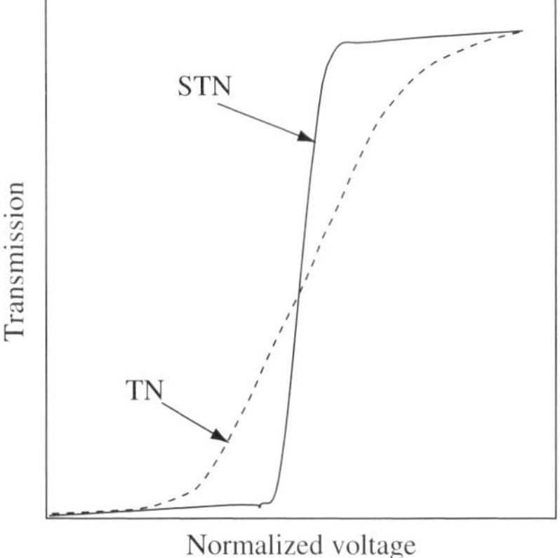

Fig. 4.3.4 Typical voltage–transmission response of an STN-LCD in comparison with a TN-LCD.

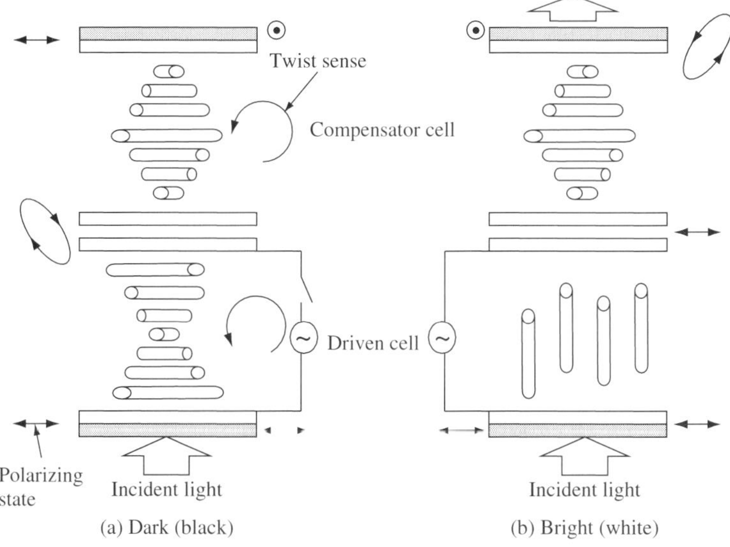

Fig. 4.3.5 Principle of the D-STN-LCD.

Accordingly, D-STN-LCDs fabricated according to these conditions (with micro colour filters on each pixel) were used to make multi-colour LCD displays. It should be noted that a film-compensated STN-LCD using optical retarder films instead of the STN-LCD liquid-crystal compensator in the D-STN-LCD can also be used.

4.4 The IPS (In-Plane Switching) Mode

Mitsuhiro Koden

The principle of the IPS (In-Plane Switching) mode is shown in Fig. 4.4.1. In typical LCDs, both substrates have electrodes and the direction of the electric field is perpendicular to the substrates. In the IPS mode, however, only one substrate has electrodes and the electric field runs parallel to the substrates. The molecular long axes of the liquid crystal molecules are thus parallel to the substrates whether an electric field is or is not applied. In other words, the switching of the liquid crystal molecules by the electric field remains within a single plane.

The strong merit of operation in the ISP mode is its wide viewing angle. Kiefer *et al.* proposed this mode in 1992 [1] and, in 1995, Oh-e *et al.* presented a wide viewing angle TFT-LCD in which the IPS mode was applied [2].

Operation in the IPS mode requires homogeneous alignment. Such homogeneous alignment is usually produced by rubbing. One question is the effect of the pretilt angle induced by rubbing on the characteristics of the display, and especially on the viewing angle. Oh-e *et al.* have investigated this subject [3]. They reported that the pretilt angle has a strong effect on the dependence of the viewing angle on the contrast ratio. When the pretilt angle is low, a high contrast ratio is obtained with a wide viewing angle. On the other hand, when the pretilt angle is high, the contrast ratio rapidly decreases as the viewing angle grows.

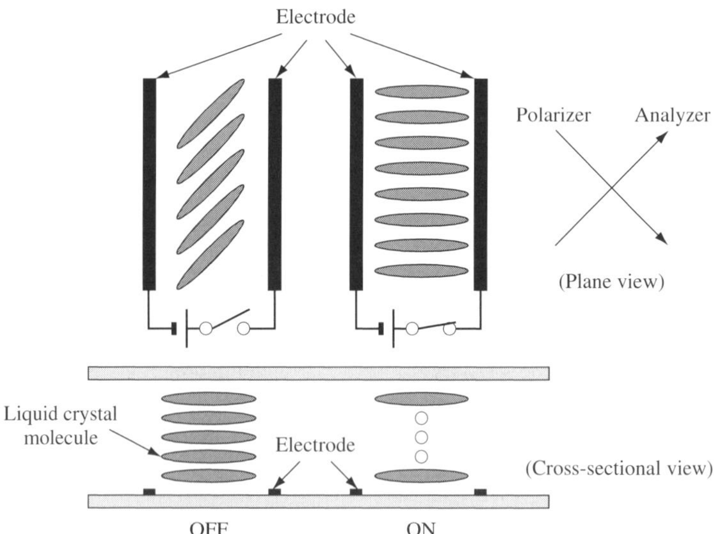

Fig. 4.4.1 The IPS mode.

4.5 Vertical Alignment (VA) Mode and Multi-domain Vertical Alignment (MVA) Mode

Kohki Takatoh

4.5.1 Vertical alignment (VA) mode

In the vertical alignment (VA) mode, nematic liquid crystalline materials with negative dielectric anisotropy align perpendicular to the surface of the alignment layers. Applying an electric field to the liquid crystalline material between the two electrodes causes the molecules to rotate into a direction parallel to the electrodes or perpendicular to the electric field. Comparison of the viewing angle dependence of the contrast ratio between the TN mode and the VA mode reveals that the viewing angle dependence of VA mode is smaller than that of TN mode (see Fig. 4.5.1).

By using an optically uniaxial polymer film or retardation film, the viewing angle dependence of the VA mode can be reduced [2]. Concerning the usage of retardation films, there are two methods, one using only the negative refractive index [3] and the other using two films [1, 4], one of positive refractive index and

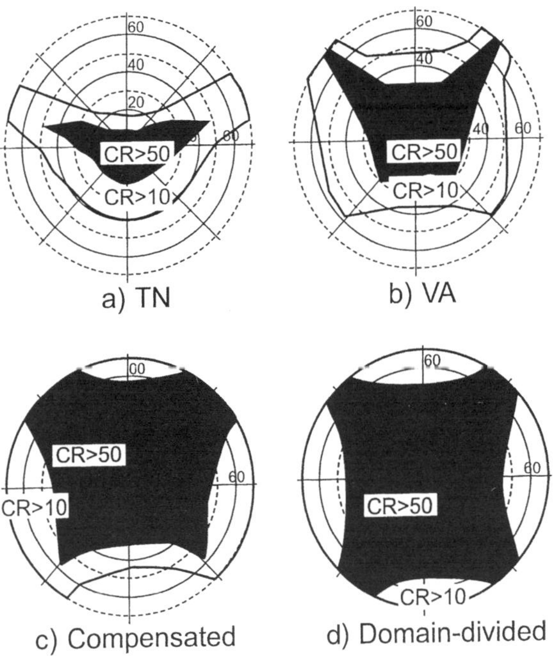

Fig. 4.5.1 Viewing angle dependence of the contrast ratio of the VA mode (a) TN mode (b) VA mode (c) VA mode with two retardation films (d) MVA mode [1].

another of negative refractive index. By using two films, the viewing angle dependence can be reduced more than in the case of using only the negative refractive index.

4.5.2 MVA method (Multi-domain vertical alignment method) [5]

To reduce the viewing angle dependence of the TN mode, the division of each pixel into dual- or multi-domains was proposed. For the vertical alignment (VA) mode, the mode called "multi-domain vertical alignment (MVA) mode" was also proposed to compensate the viewing angle dependence of each domain by those of the adjacent domains. In the case of the dual- or multi-domain TN mode, the division is carried out, for example by alternating the rubbing direction. In one pixel, the tilt direction or the direction in which the liquid crystal molecules rise on applying an electric field in any given domain is different from that in adjacent domains. On the other hand, in the case of the VA mode, the direction in which the liquid crystal molecules decline is alternated to change the direction of large viewing angle dependence in each domain. The viewing angle dependence of a domain compensates those of other domains, thereby reducing the viewing angle dependence over the whole area.

For the VA mode or MVA mode, vertical alignment is realized by using polymer materials without prior rubbing treatment.

Changing the subsequent rubbing direction in each division or pixel makes the processes complicated and increases the production cost. For the MVA mode, a methodology based on a completely different concept has been developed. The division into domains of different alignment is realized by forming "protrusions" on the alignment layers as shown in Fig. 4.5.2 [1, 3, 4, 5, 6].

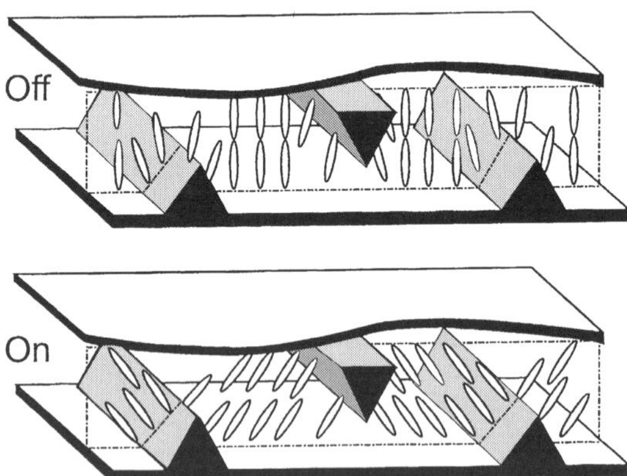

Fig. 4.5.2 The division into domains of different alignment realized by forming "protrusions" on the vertical alignment layers [1, 3, 4, 5, 6].

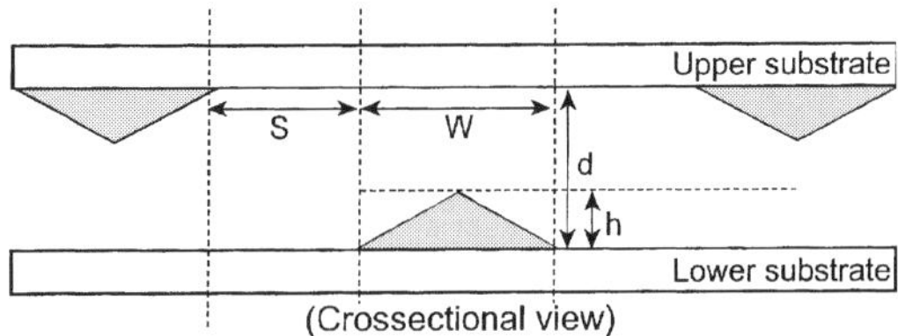

Fig. 4.5.3 Factors which determine the properties of MVA mode [5, 6].

In Fig. 4.5.2, the liquid crystal molecules in different domains are shown separated by the protrusions and the molecules decline in different directions on applying an electric field. In a given domain, not only on the protrusion, but also on the alignment layer parallel to the substrates, the liquid crystal molecules decline in the same direction. Unlike the rubbing method, this method involves no complicated procedures.

The properties of the MVA mode are determined by the factors shown below (see Fig. 4.5.3 [6])

(1) Height of the protrusions (h)
(2) Width of the protrusions (w)
(3) Interval between the protrusions (s)
(4) Dielectric constant of the LC materials ϵ_{lc} and protrusions ϵ_p.

In LCDs operating under the MVA mode as shown in Fig. 4.5.4, applying an electric field causes the LC molecules on the protrusions to begin to decline. The molecular rotation spreads in the same direction to the part parallel to the substrates. By changing the two-dimensional patterns of the protrusions, as shown in Fig. 4.5.5, the division into plural kinds of domains can be realized. This division method introduces an alignment methodology based on the new concept that the control of alignment is realized by the polymer protrusions present on the alignment layers.

Figure 4.5.6 [6] shows the transmittance dependence on the height of the protrusion, h, for various strengths of applied electric field. This suggests that a high protrusion can realize a high value of the transmittance.

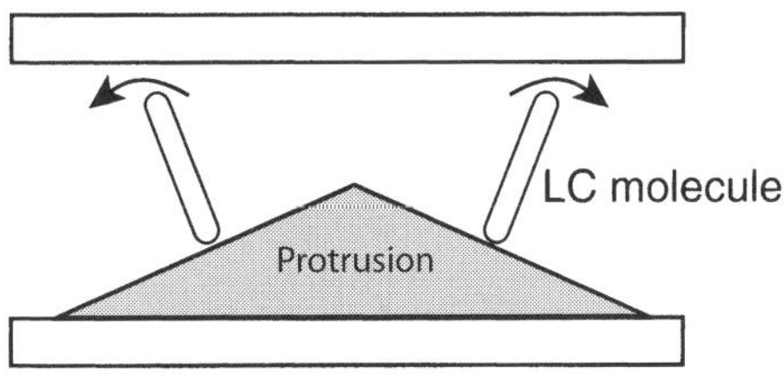

Fig. 4.5.4 Switching of LC molecules in the MVA mode with protrusions [5].

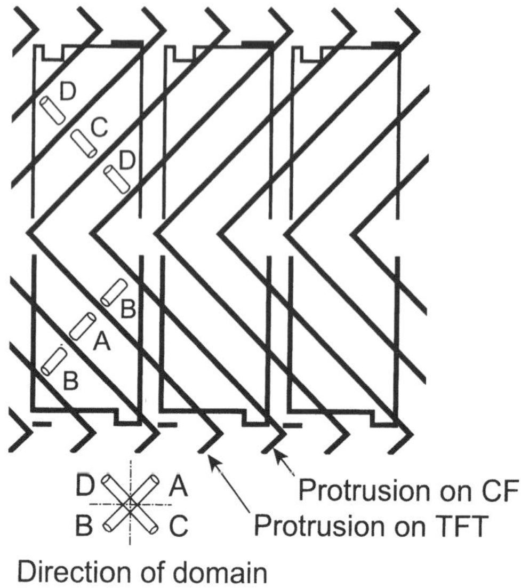

Fig. 4.5.5 Protrusion pattern of the MVA mode (four-domain type) [5]. CF = Colour filter.

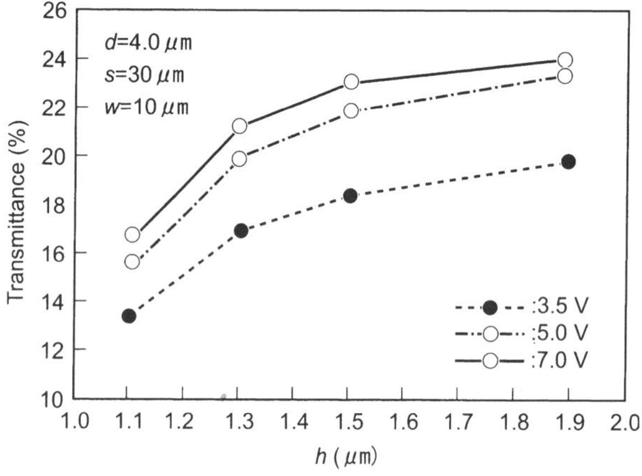

Fig. 4.5.6 Transmittance dependence on the height of the protrusion, h, for various strengths of applied electric field [6].

Figure 4.5.7 [6] shows the transmittance dependence on the interval between the protrusion, s. The larger the interval (s), the higher the transmittance becomes. However, for intervals larger than 30 μm, no dependence can be observed. This suggests that the protrusions do not influence the liquid crystalline alignment in areas far from them.

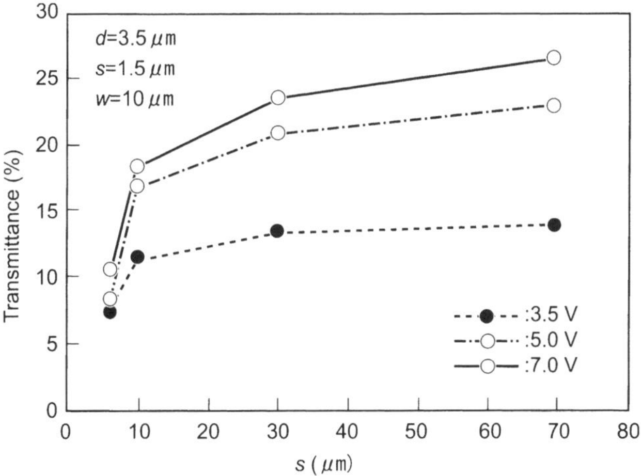

Fig. 4.5.7 [6] Transmittance dependence on the interval of the protrusion, *s*.

Figure 4.5.8 shows the difference in the molecular response to the applied electric field due to the different dielectric constants of protrusions using different polymer materials. The broken lines show the lines of electric force. As shown by these broken lines, the distribution of the electric field changes greatly depending on the relationship between the dielectric constants of the LC materials and the

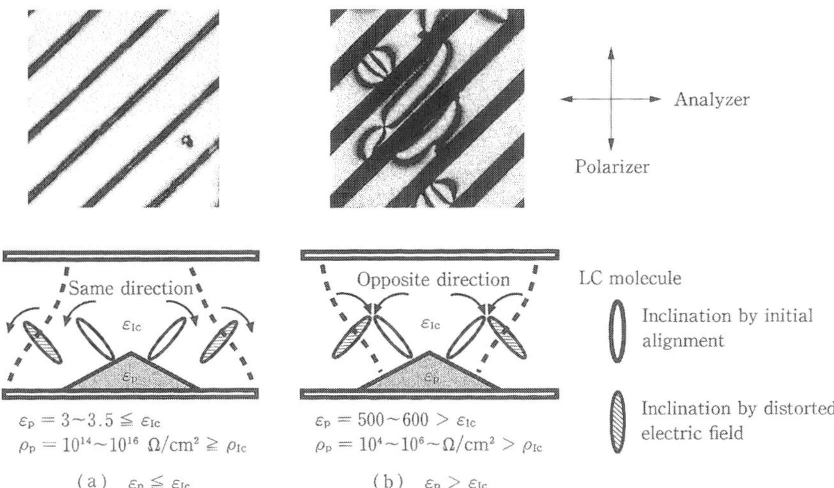

Fig. 4.5.8 Difference in the molecular response to the applied electric field due to the different dielectric constants of the protrusions using different polymer materials [6].

polymers used for the protrusions. In the case (Fig. 4.5.8 that $\epsilon_p < \epsilon_{lc}$, the lines of electric force run as lines keeping away from the protrusions. The LC molecules on either side of each protrusion rotate in opposite directions, symmetrically. The molecular declining direction is determined in one direction. As a result, homogeneous molecular orientation can be realized between the two protrusions. By microscopic observations, no defect line can be observed in the area between the protrusions.

On the other hand, in the case that $\epsilon_p > \epsilon_{lc}$, the lines of electric force run toward the protrusions. The long axes of the LC molecules with negative $\Delta\varepsilon$ rotate in the direction perpendicular to the lines of electric field. As a result, the liquid crystal molecules rotate in a complicated manner as shown in Fig. 4.5.8(b). In this case, for the LC molecules on the protrusion, the electric field is applied in the same manner as in the parts parallel to the substrates. LC molecules on the tip of the protrusions decline largely in the direction parallel to the protrusion. LC molecules on the slope of the protrusion and on the area parallel to the substrates also rotate, influenced by the movement of the LC molecules on the tips. By microscopic observations, many line defects or Schlieren brushes can be observed in the area between the protrusions.

In the case that $\varepsilon_p < \varepsilon_{lc}$, on the protrusion, the direction of the molecules without an applied electric voltage coincides with the declining direction on applying electric field. As a result, the intended division of alignment layers can be realized. In this case, the LC molecules on the tip of the protrusion again decline in the direction parallel to the protrusion. However, the extent of this LC rotation is so small that it does not influence the movement of the LC material on the slope of the protrusion and in the part parallel to the substrates.

The viewing angle dependence of the MVA-mode is wider than 160° from left to right and from up to down and the contrast is more than 10 to 1. For any viewing angle, no inversion of the grey scale can be observed. The MVA mode was commercialized in December 1997, and the specification of an MVA LCD is shown in Table 4.5.1 [5].

Table 4.5.1 Specifications of an MVA LCD [5]

Display area	15 inch diagonal
Number of pixels	1024 × 768 × RGB
Number of colours	26 million
Brightness	200 cdm^{-2}
Contrast ratio (max.)	300 : 1 or more
Response time	less than 25 ms
Viewing angle (CR > 10)	160 deg. or more (no inversion)
Driving voltage	5 V

4.6 Pi Cell

Masaki Hasegawa

In the Pi cell, also known as the Optically Compensated Bend alignment (OCB) cell, the LC is aligned to produce a bent structure in the middle of the cell [1, 3]. The pretilt angle directions at both surfaces of the substrate are opposite as shown in Fig. 4.6.1. At first, this configuration was proposed for an electrically controllable birefringence plate [1]. After this suggestion, it became clear that this configuration has many good characteristics. The upper and lower layers of the cell have a symmetric configuration. Therefore, this structure is suitable for a wide viewing angle display, because the viewing angle dependency of the upper and lower layers of the cell compensate each other [2, 3]. In addition to the optically compensated characteristics, the response time to changes in the electric field is very fast. From these characteristics, this configuration has received attention as a display which has a wide viewing angle and is suitable for moving images. Here, we will introduce some characteristics of this configuration.

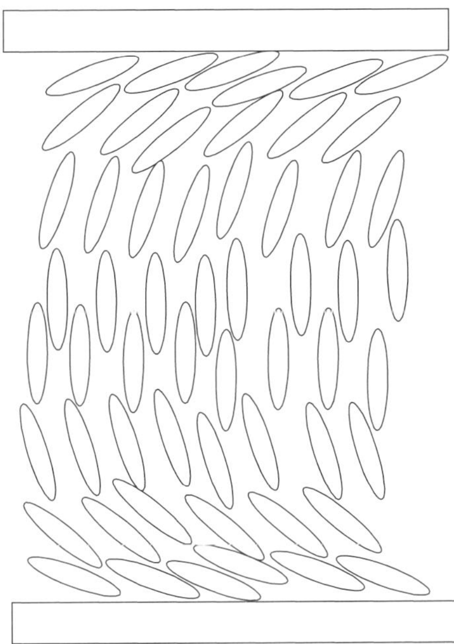

Fig. 4.6.1 Director orientation distribution of the Pi cell in the bend mode.

4.6.1 LC configuration

The initial alignment of the LC in this structure is configured as a splay arrangement as shown in Fig. 4.6.2. After the voltage is applied to align the middle of the cell perpendicular to the substrate, the LC aligns to produce the bent structure. When the applied electric field is removed, the LC alignment slowly returns to the splay configuration. However, if the electric field remains at a certain level, the LC configuration remains in the bend configuration. The LC configuration change from splay to bend is caused by the elastic energy. Figure 4.6.3 shows the relationship between the applied voltage and the Gibbs free energy. The free energy of the bend structure with a certain voltage applied is lower than that of the splay structure. Therefore, when that voltage is applied to the cell, its configuration changes to the bend structure.

When the Pi cell is used for a display, it requires a compensation film to create the black state. Because of its symmetrical tilt alignment near the substrate, the Pi cell has a biaxial birefringence in the black state. Several approaches were tried to develop a compensation film for the Pi cell. To compensate for the birefringence of the black state by using only one film, biaxiality is required by the film [2]. A combination of an a-plate and c-plate was used [7] to reduce the light leakage in the black state. To compensate for each side of the distributed LC molecules, discotic films were used as the compensation films as shown in Fig. 4.6.4 [4, 5, 8, 9, 10].

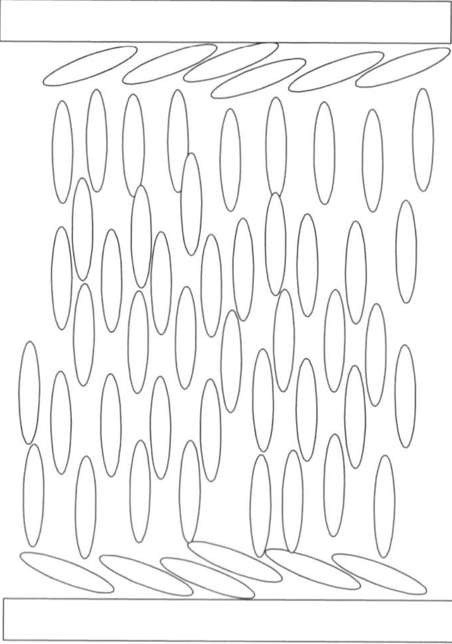

Fig. 4.6.2 Initial director orientation distribution of the Pi cell in the splay mode.

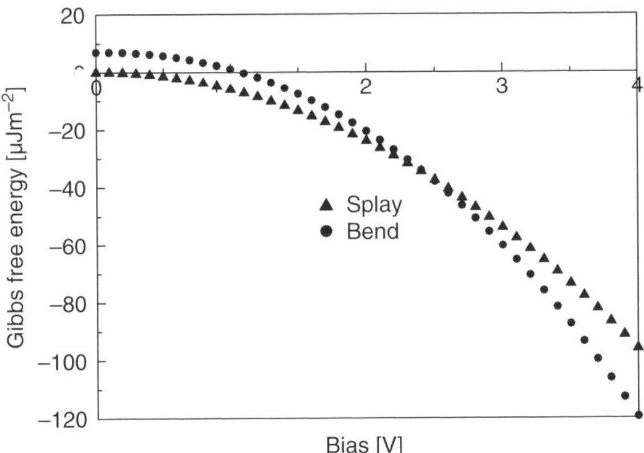

Fig. 4.6.3 Comparison of the free energy between the splay and the bend distribution.

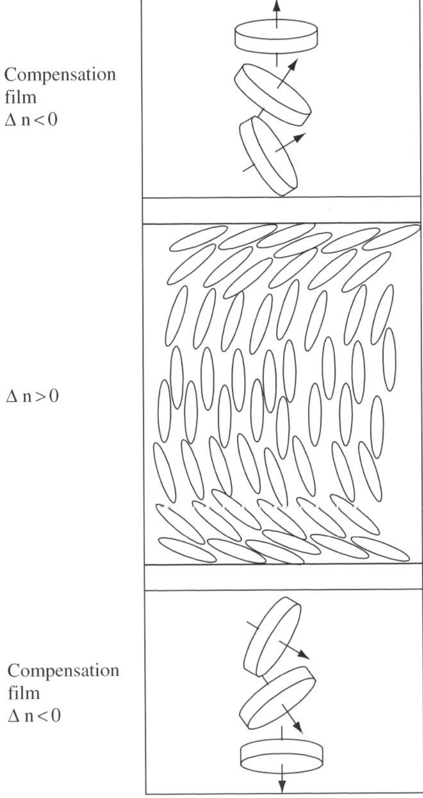

Fig. 4.6.4 Configuration of the compensation films and the Pi cell. Reproduction by permission from [4].

4.6.2 Dynamics

The response to the applied electric field of the typical twisted nematic mode is not rapid. Its rise and decay times in response to the applied voltages are in the range of several 10 s of msec. However, the response of the Pi cell is very rapid, and the summation of its rise and decay times in response to the applied voltage is less than 10 msec, even for grey levels. The reason for this fast response of the Pi cell is believed to be its back flow [11]. In the relaxation from the twisted nematic display mode, the flow, near the centre of the cell disturbs the relaxation of the centre as shown in Fig. 4.6.5. However, in the case of the Pi cell, the flow of the relaxation accelerates the relaxation of the centre part of the cell as shown in Fig. 4.6.6 [11]. Another reason for this quick response is that the area of the alignment change caused by the applied voltage is near the surface and it is highly deformed because of the bend configuration. Based on this quick response characteristic, it was suggested that the field sequential display mode could be used for eliminating colour filters [6].

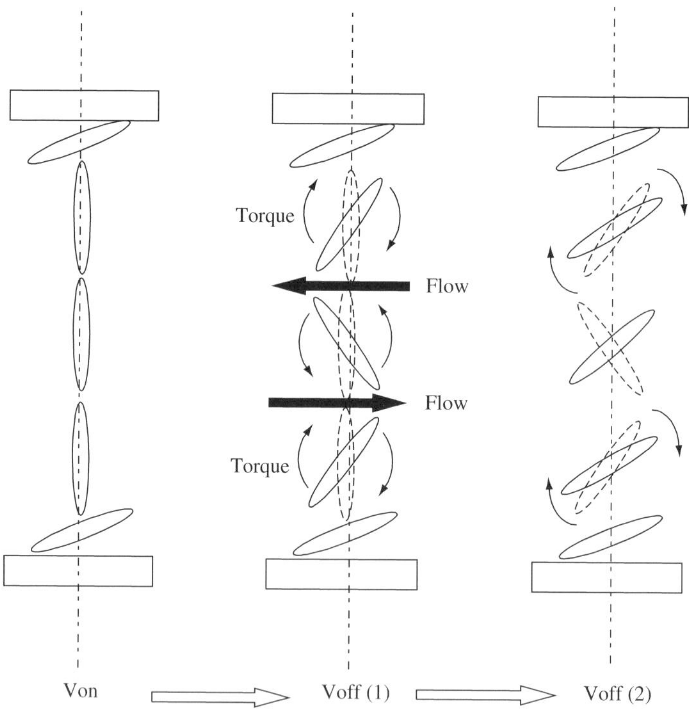

Fig. 4.6.5 Relaxation state of the TN mode: The flow prevents the relaxation motion in the centre. Reproduction by permission from [11].

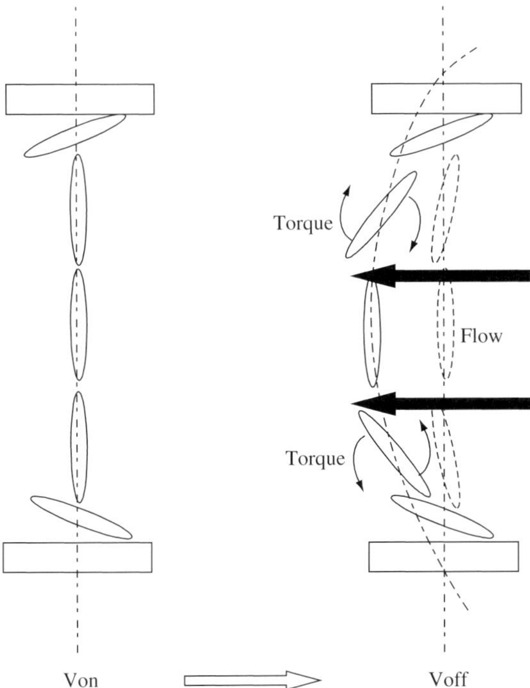

Fig. 4.6.6 Relaxation state of the bend distribution: The flow accelerates the relaxation motion in the centre. Reproduction by permission from [11].

4.7 Multi-domain Mode

Kohki Takatoh

4.7.1 Improvement of the viewing angle dependence of TN-LCDs

When TN-LCDs are viewed, the contrast of the displayed images varies depending on the angle from which the panels are observed. This phenomenon is called viewing angle dependence of contrast. The viewing angle dependence occurs because the anisotropy of the refractive index of the LC material depends on the angle of observation. The viewing angle dependence is particularly marked in grey scale expression where the LC molecules align in the direction oblique to the substrate plane. To reduce the viewing angle dependency, methods involving the division of each pixel into more than two parts were proposed. These parts are aligned so as to compensate the viewing angle dependence for the various parts of each pixel. To realize such structures in one pixel, various methods have been proposed.

4.7.2 Mechanism of viewing angle dependence of TN-LCDs

Figure 4.7.1 shows the LC molecular configuration of TN-LCDs. Parts (a) and (b) show the cases without and with an applied voltage, respectively. The applied voltage in the case of (b) is smaller than the saturation voltage. In the case of (a), the LC molecules align nearly parallel to the substrates. In this case, the change in contrast, depending on the viewing angle, is limited, or alternatively the viewing angle dependence is small.

In the case of (b), the LC molecules tilt relative to the substrate plane. This shows the LC configuration for grey scale expression. During the passage of light through this structure, the light transmittance changes greatly depending on the light direction, and so the viewing angle dependence of the contrast becomes large. The change is largest in the direction shown by the outlined arrow in Fig. 4.7.1. This is the direction in which the change of the anisotropy of the refractive index from the front direction is largest and the direction to which LC molecules mainly incline. This phenomenon can be compared with the following situation. The length of a pencil on a desk does not change even if it is viewed from several directions. However, if we incline the pencil at a certain angle from the desk, it appears to be a long pencil, a short pencil, or a small circle depending on the observation angle. To the viewer, this phenomenon is not marked when LC molecules align perpendicular to the substrates. Therefore the viewing angle dependence is small when the applied electric voltage is larger than the saturated voltage. As a result, the viewing angle dependence is especially large in the case of grey scale expression.

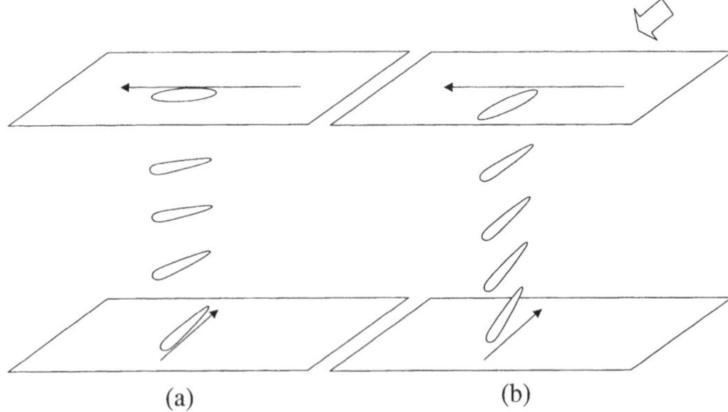

Fig. 4.7.1 LC molecular configuration of TN-LCDs. (a) and (b) show the cases without and with an applied voltage, respectively. The white outlined arrow in (b) shows the direction of the largest viewing angle dependence.

4.7.3 Reduction of the viewing angle dependence by the multi-domain mode

In Fig. 4.7.2, an example of viewing angle dependency for the transmittance of a TN-LCD is shown. Part (a) shows the dependence in the vertical direction and (b) shows the dependence in the horizontal direction.

In typical commercial TN-LCDs, the largest viewing angle dependency is in the vertical direction. The viewing angle dependence shown on the right side of (a) is that from the lower direction in commercial products, such as LCDs for notebook-type computers. This direction corresponds to the outlined arrow direction in Fig. 4.7.1. The lines from 2 to 7 are those for grey scale expression. From these lines, it can be confirmed that the viewing angle dependency in one direction, the right side of Fig. 4.7.2(a), is especially large. For the lines from 2 to 7, the transmittance observed from this side is smaller than from the front. The images observed from this side are darer than those observed from the front. On the other hand, the transmittance in the left part is larger than that observed from the front, and the images are brighter than those observed from the front. This phenomenon reduces the direction range from which the images can be observed properly.

To minimize the viewing angle dependence by reducing this phenomenon, methods involving the division of each pixel into multiple parts have been proposed [1]. These methods are called multi-domain modes. In these modes, the viewing angle dependence in one domain can be compensated by those in the other domains as shown in Fig. 4.7.3. Figure 4.7.4 [11] (a) shows a typical example of the viewing angle dependency of the transmittance for multi-domain LCDs.

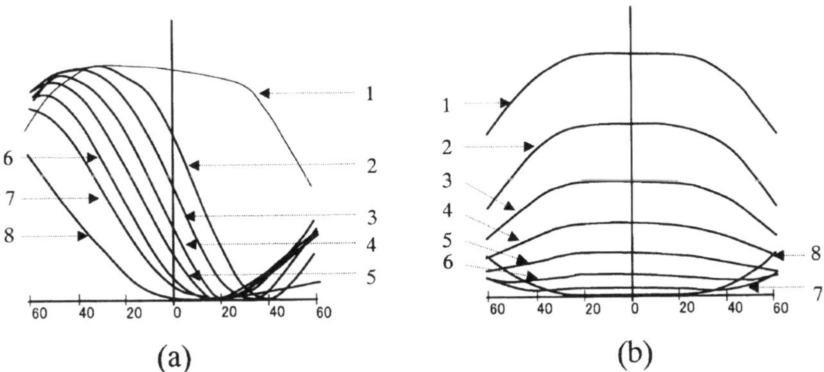

Fig. 4.7.2 Example of viewing angle dependency of the transmittance of a TN-LCD. (a) Shows the dependences in the vertical direction and (b) show the dependences in the horizontal direction. The line 1 shows the case without applied voltage. The lines from 2 to 7 are those for grey scale expression. The line 8 is the one for black expression with applied voltage almost equal to the saturated voltage.

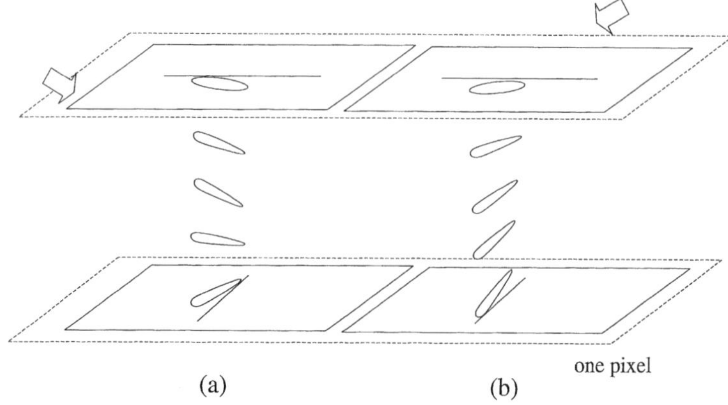

Fig. 4.7.3 LC molecular configuration of the multi-domain mode. The outlined arrows show the directions of the largest viewing angle dependence. The viewing angle dependence of each domain can be compensated by that of another domain.

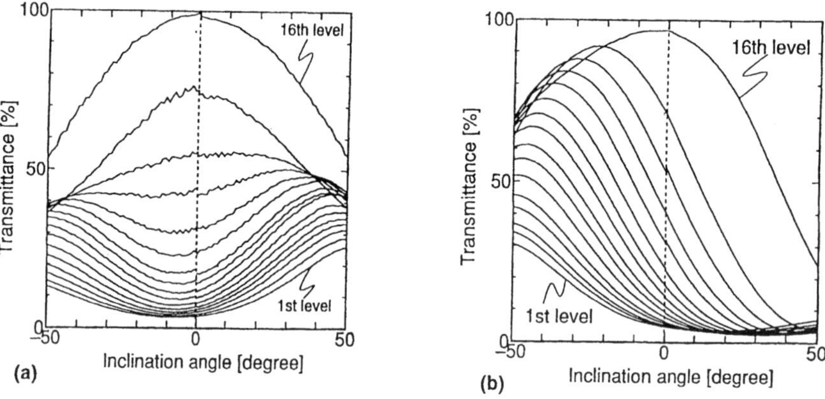

Fig. 4.7.4 Viewing angle dependence of the transmittance for LCDs having multi-domain mode (a) and for a typical TN-LCD (b) [11].

Figure 4.7.4(b) shows an example of the viewing angle dependency of the transmittance for typical TN-LCDs. In part (a), the strong viewing angle dependency in part (b) is not observed. Moreover, each curve shows symmetrical properties.

By rubbing the surfaces of two regions in different directions, the LC configuration in Fig. 4.7.3 can be realized. However, in these methods, the number of the processes and the production cost increase greatly. To realize the multi-domain mode with plural domains of different LC configuration without serious increase in production cost, a method involving the addition of only one UV photoradiation process to the conventional production procedure was proposed. In this case, two regions having different pretilt angles on the alignment layers are formed by UV photoradiation.

4.7.4 Formation of two kinds of regions possessing different alignment directions in one pixel

To form two regions having different alignment directions in one pixel, methods using photoresist materials [2, 3] and those using polarized light radiation, as described in chapter 3–2, were proposed [4, 5, 6, 7, 8]. Figure 4.7.5 shows the processes using photoresist materials [9]. First of all, a rubbing treatment is curried out on the whole surface of the polyimide alignment layer. The rubbed surface is then coated with photoresist. UV light is irradiated on the resist layer through a photomask to some region in each pixel. The irradiated resist layer is developed, and in the case of positive-type resist materials, the irradiated parts are removed. On the other hand, in the case of negative type resists, irradiated parts remain. For resist material selection, it is especially important that an alignment layer surface is not damaged in the development and removal processes. From this point of view, the positive type which can be developed and stripped off by aqueous solutions is more appropriate than the negative type. In the case of negative resists, a dry etching process is required for their removal [4], but this process results in serious damage to the alignment layer surface. On the alignment layer

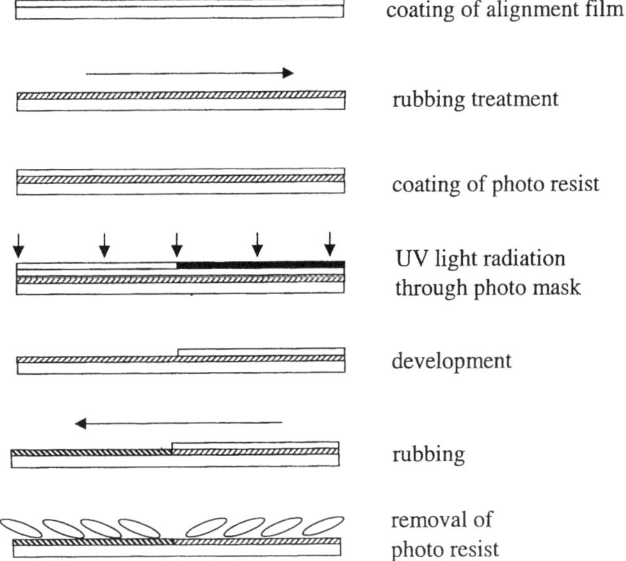

Fig. 4.7.5 Processes using photoresist materials for formation of two kinds of region possessing different alignment directions in one pixel [9].

on which a part of each pixel is covered by the resist layer, a rubbing process is now carried out in the direction opposite to the initial rubbing direction. After stripping off the resist layer, an alignment layer with two regions of different directions of rubbing in each pixel is formed. In this method, the methodology of semiconductor devices is applied. Therefore, this method is suitable for adoption in a TFT-LCD production line. However, this method is more expensive than that in which the pretilt angle is controlled by photo irradiation, because of the increase in the number of processing steps.

In the above method, both of the alignment layers on the two substrates have to be treated, although a modified version of this method has been proposed in which only one alignment layer is treated by these processes. In the method, two kinds of polyimide of different pretilt angle magnitudes are applied (Fig. 4.7.6) [10, 11]. The LC molecular configuration is determined by the alignment direction on the high pretilt angle alignment layer and the twisting direction of LC materials. The mechanism is discussed in detail below.

Another method for producing two regions of different alignment direction in one pixel is by the polarized light irradiation of polyimide layers or photocurable polymer layers as discussed in chapter 3–2 [4, 5, 6, 7, 8]. In both methods, the direction of the pretilt angle can be controlled by UV light irradiation from an oblique direction. By these methods, an alignment layer having the same function as the one fabricated by using photoresist can be realized. Polarized light is irradiated through a photomask to form two regions of different alignment directions in

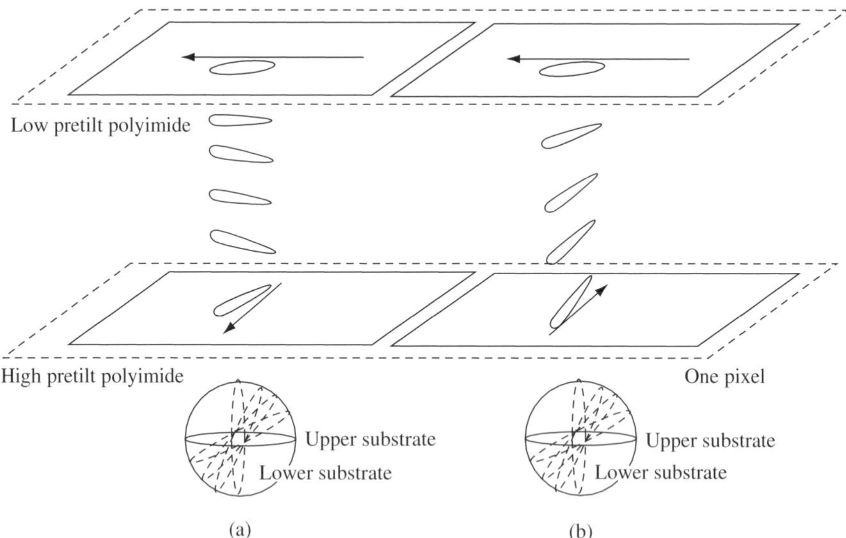

Fig. 4.7.6 LC molecular configuration of the multi-domain mode formed by two kinds of polyimide with different pretilt angle magnitudes. Part (a) shows the "splayed" TN configuration and part (b) shows a conventional TN configuration.

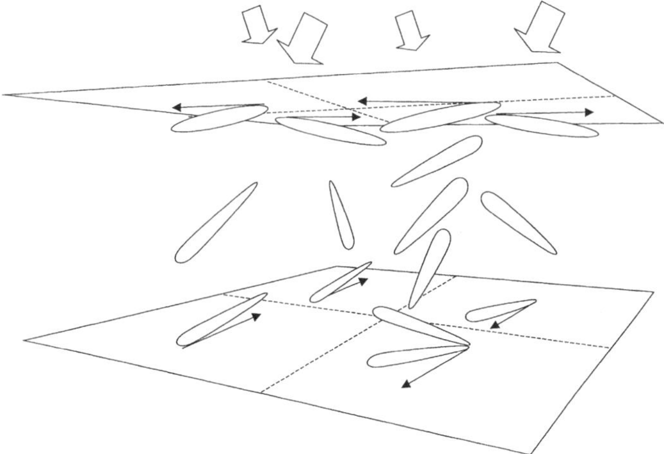

Fig. 4.7.7 Four kinds of LC configuration formed by alignment layers with two regions with different pretilt directions.

one pixel. By combination of two alignment layers with two such regions, four kinds of configurations can be formed [4] as shown in Fig. 4.7.7. In the figure, four large outline arrows show the directions of the largest viewing angle of each domain. The viewing angle dependences of the four regions can be mutually compensated.

4.7.5 Formation of two kinds of region with different pretilt angles in one pixel

The methods involving control of pretilt angle magnitude are similar to those involving control of alignment direction and the viewing angle dependency of the LC materials within one pixel is mutually compensated.

In Fig. 4.7.8, two kinds of LC configuration determined by the fixed pretilt angle directions on both alignment layers are shown. In these two LC configurations, the twisting directions are different. Configuration (b) is more splayed and is more unstable than configuration (a). As a result, a LC material forms configuration (a) spontaneously. Here it is noted that in TN-LCDs, optically active materials called "chiral reagents" are added to reduce the occurrence of a defect called "twist reverse". The optically active materials determine the twisting direction of the LC materials. The LC configuration in Fig. 4.7.8(b) is formed by the addition of optically active material that induces a twist in a clockwise direction from upper substrate to lower substrate. This configuration is splayed to a great extent and is thermodynamically unstable. Note that, to form this configuration, the relationship between the twist direction caused by optically active materials and the one caused by the combination of tilt angle directions on two alignment

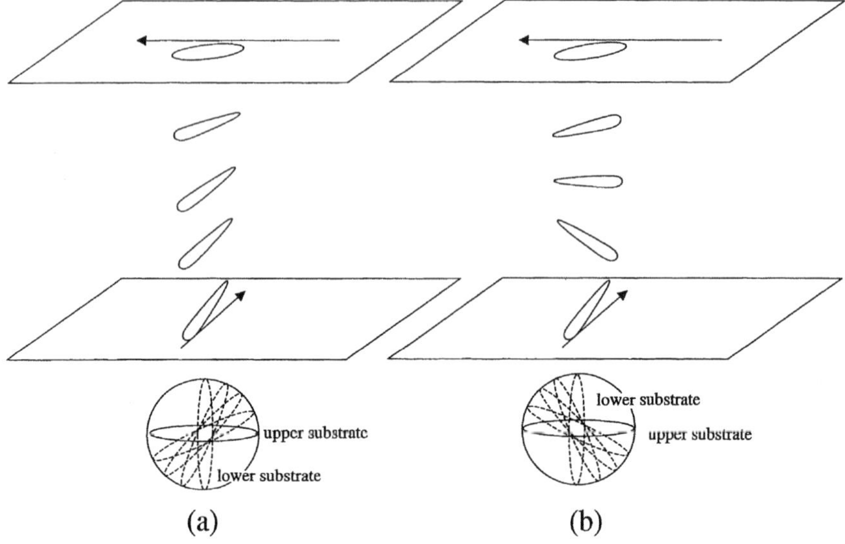

Fig. 4.7.8 Two kinds of LC configuration determined by the fixed pretilt directions on both alignment layers. Configuration (b) is more splayed and is more unstable than configuration (a).

layers is opposite to that in the case of usual TN-LCDs, in which the optically active materials are used to stabilize the configuration (a). In the case of usual TN-LCDs, the twist direction caused by the optically active materials should be equal to the one caused by the combination of tilt angle directions on two alignment layers. On the other hand, for the formation of the LC configuration in Fig. 4.7.8(b), the optically active material possessing the twist direction opposite to the one caused by the combination of tilt angle directions must be selected. The LC configuration in Fig. 4.7.8(b) is called the "splayed-twist" state.

In the multi-domain mode, this "splayed-twist" state is used as shown below. However, the "splayed-twist" state is unstable and changes into other states. For example, when the chirality decreases on reducing the chiral reagent concentration, the twisting direction becomes the opposite to form the usual TN configuration without splayed structure. This configuration is called the "reverse twist state". Furthermore, the occurrence of a configuration called the "reverse tilt state" is also reported. In this state, the LC material behaves as if the directions of the pretilt angles are reversed.

It is reported that the splayed-twist state can be stabilized by the conditions shown below [18–21].

(1) Short chiral pitch
(2) Small angle between the directions used in the rubbing process (70° is better than 90°)

(3) Wide cell gap
(4) Large K_{22}/K_{11}.

Generally speaking, with a low applied voltage the "splayed-twist" state is stable, but with a high voltage, the twist direction tends to be reversed to form the "reverse-twist" state. Now, let us think about the two kinds of configuration, (a) and (b), shown in Fig. 4.7.9. Part (b) shows the "reverse-twist" state, of which the configuration is the same as for a usual TN configuration. Even if the pretilt angle on one or two substrates varies, the change of the viewing angle dependence is limited. The direction of the viewing angle dependence does not change. So, by changing the magnitude of the pretilt angles on one alignment layer of the "reverse-twist" state, the viewing angle compensation between two domains cannot be realized.

On the other hand, in the case of the "splayed-twist" state shown in part (a) of Fig. 4.7.9 the change of the pretilt angle on one alignment layer can alter the direction of the viewing angle dependence because of the splayed structure. As a result, by changing the pretilt angle on either the upper or lower alignment layer, the direction and the magnitude of the viewing angle dependence can be controlled.

Figure 4.7.10 shows the LCD structure of the multi-domain mode in which each domain forms the "splayed-twist" state. In the case of (a), the pretilt angle on the upper alignment layer is larger than that on the lower layer and the LC configuration near the upper alignment layer becomes predominant. As a result, the domain shows the viewing angle dependence induced by the configuration near the upper alignment surface. On the contrary, in the case of (b), the pretilt angle on the lower alignment

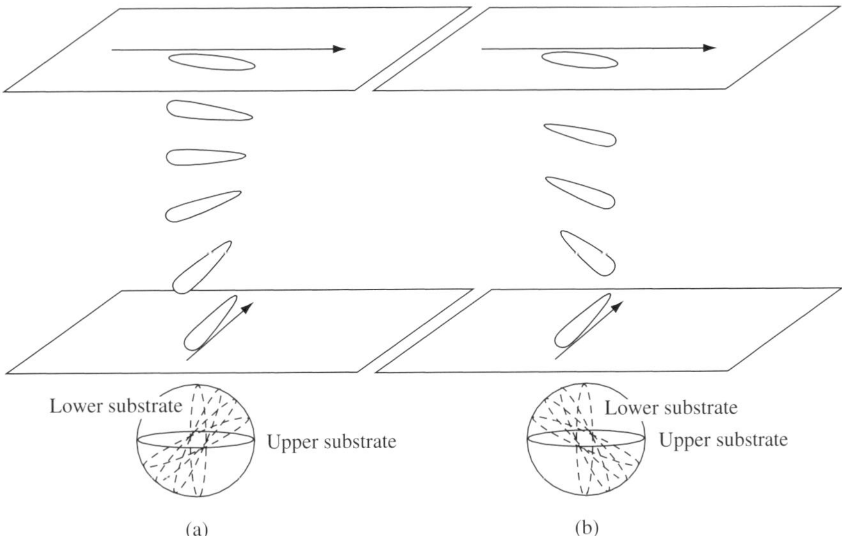

Fig. 4.7.9 LC molecular configuration of "splayed-twist" state (a) and "reverse-twist" state (b).

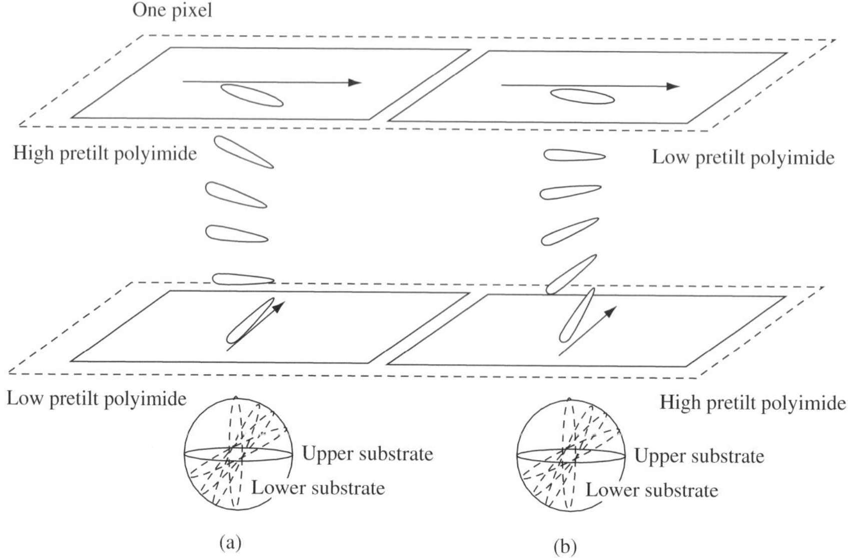

Fig. 4.7.10 LCD structure with two kinds of LC molecular configuration realized by a combination of a low pretilt angle alignment layer and a high pretilt angle alignment layer.

layer is larger than that on the upper layer and the domain shows the viewing angle dependence induced by the configuration near the lower alignment surface.

By forming two domains, (a) and (b), in one pixel, the viewing angle dependence of each domain can be compensated to realize LCDs showing a wide viewing angle.

4.7.5.1 Control of pretilt angles by using two kinds of polyimide [14, 15]

By using two kinds of polymer materials, two regions in which the magnitudes of the pretilt angles are different can be formed. In Fig. 4.7.11, the process for forming alignment layers with two regions having different pretilt angles in one pixel is shown. In this method, the alignment layer surfaces need not be treated by processes such as coating, development and removal of resist material after the rubbing treatment. These processes may result in deterioration of the alignment surfaces. It has been reported that [14] when the difference in pretilt angle magnitudes for the two alignment layers is larger than 2°, two regions of different configuration can be formed.

4.7.5.2 Control of pretilt angle by photoradiation [16, 17]

The magnitude of the pretilt angle depends on the magnitude of the polarity of the alignment layer surface. The higher the polarity of the alignment layer surface, the smaller the pretilt angle [17].

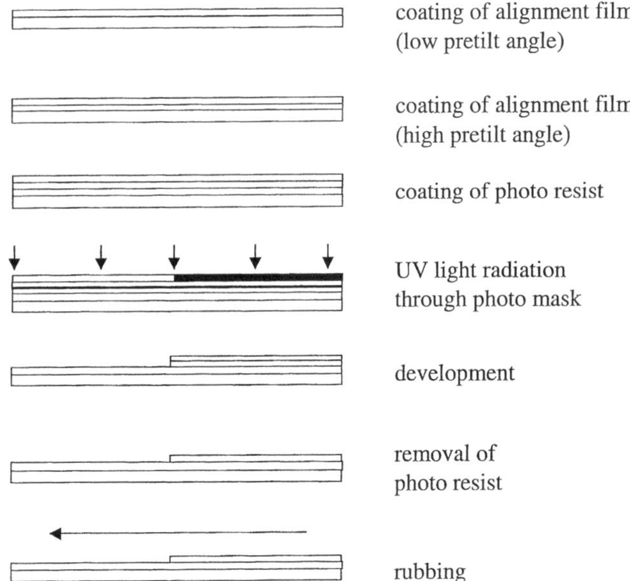

Fig. 4.7.11 The processes required to form alignment layers with two regions having different pretilt angles in one pixel.

By UV light radiation, polyimide surfaces are oxidized to generate polar groups such as hydroxy groups to increase the surface polarity. As a result, UV light irradiation of the alignment layer reduces the magnitude of the pretilt angle.

By UV light radiation through a photomask, an alignment layer with two kind of region showing large and small pretilt angle magnitudes can be realized.

Now, the real process used will be introduced [17]. Deep UV light (240–600 nm, mainly 254 nm) is irradiated onto the polyimide layer surface through a photomask. By one irradiation, two regions having different pretilt angle magnitudes are produced. In practical displays, each pixel is divided into two regions. In one region, two kinds of alignment layers with/without UV light irradiation must be combined as the upper/lower alignment layers. In the example shown in Ref. 17, deep UV radiation reduces the pretilt angle magnitude from 4–10° to less than 1°. On the surface in which two kinds of region are formed, a rubbing treatment is carried out in the conventional manner. Using alignment layers formed by this method, LCDs with two kinds of region having different configurations can be obtained. This is quite a simple method in which only one UV radiation is added to the usual LCD production process. Furthermore, it is a well-established process for TFT (thin film transistor) production. Therefore, in the practical production line, this method is adopted for multi-domain LCD production.

4.8 Polymer Dispersed Liquid Crystals (PDLC) [1]

Ray Hasegawa

PDLC are nematic liquid crystals randomly dispersed as microdroplets in polymer films. PDLC films can be switched from an opaque (light scattering) state to a transparent state and do not require polarizers and an alignment film. The light scattering properties of PDLC are brought about by the difference in refractive index between the liquid crystal and the polymer. Typically, $n_e \gg n_o \approx n_p$, where n_e and n_o are the extraordinary and ordinary refractive indices of the liquid crystal, and n_p is the refractive index of the polymer. Typical operation using a nematic liquid crystal with positive electric anisotropy is shown in Fig. 4.8.1. In the absence of an electric field, the directors of the liquid crystal in each droplet are randomly oriented by the interaction between the liquid crystal molecules and the polymer surface. In this state, the effective refractive index of the liquid crystal droplets is mismatched with that of the polymer matrix, resulting in an opaque appearance. In the presence of an electric field, the directors in the droplets are reoriented along the direction of the applied electric field, and the droplets appear to have n_o. Consequently, the refractive indices of the droplets and the polymer matrix are matched and the PDLC film becomes transparent.

There are three principal methods of fabricating PDLC (to disperse liquid crystals in polymer matrix); phase separation, encapsulation and permeation. In the phase separation method, liquid crystal materials, prepolymer (monomer or oligomer) and photoinitiator (curing agent) are mixed, and then polymerization is brought about by heating [2] or UV irradiation [3]. During the polymerization, the liquid crystal material phase separates from the solution, and liquid crystal droplets and a polymer matrix are formed. By measurement of the nematic-isotropic transition temperature, the purity of the liquid crystal in the droplets can be checked. In the case that the polymerization-induced phase separation is not complete and unreacted prepolymer is dissolved in the liquid crystal droplets, the

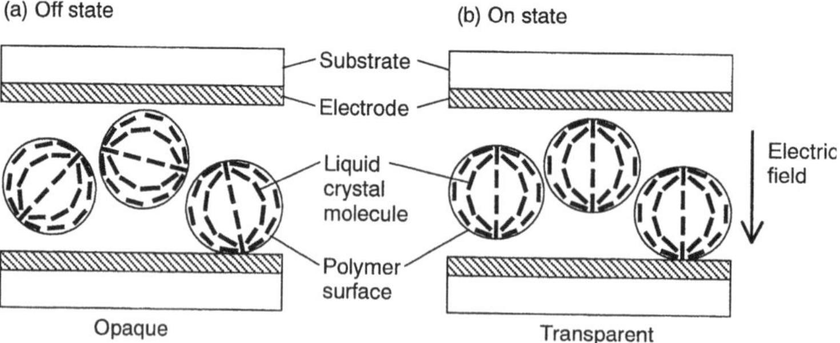

Fig. 4.8.1 Typical operation of a PDLC system.

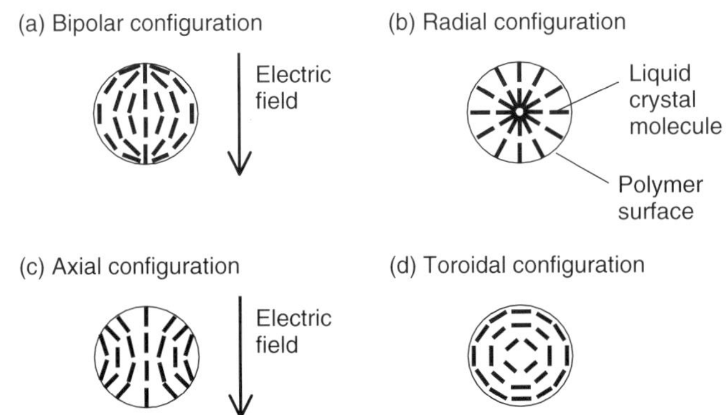

Fig. 4.8.2 Schematic illustrations of typical director configurations in PDLC droplets.

nematic-isotropic transition temperature varies from that of the pure liquid crystal material and the temperature range of the transition becomes wide.

Encapsulation [4] is brought about by emulsifying a liquid crystal in a waterborne polymer such as polyvinyl alcohol (PVA); the emulsion is then coated on a substrate and dried. The permeation method [5] is used for basic studies and the liquid crystals are permeated into the pores of a prefabricated polymer matrix.

The electro-optical properties of PDLC depend not only on the chemical nature of liquid crystal and polymer, but also on the morphology of the polymer matrices. Figure 4.8.2 shows a schematic illustration of the nematic director configurations in spherical cavities. Case (a) is bipolar and (b) has a radial configuration. These configurations are the most commonly observed nematic director configurations resulting from parallel and perpendicular boundary conditions, respectively. In the case where the electric anisotropy is positive, the effect of an electric field on the bipolar configuration is to align the symmetry axis parallel to the field with little distortion of the director configuration within the droplets. The effect of an electric field on the radial configuration is to induce a configurational transition to the axial configuration shown in Fig. 4.8.2(c). The toroidal configuration of Fig. 4.8.2(d) was found to be relatively stable for parallel anchoring conditions when $K_{33}/K_{11} < 1$ [6].

Since PDLC displays do not require polarizers and alignment layers, they offer the advantages of high-brightness and ease of the fabrication. Using these advantages, PDLC films are applied to switchable windows for privacy or solar control and in projection type displays. Although the PDLC display is usually opaque (scattering) and transparent switching, black and white switching can be obtained in a projection system with a lens and a pinhole. The PDLC projection display combined with thin film transistors (TFT) has been under development [7, 8], but in order to commercialize TFT-driven, PDLC it is necessary to reduce the driving voltage and get rid of hysteresis in the transmittance-applied voltage dependence.

References

4.1 Summary of Molecular Alignment and Device Applications

[1] L. T. Creagh and A. R. Kmetz, Mol. Cryst. Liq. Cryst., **24** (1973) 59.

[2] G. H. Heilmeier, L. A. Zanoni and L. A. Barton, Proc. IEEE, **56** (1968) 1162.

[3] M. Schadt and W. Helfrich, Appl. Phys. Lett., **18** (1971) 127.

[4] M. F. Schiekel and K. Fahrenschon, Appl. Phys. Lett., **19** (1971) 391.

[5] G. Assouline, M. Hareng and E. Leiba, Electron. Lett., **7** (1971) 699.

[6] S. Matsumoto, M. Kawamoto and K. Mizunoya, J. Appl. Phys., **47** (1976) 3842.

[7] G. H. Heilmeier and L. A. Zanoni, Appl. Phys. Lett., **13** (1968) 91.

4.2 Twisted Nematic (TN)

[1] "The Optics of Thermotropic Liquid Crystals", edited by S. J. Elston and J. R. Sambles, T&F (1998) p. 289.

[2] "Electronic Display Devices", edited by S. Matsumoto, John Wiley and Sons (1990) p. 30.

[3] K. H. Yang, Proc. 11th Intl. Display Research Conf. (1991) 68.

[4] Y. Koike, T. Kamada, SID Digest (1992) 798.

[5] K. Sumiyoshi, K. Takatori, Y. Hirai and S. Kaneko, J. SID, **2/1** (1994) 31.

[6] M. Okamoto, Y. Yamamoto, N. Fukuoka, Y. Hisatake, T. Yamamoto, Y. Tanaka, H. Hatoh, K. Monma, K. Sunohara, R. Hasegawa, H. Nagata, H. Sugiyama, S. Shishido and K. Kojima, Proc. AM-LCD '94 (1994) 88.

[7] K. W. Lee, A. Lien, J. H. Stathis and S. H. Paek, Jpn. J. Appl. Phys., **36** (1997) 3591.

[8] A. Lien and R. A. John, SID Digest (1993) 269.

[9] Y. Toko, T. Sugiyama, K. Katoh, Y. Iimura and S. Kobayashi, SID Digest (1993) 622.

[10] M. Okazaki, K. Kawata, H. Nishikawa and M. Negoro, Polym. Adv. Technol., **11** (2000) 398.

[11] T. Nose, M. Suzuki, D. Sasaki, M. Imai and H. Hayama, SID Digest (2001) 994.

[12] C. H. Gooch and H. A. Tarry, J. Phys. D, Appl. Phys., **8** (1975) 1575.

[13] F. M. Leslie, Mol. Cryst. Liq. Cryst., **12** (1970) 57.

[14] H. J. Deuling, Mol. Cryst. Liq. Cryst., **27** (1975) 81.

[15] D. W. Berreman, J. Opt. Soc. Am., **63** (1973) 1374.

[16] E. Jakeman and E. P. Raynes, Phys. Lett., A**39** (1972) 69.

4.3 Super Twisted Nematic (STN)

[1] P. M. Alt and P. Pleshko, IEEE Trans. Electron Devices **ED-21**, 146–155 (1974). "Scanning Limitations of Liquid-Crystal Displays".

[2] T. J. Scheffer and J. Nehring, Appl. Phys. Lett., **45**, 1021–1023 (1984). "A new, highly multiplexable liquid crystal display".

[3] V. G. Chigrinov, V. V. Belyaev, S. V. Belyaev and M. F. Grebenkin, Sov. Phys. JETP, **50**, 994–999 (1979).

[4] K. Katoh, Y. Endo, M. Akatsuka, M. Ohgawara and K. Sawada, Jpn. J. Appl. Phys., **26**, L1784 (1987).

4.4 The IPS (In-Plane Switching) Mode

[1] R. Kiefer, B. Weber, F. Windscheid and G. Baur, Proc. Japan Display '92, 547 (1992).

[2] M. Oh-e, M. Ohta, S. Aratani and K. Kondo, Proc. Asia Display '95, 577 (1995).

[3] M. Oh-e, M. Yoneya, M. Ohta and K. Kondo, Liq. Cryst., **22**, 391 (1997).

4.5 Vertical Alignment (VA) Mode and Multi-domain Vertical Alignment (MVA) Mode

[1] K. Koike, S. Takaoka, T. Sasaki, H. Chida, H. Tsuda, A. Takeda and Ohmuro, Proc. AM-LCD 97, 25 (1997).

[2] K. Ohmuro, S. Kataoka, T. Sasaki and Y. Koike, SID 97 DIGEST, 845 (1997).

[3] S. Yamauchi, M. Aizawa, J. F. Clerc, T. Uchida and J. Duchen, SID 89 DIGEST, 378 (1989).

[4] S. Ohmuro, S. Kataoka, T. Sasaki and Y. Koike, SID 97 DIGEST, 845 (1997).

[5] A. Takeda, S. Kataoka, T. Sasaki, H. Chida, H. Tsuda, K. Ohmuro and K. Koike, SID 98 DIGEST, 1077 (1998).

[6] A. Takeda, EKISHO, **3** (2) 117 (1999) (in Japanese).

4.6 Pi Cell

[1] P. J. Bos and L. R. Koehler/Beran, Mol. Cryst. Liq. Cryst., **113** (1984) 329.

[2] T. Miyashita, P. Vetter, M. Suzuki, Y. Yamaguchi and T. Uchida: Proc. Eurodisplay (1993) 149.

[3] Y. Yamaguchi, T. Miyashita and T. Uchida, SID '93 Digest (1993) 277.

[4] H. Mori, Y. Itoh, Y. Nishiura, T. Nakamura and Y. Shinagawa, SID '97 Digest (1997) 941.

[5] H. Mori and P. J. Bos, 1997 Int. Display Res. Conf. (1997) M-88.

[6] T. Uchida, K. Saitoh, T. Miyashita and M. Suzuki, SID '97 Digest (1997) 37.

[7] J. Chen, K.-H. Kim, J.-J. Jyu, J. H. Souk, J. R. Kelly and P. J. Bos, SID '98 Digest (1998) 315.

[8] H. Mori and P. Bos, SID '98 (1998) 830.

[9] H. Nakamura, K. Miwa, M. Noguchi, Y. Watanabe, J. Mamiya, J. Watanabe, Y. Nishiura and Y. Shinagawa, SID '98 (1998) 143.

[10] H. Mori, E. C. Garatland Jr., J. R. Kelly and P. J. Bos, Jpn. J. Appl. Phys., **38** (1999) 135.

[11] S. Onda, T. Miyashita and T. Uchida, Mol. Cryst. Liq. Cryst., **331** (1999) 383.

4.7 Multi-domain Mode

[1] K. H. Yang, Proceedings of Int. Display Research Conf., **68** (1991).

[2] M. Okamoto, Y. Yamamoto, N. Fukuoka, Y. Hisatake, T. Yamamoto, Y. Tanaka, H. Hatoh, K. Monma, K. Sunohara, R. Hasegawa, H. Nagata, H. Sugiyama, S. Shishido and K. Kojima, Proceeding of AM-LCD '94, 88 (1994).

[3] S. H. Jamal, J. R. Kelly and J. L. West, Jpn. J. Appl. Phys., **34**, Part 2, 10B, L1368 (1995).

[4] M. Schadt, H. Seiberle and A. Schuster, Nature, **381**, 212 (1996).

[5] M. Schadt and H. Seiberle, SID 97 DIGEST, **397** (1997).

[6] H. S. Soh, J. W. Wu, M. S. Nam, Y. J. Choi, Jungha Kim, K. J. Kim, J. H. Kim and S. B. Kwon, Proceedings of Euro Display 96, 579 (1996).

[7] S. H. Ahn, Y. S. Ham, W. Kim, W. S. Kim and S. B. Kwon, Proceedings of Euro Display 96, 377 (1996).

[8] M. S. Nam, J. W. Wu, Y. J. Choi, K. H. Yoon, J. H. Jung, J. Y. Kim, K. J. Kim, J. H. Kim and S. B. Kwon, SID 97 DIGEST, 933 (1997).

[9] M. Nishikawa, T. Shimokawa, M. Endo, Y. Matsuki and N. Bessho, Proceedings of Asia Display '95, 357 (1995).

[10] K. Takatori, K. Sumiyoshi, Y. Hirata and S. Kaneko, Proceedings of Japan Display '92, 591 (1992).

[11] K. Sumiyoshi, K. Takatori, Y. Hirai and S. Kaneko, J. SID, **2/1**, 31 (1994).

[12] K. Kawata, K. Takatoh, M. Hasegawa and M. Sakamoto, Liq. Cryst., **16** (6), 1027 (1994).

[13] M. Schadt, K. Schmitt, V. Kozinkov and V. Chigrinov, Jpn. J. Appl. Phys., 31(7), 2155 (1992).

[14] Y. Koike, T. Kamada, K. Okamoto, M. Ohashi, I. Tomita and M. Okabe, SID 92 DIGEST, 798 (1992).

[15] M. Nishikawa, T. Shimokawa, M. Endo, Y. Matsuki and N. Bessho, Proceedings of Asia Display '95, 357 (1995).

[16] A. Lien, R. A. John, M. Angelopoulos, K. W. Lee, H. Takano, K. Tajima and A. Takenaka, Appl. Phys. Lett., 67 (21), 3108 (1995).

[17] K.-W. Lee, A. Lien, J. H. Stathis and S.-H. Paek, Jpn. J. Appl. Phys., 36, Part 1, 6A, 3591 (1997).

[18] K. Takatori, K. Sumiyoshi and S. Kaneko, Tech. Rep. IEICE, EID94-136, ED94-164, SDM94-193 (in Japanese).

[19] Y. Saito, H. Takano, C. J. Chen and A. Lien, SID96 DIGEST 662 (1996).

[20] J. Bruinik, ASIA Display '95, 123 (1995).

[21] Y. Saitoh, H. Takano, C. J. Chen and A. Lien, Jpn. J. Appl. Phys., 36, 7216 (1997).

4.8 Polymer Dispersed Liquid Crystals (PDLC)

[1] "The Optics of Thermotropic Liquid Crystals", edited by S. J. Elston and J. R. Sambles, T&F (1998) p. 233.

[2] J. L. West, Mol. Cryst. Liq. Cryst., **157** (1988) 427.

[3] F. G. Yamagishi, L. J. Miller and C. I. Van Ast, Proceedings SPIE, **1080** (1989) 24.

[4] P. S. J. Drziac, J. Appl. Phys., **60** (1986) 2142.

[5] G. P. Crawford, L. M. Steele, R. Ondris-Crawford, G. S. Iannacchione, C. J. Yeager, J. W. Doane and D. J. Finotello, J. Chem. Phys., **96** (1992) 7788.

[6] P. S. J. Drziac, Mol. Cryst. Liq. Cryst., **154** (1988) 239.

[7] Z. Yaniv, J. W. Doane, J. L. West and W. Tamura-Lis, SID Digest (1989) 572.

[8] M. Kunigida, Y. Hirai, Y. Ooi, S. Niiyama, T. Asakawa, K. Masumo, H. Kumai, M. Yuki and T. Gungima, SID Digest (1990) 227.

Chapter 5

Alignment Phenomena of Smectic Liquid Crystals

5.1 Layer Structure and Molecular Orientation of Ferroelectric Liquid Crystals

Nobuyuki Itoh

5.1.1 Introduction

At first in this chapter, ferroelectric liquid crystals (FLCs) and their most interesting application as surface stabilized ferroelectric liquid crystals (SSFLCs) are briefly explained.

Molecular orientational models of SSFLCs have previously been studied, so the orientational states of SSFLCs and their optical properties are fully understood. For example, the smectic layer structures of various SSFLCs have been studied by using high-resolution X-ray diffraction and the relationship between the layer tilt angle and the optical molecular tilt angle has been confirmed.

The molecular orientational states of SSFLCs are classified by the optical viewing conditions and the relationship between the directions of bend of the layer structure and the surface pretilt angle. The molecular orientational models of the states have been considered and illustrated with regard to the experimental results, and useful information has been obtained from optical simulations using the models. The influence of the surface pretilt angle on the orientational and the optical properties of SSFLCs has been described.

5.1.2 Ferroelectric SmC* liquid crystals

The chiral smectic C (SmC*) phase is well known as a liquid crystal phase that is ferroelectric. Smectic A (SmA) and smectic C (SmC) are the fundamental smectic liquid crystal phases. As shown in Fig. 5.1.1, the director **n** is parallel to the smectic layer normal **z** in the SmA phase and it is an optically and dielectrically uniaxial

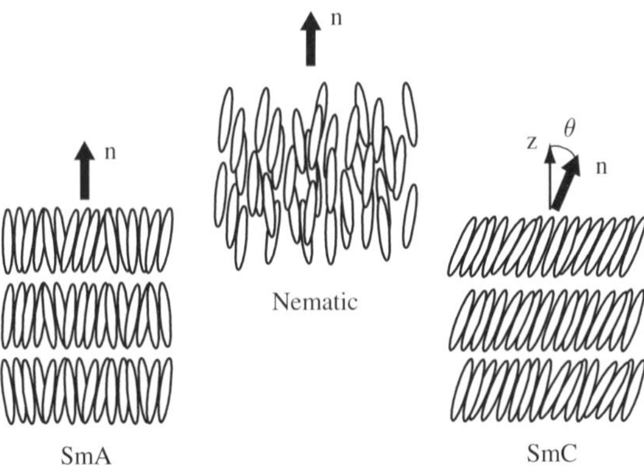

Fig. 5.1.1 Molecular orientation in some liquid crystal phases.

phase; on the other hand, the director **n** tilts at a tilt angle θ from the smectic layer normal in the SmC phase and it is a biaxial phase. In the SmC* phase, which is an SmC phase consisting of chiral molecules, but not racemic, the helical structure has a constant tilt angle θ, but with a slightly different azimuthal angle Φ in each layer. It appears as shown in Fig. 5.1.2 because of the chirality of molecules. Although only four layers in Fig. 5.1.2 are shown to express a pitch, more than several hundred layers are required to form an actual pitch. The FLC molecules lie on the smectic cones and move around them.

The spontaneous polarization is originated in the C_2 symmetry of the SmC* phase perpendicular to the molecular long axis. The molecular structure and the phase sequence of the world's first FLC, DOBAMBC (*p*-decyloxybenzylidene-*p*′-amino-2-methylbutylcinnamate) synthesized by Meyer *et al.* [1], is shown in Fig. 5.1.3.

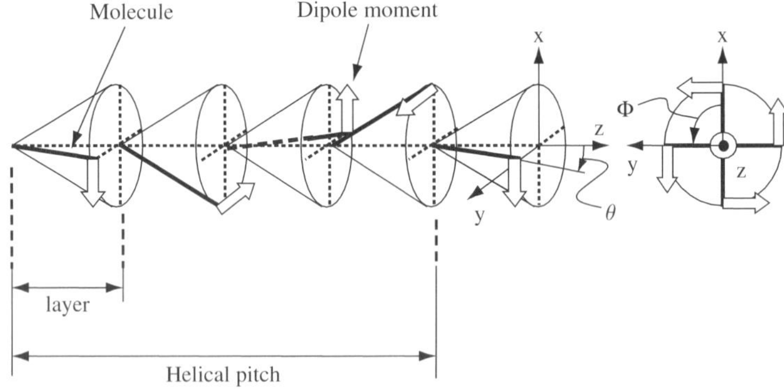

Fig. 5.1.2 Molecular orientation in the SmC* phase.

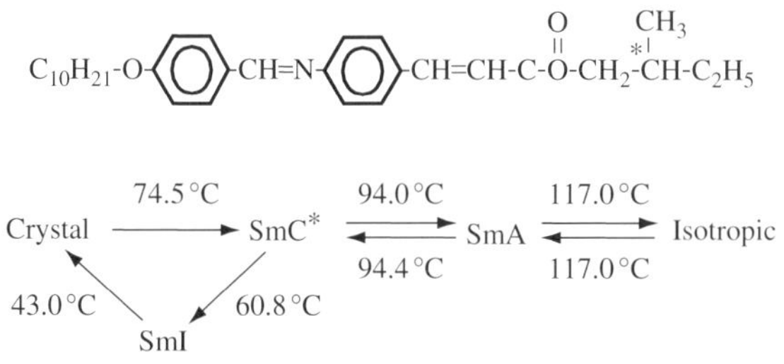

$$C_{10}H_{21}\text{-O-}\langle\bigcirc\rangle\text{-CH=N-}\langle\bigcirc\rangle\text{-CH=CH-}\overset{\overset{\text{O}}{\|}}{\text{C}}\text{-O-CH}_2\text{-}\overset{\overset{\text{CH}_3}{*|}}{\text{CH}}\text{-C}_2\text{H}_5$$

Crystal —74.5 °C→ SmC* ⇄ (94.0 °C →, ← 94.4 °C) SmA ⇄ (117.0 °C →, ← 117.0 °C) Isotropic

SmC* —60.8 °C→ SmI —43.0 °C→ Crystal

Fig. 5.1.3 Molecular structure and phase sequence of DOBAMBC.

5.1.3 Surface-Stabilized Ferroelectric Liquid Crystals (SSFLCs)

5.1.3.1 Device structure

The free rotation around the molecular long axis is hindered by chiral parts such as those at *C (Fig. 5.1.3). The origin of the spontaneous polarization is this hindered rotation. The helical structure arises from the chirality of the molecules and the flexoelectricity, but it is not essential for the spontaneous polarization. Therefore the spontaneous polarization remains if the structure is unwound. Clark and Lagerwall [2] reported that helix unwinding is produced by the surface-pinning effect when the cell spacing is less than 2 μm. This is called the surface stabilized state. The cell spacing is quite small compared with that of conventional nematic devices. Figure 5.1.4 shows the helix unwinding state. There are two domains which are tilted at $\pm\theta$ from the smectic layer normal before an electric field is applied. The molecules can switch between the two domains when opposite electric fields are applied. Bright and dark states, which are switchable alternately by using crossed polarizers, are obtained.

5.1.3.2 Technical merits

(a) Fast electro-optic response

SSFLCs have a very short switching time (in the order of microseconds) because of the direct coupling force of the spontaneous polarization and an electric field. This response is about a thousand times faster than that of conventional nematic devices, which utilize the dielectric anisotropy of the molecule.

(b) Memory effect

Both the dark and bright states of the SSFLCs shown in Fig. 5.1.4 are stable after an electric field is removed. This memory effect is a result of the surface anchoring effect and the fact that both states are elastically equivalent. The memory effect is very important for high multiplexing of displays in order to avoid the drawback of crosstalk, and to obtain an optical storage device.

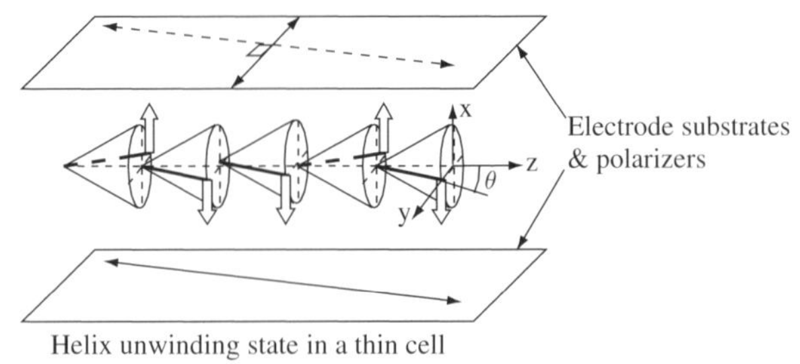

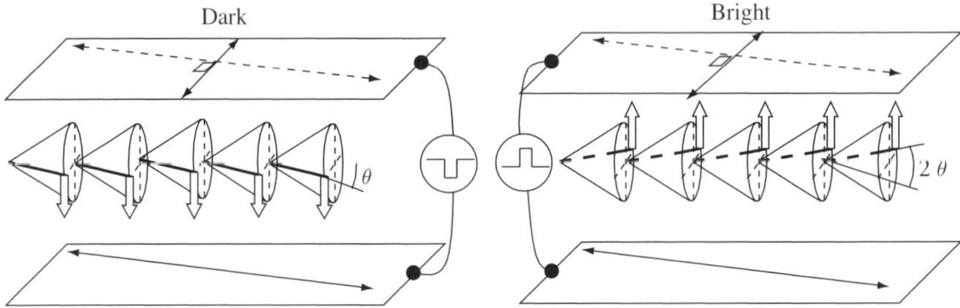

Fig. 5.1.4 Principle device structure of a SSFLC.

(c) Wide viewing angle
A wide viewing angle is derived from the in-plane molecular motions (as shown in Fig. 5.1.4). There is a slight dependence of effective birefringence on viewing angular under the in-plane molecular motion. Wide viewing angle is especially important for applications like a large-area display system viewed from multi directions by a number of persons.

5.1.4 Smectic layer structure study

5.1.4.1 Smectic layer structures in various alignments

The smectic layer structure strongly depends on the surface alignment as shown in Fig. 5.1.5. The smectic layer is formed so that it is parallel to the substrates in a homeotropic aligning cell, and the schlieren texture is observed by polarizing optical microscopy [3]. An oblique layer structure is formed in a planar homogeneous cell with antiparallel aligning treatment [4], but Rieker *et al.* found by X-ray measurements that the smectic layer structure in most SSFLCs using parallel alignment conditions involve a "chevron" shape, not a simple "bookshelf structure" [5]. Although a few unique FLC materials [6, 7] do show a quasi-bookshelf structure, the chevron layer structure is essential for SSFLCs using parallel alignment [8]. Although the theoretical understanding is not clear so far, it is reported that a uniform mono-domain can be easily obtained by using not antiparallel alignment conditions, but parallel conditions. The parallel alignment is most important and useful for practical applications.

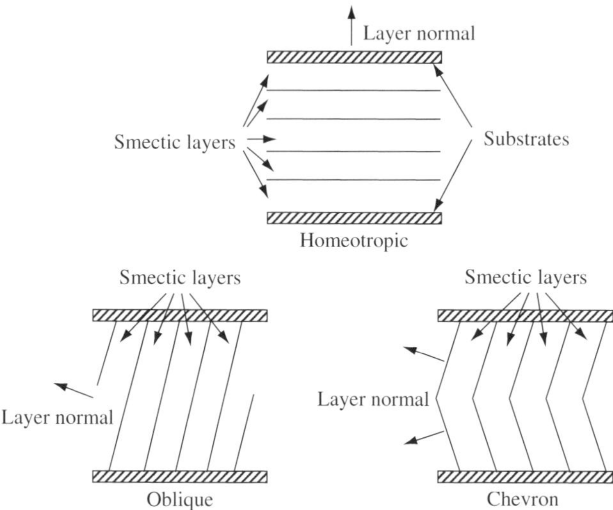

Fig. 5.1.5 Smectic layer structures in various alignment cells.

5.1.4.2 Relationship between molecular tilt angle and chevron layer tilt angle

The origin of the chevron layer structure is explained by the discrepancy between the layer spacing of the SmC* phase and that of the high temperature phase, which is usually the SmA phase. The layer spacing in the SmA phase, d_A is fixed at the surfaces, and it decreases to d_C in the SmC* phase because the molecules decline at the tilt angle θ from the smectic layer normal as shown in Fig. 5.1.6. This figure leads to a simple relation,

$$d_C = d_A \cos \delta_c, \tag{5.1.1}$$

where δ_C is the layer tilt angle in the SmC* phase.

Figure 5.1.6 and Eq. (5.1.1) indicate that the layer tilt angle is equal to the molecular tilt angle. However, it has been reported that, for a few materials, the layer tilt angle is slightly smaller than the optically measured molecular tilt angle [5, 7]. This discrepancy can be explained by recognizing the difference between the optical molecular tilt angle θ_{OPT}, resulting from the rigid central core, and the structural molecular tilt angle $\theta_{X\text{-}RAY}$ of the zigzag molecular structure overall [9, 10], including the flexible alkyl chains (Fig. 5.1.7). The liquid crystal molecular structure is assumed to be a rod as a first physical approximation. However, in the SmC* phase, the free rotation around the molecular long axis is hindered and the molecular structure is expressed by the zigzag model. Two types of zigzag model are possible as shown in Fig. 5.1.7, one is where θ_{OPT} is larger than $\theta_{X\text{-}RAY}$ and the other is where $\theta_{X\text{-}RAY}$ is larger than θ_{OPT}.

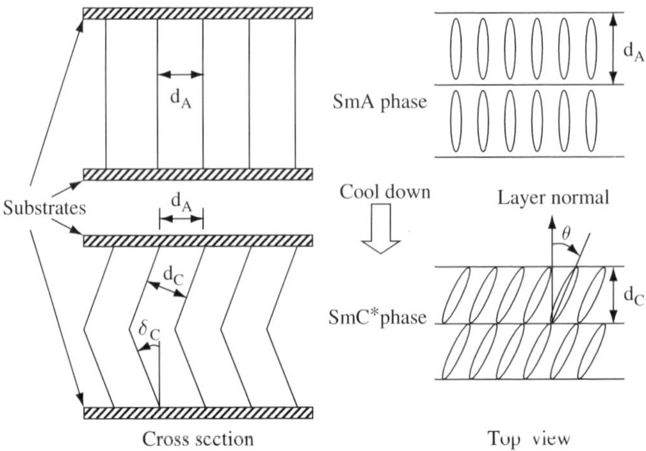

Fig. 5.1.6 Origin of the chevron layer structure.

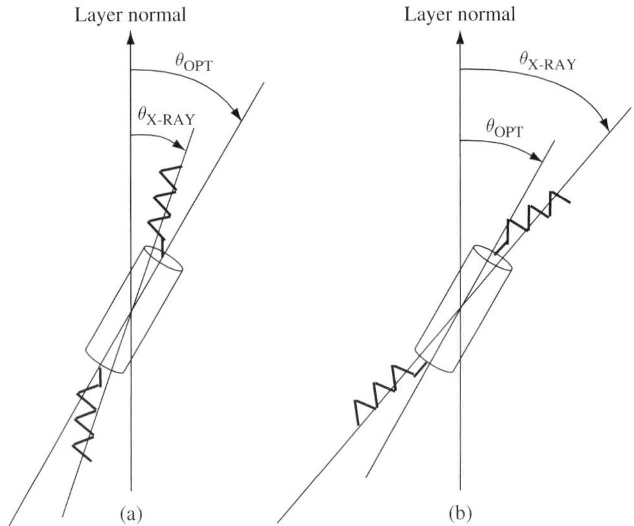

Fig. 5.1.7 Zigzag molecular structure of a SmC* material.

The chevron layer structures of several SSFLCs from various FLC materials exhibiting different optical molecular tilt angles have been precisely investigated, and a correlation between the layer tilt angle of the chevron structure and the optical molecular tilt angle was confirmed.

The layer structure was measured by using an X-ray scattering system with a goniometer at room temperature (25 °C). The measurement geometry is shown in

Fig. 5.1.8. Micro-cover glass plates about 150 μm thick were used in order to avoid X-ray absorption. The Bragg diffraction angle is 2θ. The layer tilt angle is determined as the position of strongest diffraction intensity obtained by rotating a cell through an angle β. The molecular tilt angle was measured by observing half of the angle between two extinction positions of the SSFLC cells when a sufficiently strong square wave for full switching was applied using crossed polarizers in the polarizing microscope. This method is used conventionally to measure the molecular tilt angle. It should be noted that the measured molecular tilt angle is the apparent tilt angle $\theta_{app.}$, differing from the real molecular tilt angel θ by half of the cone angle. The Cartesian coordinate system of the tilted layer structure is shown in Fig. 5.1.9. The layer is assumed to tilt in the YZ plane, where the vector **z** represents the perpendicular line of the cone. The FLC molecule is stabilized, as the spontaneous polarization vector **p** becomes parallel to the YZ plane when an electric field **E** is applied along the Y-axis. The apparent tilt angle $\theta_{app.}$ is the angle

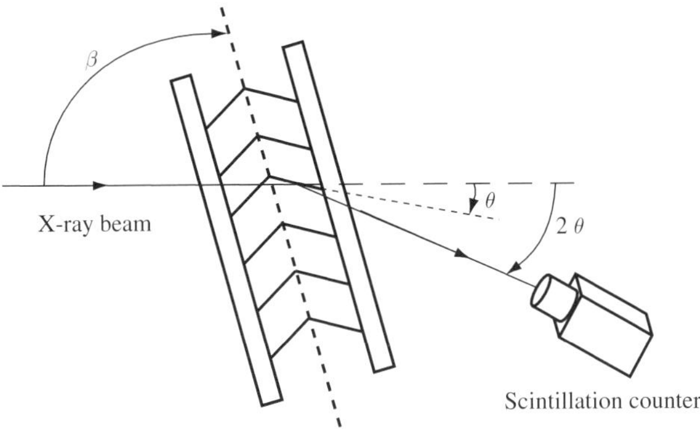

Fig. 5.1.8 Scattering geometry of the X-ray measurements.

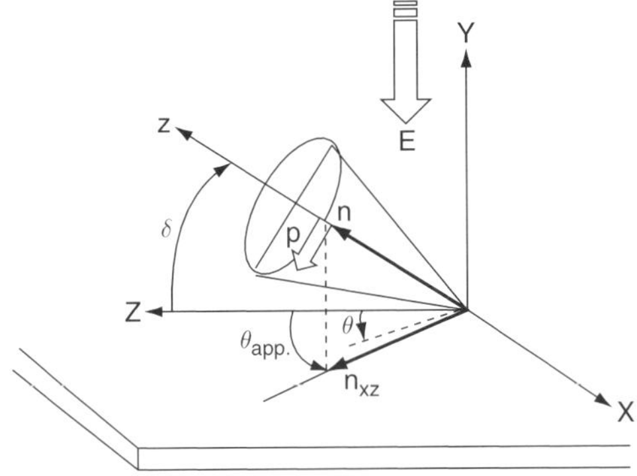

Fig. 5.1.9 Coordinate system of the tilted layer structure. $\theta_{app.}$ is the apparent molecular tilt angle.

between the layer normal Z and the projection of the director **n** on the boundary XZ plane, $\mathbf{n}_{xz}$, and is expressed as

$$\theta_{\text{app.}} = \tan^{-1}(\tan\theta \cdot \sec\delta). \tag{5.1.2}$$

The optical molecular tilt angle θ was thus determined from Eq. (5.1.2).

Some FLC materials are shown in Table 5.1.1 with the transition temperatures and the apparent tilt angles. Rieker *et al.* [5] and Ouchi *et al.* [8] reported that the layer tilt angle does not depend on the surface treatment. A high pretilt angle polyimide (PSI-A-2001) supplied by Chisso Co. Ltd., and a low pretilt angle PVA were used in order to reconfirm their reports. The rubbing direction was parallel and the cell thickness was nominally 2 µm. The pretilt angles θ_p of antiparallel treated, nominally 50 µm cells for the nematic liquid crystal E-8 (Merck Ltd) were determined from the capacitance-magnetic field (C-H) curves [11]. The pretilt angle of PSI-A-2001 was 15° and that of PVA was 0.5°. Typical X-ray diffraction profiles are shown in Fig. 5.1.10: (a) shows the profile of mixture A with PSI-A-2001 and (b) shows that of FELIX-002 with PVA. The peaks of the diffraction patterns are quite sharp, and their positions and intensities are symmetric around the surface normal where β is 90°. The typical chevron layer structure was formed in all SSFLC samples. Figure 5.1.11 shows the relationship between tilt angle θ and layer tilt angle δ. The circles and crosses marked (a) to (f) in Figs 5.1.11 and 5.1.12 represent the physical properties of the FLCs shown in Table 5.1.1 with identical letters. The dashed lines indicate the approximation that the layer tilt angle is equal to the optical tilt angle. It is clear that the layer tilt angle is slightly smaller than the optical molecular tilt angle. Figure 5.1.12 plots the ratio $\kappa(=\delta/\theta)$ versus *theta*. The variable $|\kappa|$ is around 0.9 and exhibits a tendency to decrease gradually as θ increases. A simple hypothesis is that the zigzag molecular structure of FLCs with a large molecular tilt angle is more marked than that of FLCs with a small tilt angle, as shown in Fig. 5.1.13. The difference between the aligning films was not recognized.

Table 5.1.1 Physical properties of FLC materials used for X-ray study

FLC material	Transition temperature/°C				Apparent tilt angle/° (25 °C)
	Cr • SmC*	• S_mA	• N	• I	
Mixture A(a)	• <RT •	55 •	82 •	93 •	17
CS-1014(b)	• <RT •	54 •	69 •	81 •	21
CS-1022(c)	• <RT •	60 •	73 •	85 •	25
ZLI-3654(d)	• <RT •	62 •	76 •	86 •	25
ZLI-3489(e)	• <RT •	65 •	71 •	87 •	29
FELIX-002(f)	• <RT •	70 •	77 •	87 •	33

(a) blended by SHARP Co.
(b), (c) supplied by Chisso Co. Ltd.
(d), (e) supplied by E. Merck
(f) supplied by Hoechst AG

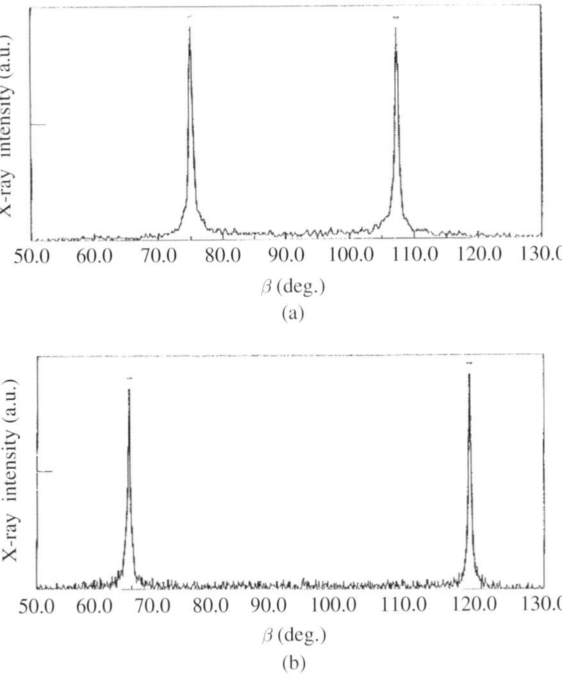

Fig. 5.1.10 X-ray scattering profiles (a) mixture A with PSI-A-2001 and (b) FELIX-002 with PVA.

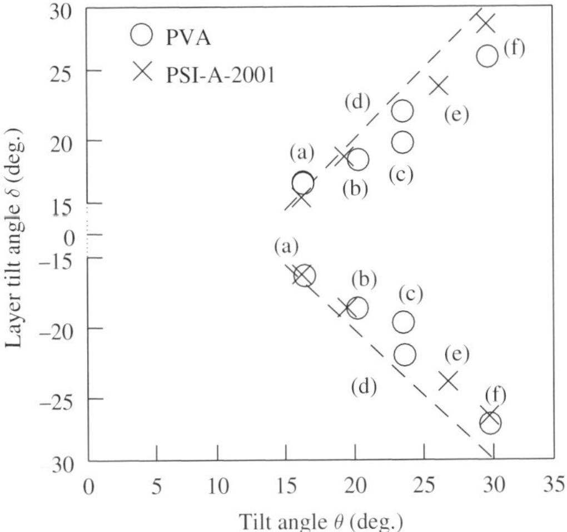

Fig. 5.1.11 Relationship between layer tilt angle and molecular tilt angle. See text for explanation of figure.

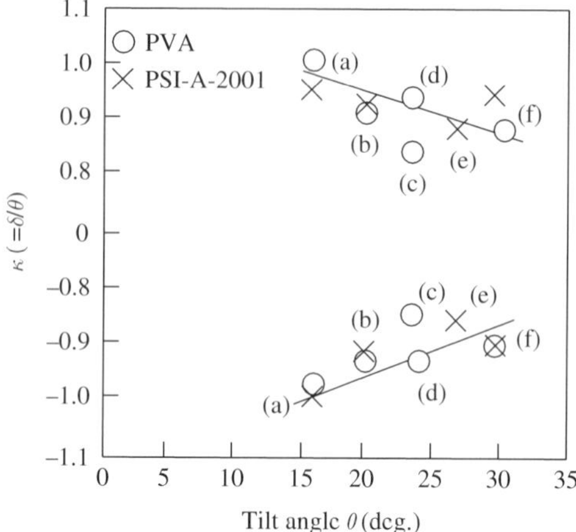

Fig. 5.1.12 Correlation between the ratio of the layer tilt angle to the molecular tilt angle, κ and the molecular tilt angle. See text for explanation of figure.

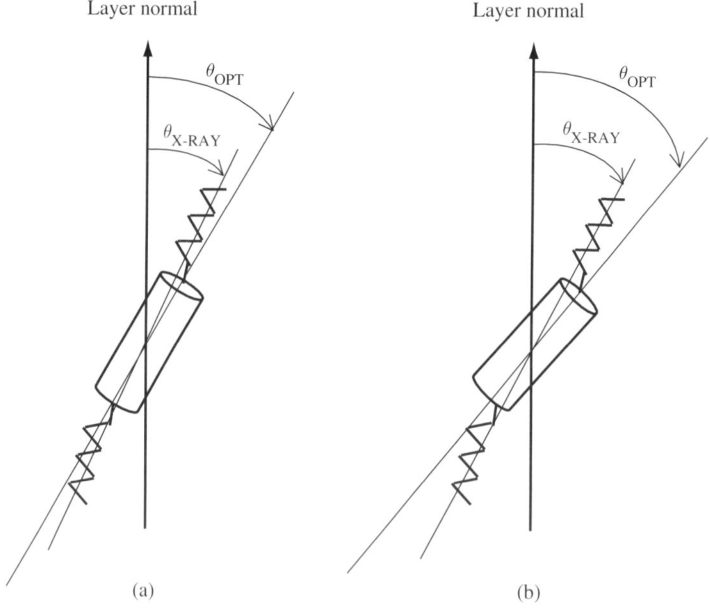

Fig. 5.1.13 Zig zag molecular structure: (a) a material showing a small molecular tilt angle; (b) a material showing a large molecular tilt angle.

5.1.5 Molecular orientational states and optical properties

The molecular orientational states of the SSFLCs have been analyzed by polarizing microspectroscopy and optical simulation. The X-ray studies indicated that the chevron layer structure is determined by the bulk properties of the FLC, but the molecular orientation in the smectic layer is strongly influenced by the surface properties. The effect of surface pretilt angle on the molecular orientation and the optical properties of SSFLCs have been studied by the optical simulation based on the molecular orientational models.

5.1.5.1 Classifications of molecular orientational states

Two basic classifications of the molecular orientational states of SSFLCs are used. One classification is based on the optical viewing behaviour when SSFLCs are placed between crossed polarizers. The uniform (U) and twisted (T) states are defined according to this classification [12, 13]. The uniform state shows extinction positions, but the twisted state shows only positions of colouration without any extinction. Another classification is based on the relationship between the tilting direction of the chevron layer structure and the direction of the surface pretilt. The C1 and C2 states are defined by this classification [14]. The C1 and C2 states are easily distinguished because the tilting direction of the chevron layer structure, confirmed by the direction of the zigzag defects [15, 16], and the direction of the surface pretilt are consistent with the rubbing direction [17]. The zigzag defects are caused by discontinuities in the chevron layer structure. Figures 5.1.14(a) and (b) show the C1 and C2 states in relation to the zigzag defect; (c) shows the smectic layer models of the C1 and C2 states.

Four states, C1U (C1-uniform), C1T (C1-twisted), C2U (C2-uniform), and C2T (C2-twisted), were found in SSFLC cells by investigating many samples aligned by various alignment films involving parallel rubbing [18]. These four states are expressed by a combination of the above two classifications.

5.1.5.2 Observed phenomena and considerations

FLC materials used for study are shown in Table 5.1.2 with apparent tilt angles, layer tilt angles, and tilt angles. Baking temperatures and pretilt angles of the aligning films are listed in Table 5.1.3. It is noted that all aligning films are polyimide, and as it is known that the baking temperature of polyimide affects the pretilt angle [19], the baking temperatures are also listed in Table 5.1.3. The SSFLC cells were mounted on a hot stage and positioned at extinction between the crossed polarizers of a polarizing microscope. Memory angle θ_m was defined as the half angle between the two extinction positions when no field was applied. A spectroscope was attached to the polarizing microscope in order to measure the transmitted light from small areas such as those inside zigzag defects.

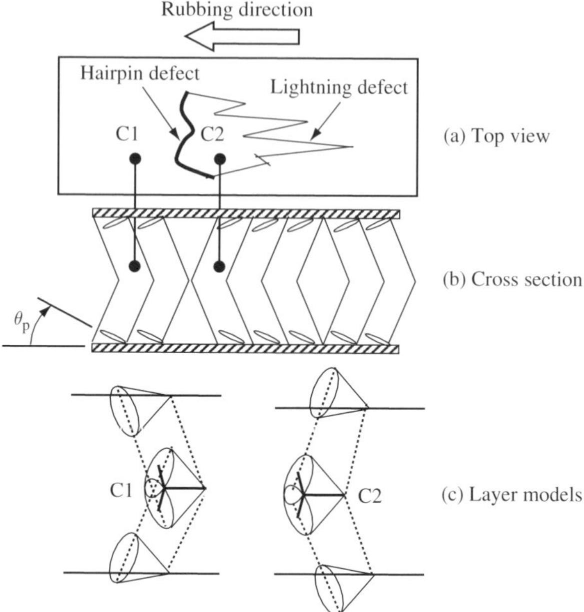

Fig. 5.1.14 C1 and C2 states, distinguished by the relationship between the direction of the chevron layer structure and the direction of the surface pretilt, as shown in (a) and (b). The tilting direction of the chevron layer structure is confirmed by the direction of the zigzag defects, as shown in (b). (c) Smectic layer models of the C1 and C2 states.

Table 5.1.2 Tilt angles and layer tilt angles of FLC materials used for a molecular orientational study at 25 °C

FLC material	Apparent tilt angle/°	Layer tilt angle/°	Tilt angle/°
CS-1014(i)	21.0	18.0	20.0
SF-1212(ii)	9.5	9.0	9.4
SCE-8(iii)	22.0	19.5	21.0

(i) supplied by Chisso Co. Ltd.
(ii) blended by SHARP Co.
(iii) supplied by E. Merck

Figure 5.1.15 shows the polarized optical micrographs of the cell made from FLC CS-1014 and the high pretilt angle aligning film PSI-A-2001 ($\theta_p = 15°$). Figure 5.1.16 shows the micrographs of the cell made from CS-1014 and the low pretilt angle aligning film PI-X ($\theta_p = 3°$). Both micrographs were taken in the crossed

Table 5.1.3 Baking temperatures and pretilt angles of aligning films used for molecular orientational study

Aligning film	Baking temperature /°C	Pretilt angle/°
PI-X	200	3
PSI-A-2101	200	6
PSI-A-X018	250	10
PI-Z	250	10
PSI-A-2001	200	15
PSI-A-X021	250	20

All aligning films were supplied by Chisso Co. Ltd.

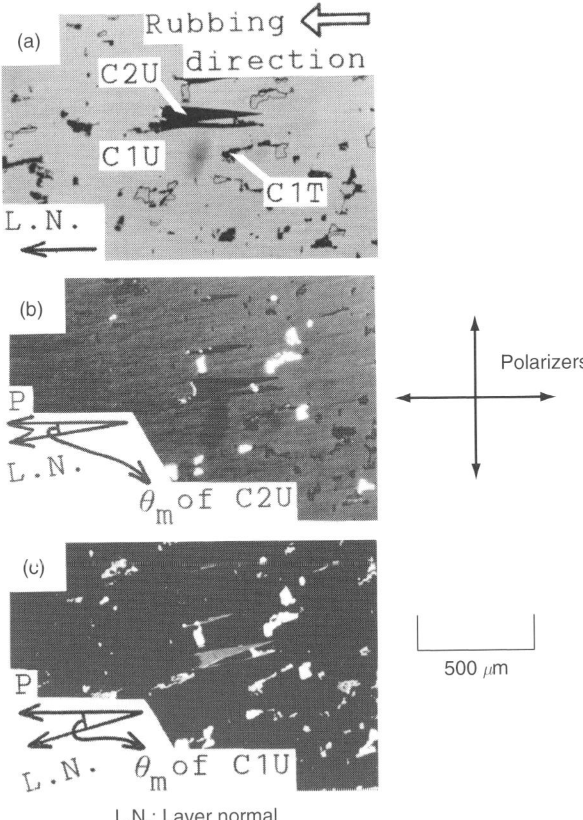

Fig. 5.1.15 Polarized optical micrographs of the sample of CS-1014 with high pretilt angle aligning film PSI-A-2001. See text for explanation. L. N. denotes the layer normal.

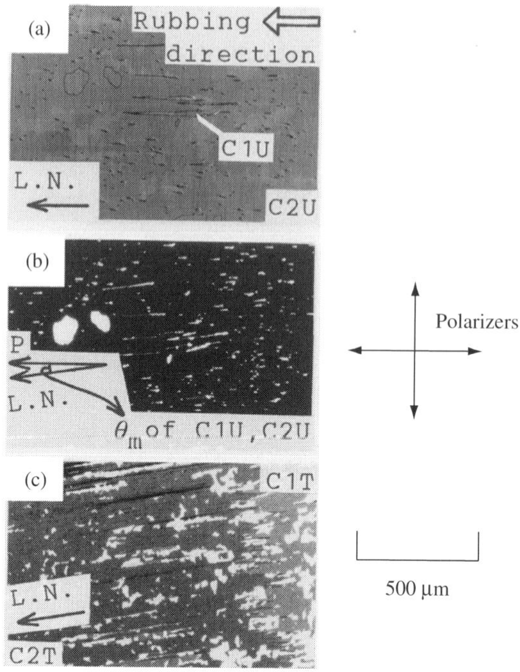

Fig. 5.1.16 Polarized optical micrographs of the sample of CS-1014 with low pretilt angle aligning film PI-X. See text for explanation. L. N. denotes the layer normal.

polarizers position at 25 °C. The C1 and C2 states can be seen on each side of the zigzag defects. The layer normal is parallel to the polarizer in Figs 5.1.15(a) and 5.1.16(a). Figures 5.1.15(b) and 5.1.16(b) show the viewing states when the cells are rotated from the positions shown in Figs 5.1.15(a) and 5.1.16(a), respectively. Only the C2 state shows the extinction position (C2U) in Fig. 5.1.15(b), and the C2 state showed only the C2U state everywhere in this PSI-A-2001 cell. Both the C1 and C2 states show the extinction position (C1U and C2U) in Fig. 5.1.16(b). Figure 5.1.15(c) shows the viewing state when the cell is rotated further from the position of that in Fig. 5.1.15(b). Only the C1 state showed extinction (C1U) at this position, and the small C1 state without any extinction position (C1T) can be seen. Figure 5.1.16(c) shows another area of the cell. Both the C1 and C2 states showed only colouration positions without any extinction (C1T and C2T). From Figs 5.1.15 and 5.1.16, it is found that the four states, C1U, C1T, C2U and C2T, can appear in SSFLCs treated by parallel rubbing and that SSFLCs with high pretilt angle aligning films show only one state with extinction positions in the C2 state. This type of the C2 state is defined as a special case of the C2U state, called the high pretilt C2U state. On the other hand, both the C1U and C1T states can appear regardless of the pretilt angle.

The C1U and C2U states are interesting from the practical point of view because they give the possibility of achieving a high contrast.

The memory angles of the C1U and C2U states are shown schematically in Figs 5.1.15 and 5.1.16. The memory angle of the C1U state is wider than that of the C2U state in the high pretilt cell (Figs 5.1.15(b) and 5.1.15(c)). However, the difference between the memory angles of the C1U and C2U states is very small in the low pretilt cell (Fig. 5.1.16(b)).

It is known that memory angle depends on surface pretilt angle. The temperature dependences of the memory angles of the C1U and C2U states were measured for all cells, and are shown in Figs 5.1.17(a)–(c). The solid lines represent the

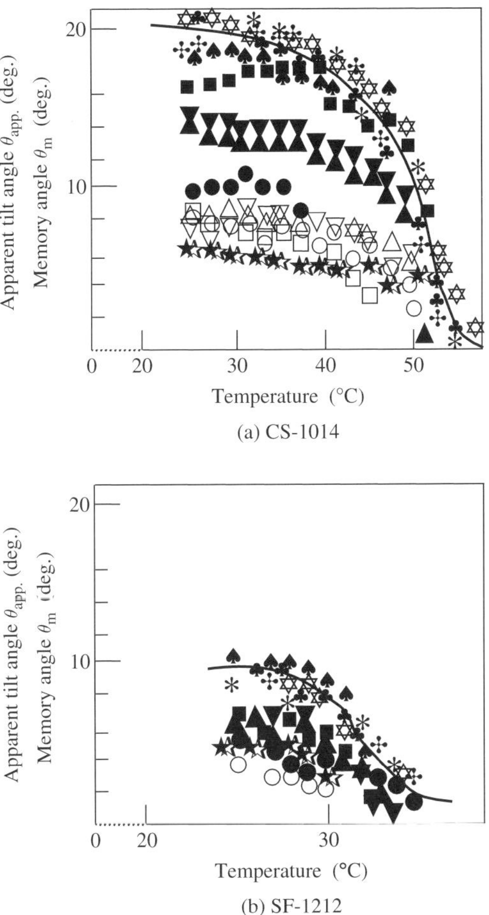

Fig. 5.1.17 (Continued)

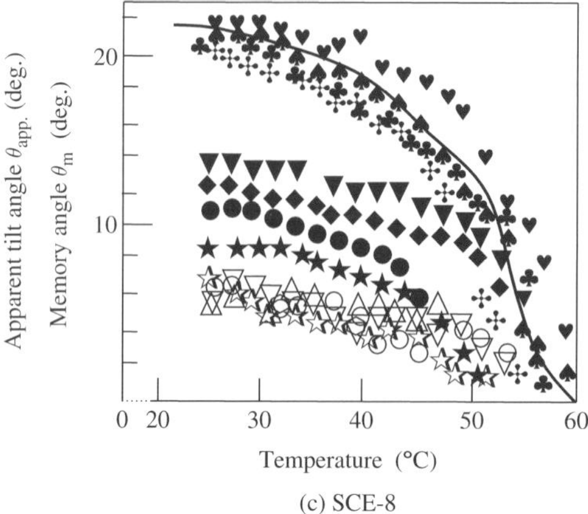

Fig. 5.1.17 Temperature dependence of apparent tilt angle (♠, ✠, ♣, ♥, ∗, ✡), the memory angle of the C1U state (★, ●, ▲, ◆, ▼, ■) and the memory angle of the C2U state (☆, ○, △, ◇, ▽, □) with various aligning films PI-X (♠, ★, ☆), PSI-A-2101 (✠, ●, ○), PSI-A-X018 (♣, ▲, △), PI-Z (♥, ◆, ◇), PSI-A-2001 (∗, ▼, ▽) and PSI-A-X021 (✡, ■, □) for (a) CS-1014, (b) SF-1212 and (c) SCE-8. Solid lines are the apparent tilt angle behaviours.

apparent tilt angle behaviour. There is little difference in the apparent tilt angles of cells with the same material, indicating the reliability of measurements. Some cells made of FLC SF-1212 and aligning films, except for low pretilt angle PI-X and PSI-A-2101, showed only the C1 state with slight zigzag defects. According to Kanbe *et al.* [14], it is difficult for the C2 state to appear if the tilt angle is small or the surface pretilt angle is high. The memory angle of the C2U state is almost independent of the surface pretilt angle, while the memory angle of the C1U state strongly depends on the surface pretilt angle.

The optical properties of SSFLC cells made of CS-1014 and PSI-A-2001 were measured in order to analyze the molecular orientations. The wavelength dependences of the memory angles of the C1U and C2U states are shown in Fig. 5.1.18. The C1U and C2U states exhibit the opposite wavelength dispersion with respect to each other. The transmission spectra of both memory states, the bright and dark states, prepared by applying a pulsed electric field to sample cells are shown in Fig. 5.1.19. The transmission spectra indicate that a high contrast ratio is obtained not under the C2U state, but under the C1U state. The wide memory angle of the C1U state contributes to a high contrast ratio. The transmitted light intensity was calculated theoretically in order to discuss the above optical properties.

Molecular orientational models C1U, C1T, C2U and C2T and the high pretilt angle C2U states are shown in Fig. 5.1.20. The molecules are almost uniformly tilted to one side from the layer normal in both the C1U and C2U models. The

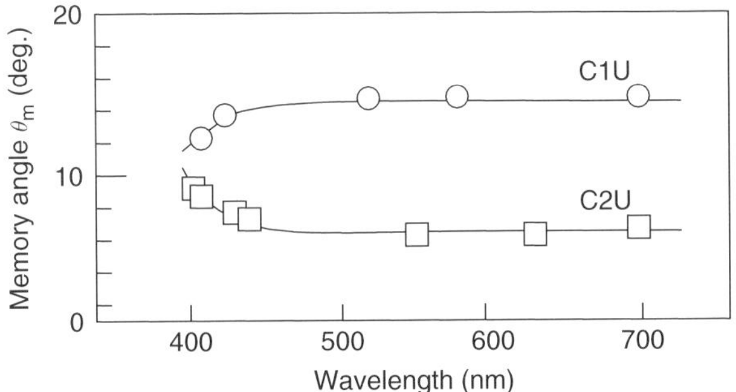

Fig. 5.1.18 Wavelength dependence of the memory angles of the C1U state (○) and the C2U state (□).

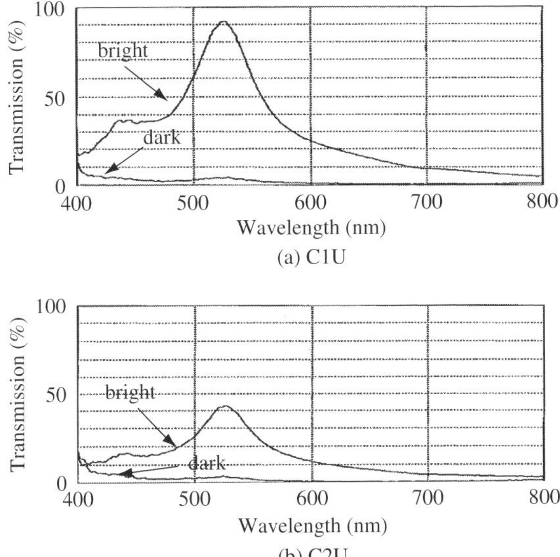

Fig. 5.1.19 Transmission spectra of (a) C1U and (b) C2U memory states. Upper curves represent the bright spectra and lower curves represent the dark spectra in each figure.

C1T and C2T models are the half splayed states [20, 21], and the boundary surfaces in the C2 state do not have wide regions wherein the molecules can exist in a stable state (Fig. 5.1.14). The c-directors at the surfaces are almost perpendicular to the substrate in the high pretilt angle C2U model. The molecules at the surfaces can move easily in the C2 state if the pretilt angle is low. These states are assumed to switch between two elastically equivalent states that produce the stable memory effect.

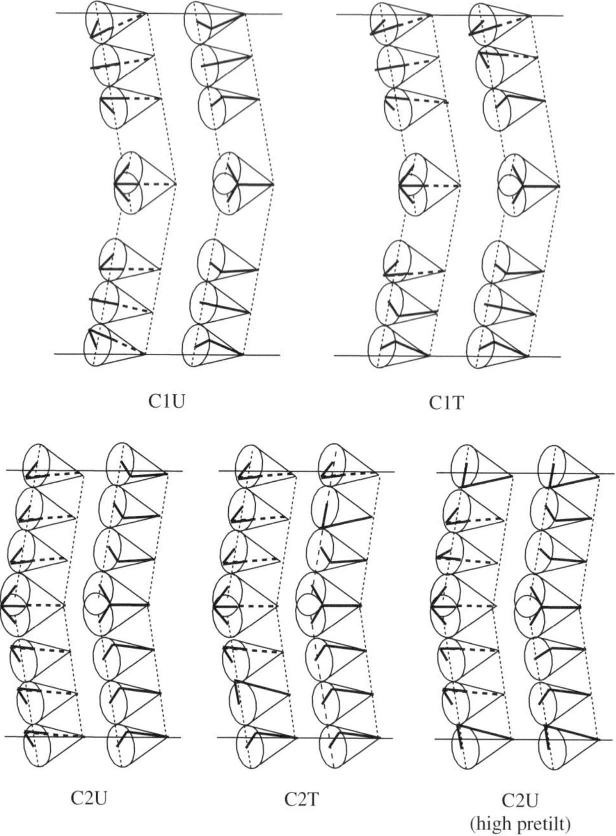

Fig. 5.1.20 Molecular orientational models of SSFLCs with a chevron layer structure.

Figure 5.1.21 shows the coordinate systems of the director in the chevron layer structure, where **n** is the director, **c** is the c-director, and **p** is the spontaneous polarization vector. It is assumed that the tilt angle θ and the layer tilt angle δ are constant, and the azimuthal angle Φ depends only on the cell thickness, direction Y. The director is expressed as

$$\mathbf{n}(\mathrm{x,y,z}) = (\sin\theta\cos\Phi,\ \sin\theta\sin\Phi,\ \cos\theta), \tag{5.1.3}$$

and

$$\mathbf{n}(\mathrm{X,Y,Z}) = \begin{bmatrix} \sin\theta\cos\Phi \\ \sin\theta\sin\Phi\ \cos\delta - \cos\theta\sin\delta \\ \sin\theta\sin\Phi\ \sin\delta + \cos\theta\cos\delta \end{bmatrix} \tag{5.1.4}$$

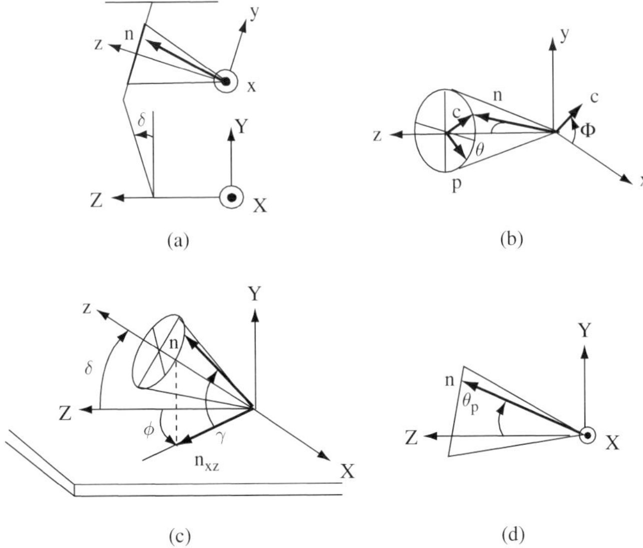

Fig. 5.1.21 Coordinate systems used for simulations. Y and Z represent the cell thickness direction and the smectic layer normal, respectively, and z is the perpendicular line of the cone. (a) Chevron layer system. (b) Cone system. (c) Director tilt angle γ and director twist angle ϕ. (d) Definition of the surface pretilt angle.

The transmitted light intensity was calculated by the Berreman 4×4 matrix method [22]. The director tensor is determined by the director tilt angle γ and the director twist angle ϕ, which are expressed as

$$\gamma = \sin^{-1}(\sin\theta \sin\Phi \cos\delta - \cos\theta \sin\delta), \tag{5.1.5}$$

$$\phi = \tan^{-1}\left[\frac{\sin\theta\cos\Phi}{\sin\theta\sin\Phi\sin\delta + \cos\theta\cos\delta}\right]. \tag{5.1.6}$$

The pretilt angle θ_p is simply defined in Fig. 5.1.21(d). The c-director pretilt Φ_0 is expressed by Eq. (5.1.7):

$$\Phi_0 = \sin^{-1}\left[\frac{\tan\delta}{\tan\theta} + \frac{\sin\theta_p}{\sin\theta\cos\delta}\right]. \tag{5.1.7}$$

It is assumed that Φ_0 at the bottom surface is $\pi/2$ and Φ_0 at the top surface is $-\pi/2$ for a high pretilt angle C2U. The azimuthal angle at the chevron interface Φ_{IN} is expressed as

$$\Phi_{IN} = \sin^{-1}(\tan\delta / \tan\theta). \tag{5.1.8}$$

For simplicity, ignoring the effect of the polarization electric field on the elastic deformation [23], it is assumed that Φ changes with Y at a constant rate. According to Kawaida *et al.* [24], the wavelength dispersion of the refractive indices was taken into account in the calculation of transmission. The wavelength dependences of the memory angles and the transmission spectra of CS-1014 ($\theta = 20.0^\circ$ and $\delta = 18.0^\circ$) and PSI-A-2001 ($\theta_p = 15^\circ$) were calculated.

The calculated wavelength dependences of the memory angles are shown in Fig. 5.1.22. Both the C1U and C2U models can predict the same dispersion as that measured. Figure 5.1.23 shows the calculated transmission spectra. Both the C1U and C2U models show slight transmission in short wavelength regions of the dark state, and show a peak transmission around 500 nm in the bright states. These calculations are almost consistent with the experimental results in Figs 5.1.18 and 5.1.19, indicating the approximate validity of the orientational models. The dependence of memory angle on surface pretilt angle was simulated by using the models (Fig. 5.1.24) and also measured (Fig. 5.1.17) at 25 °C. Solid lines indicate the simulation results and symbols express the measured results in Fig. 5.1.24. The calculated memory angle corresponds to a wide visible wavelength.

In Fig. 5.1.25, the director twist angle ϕ is plotted as a function of the cell thickness direction Y at various surface pretilt angles, where ϕ_{IN} represents the director twist angle at the chevron interface and is expressed as

$$\phi_{IN} = \cos^{-1}(\cos\theta/\cos\delta). \tag{5.1.9}$$

These director profiles were calculated for CS-1014. The simulated memory angles of each director profile are shown in Fig. 5.1.25. In Fig. 5.1.26, the calculated transmission spectra for each director profile are shown. These figures indicate that memory angle and the transmission intensity in the bright state depend on the surface pretilt angle.

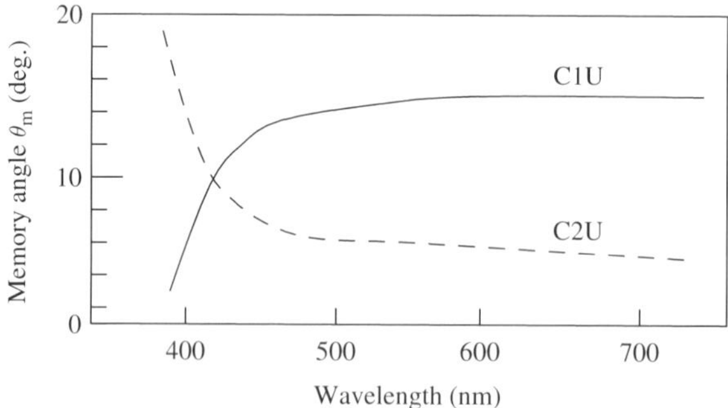

Fig. 5.1.22 Calculated wavelength dependences for the memory angles of the C1U and the C2U states.

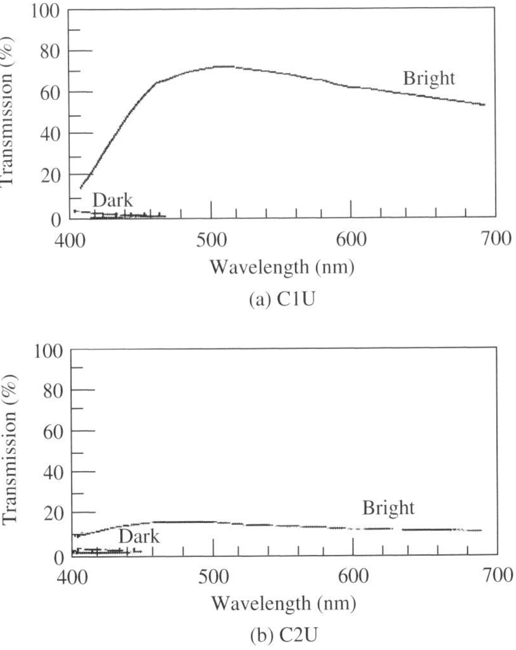

Fig. 5.1.23 Calculated transmission spectra of (a) C1U and (b) C2U states.

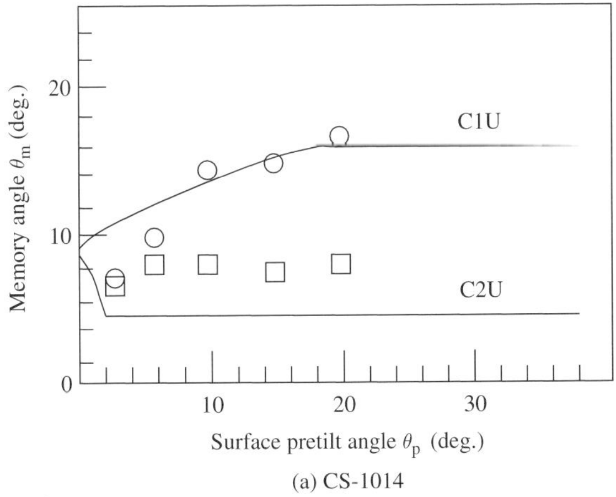

Fig. 5.1.24 (Continued)

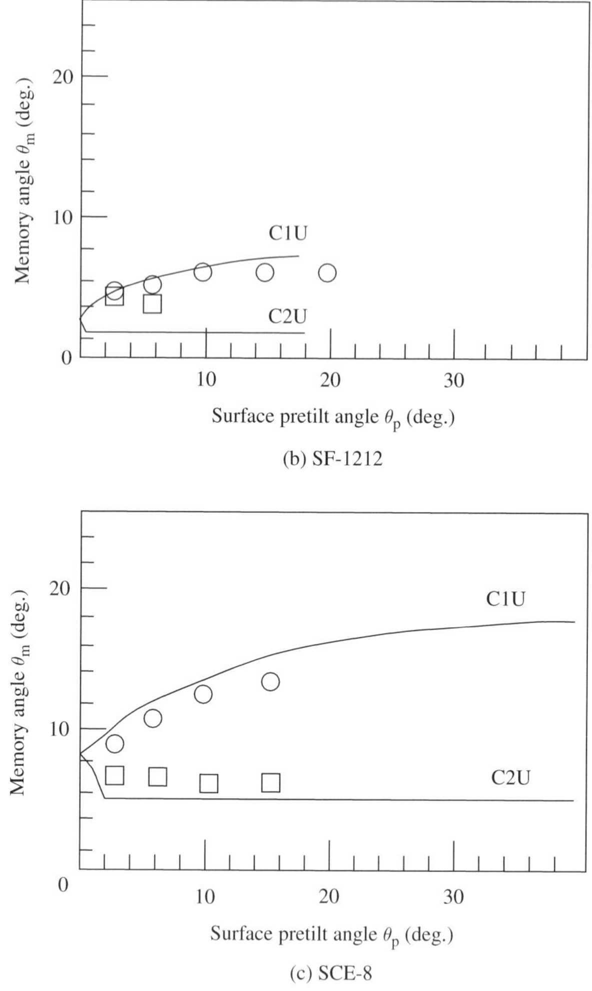

Fig. 5.1.24 Relationships between the surface pretilt angle and the memory angle of the C1U state (○) and the C2U (□) state at 25°C for (a) CS-1014, (b) SF-1212 and (c) SCE-8. Solid lines indicate simulation results by using the models.

It is shown that memory angle θ_m is approximately equal to half of the total director twist angle in the cell and determined by the following equation,

$$\theta_m \fallingdotseq (\phi_0 + \phi_{IN})/2, \tag{5.1.10}$$

where ϕ_0 is the surface director twist. The memory angle of the C1U state is made large by increasing the surface pretilt angle and saturates at $\theta_{msat.}$ given by Eq. (5.1.11),

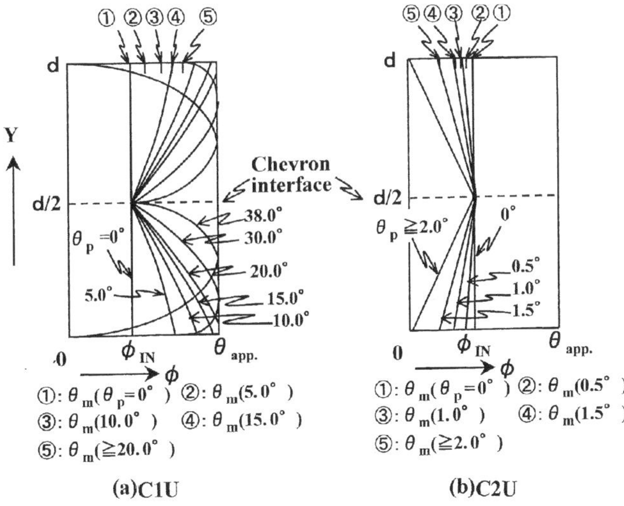

Fig. 5.1.25 Calculated director profiles and the memory angles for various surface pretilt angles of the (a) C1U and (b) C2U states. Memory angles are denoted by encircled numerals: In (a) ①: $\theta_m(\theta_p = 0°)$, ②: $\theta_m(\theta_p = 5.0°)$, ③: $\theta_m(\theta_p = 10.0°)$, ④: $\theta_m(\theta_p = 15.0°)$ and ⑤: $\theta_m(\theta_p \geq 20.0°)$ In (b) ①: $\theta_m(\theta_p = 0°)$, ②: $\theta_m(\theta_p = 0.5°)$, ③: $\theta_m(\theta_p = 1.0°)$, ④: $\theta_m(\theta_p = 1.5°)$ and ⑤: $\theta_m(\theta_p \geq 2.0°)$

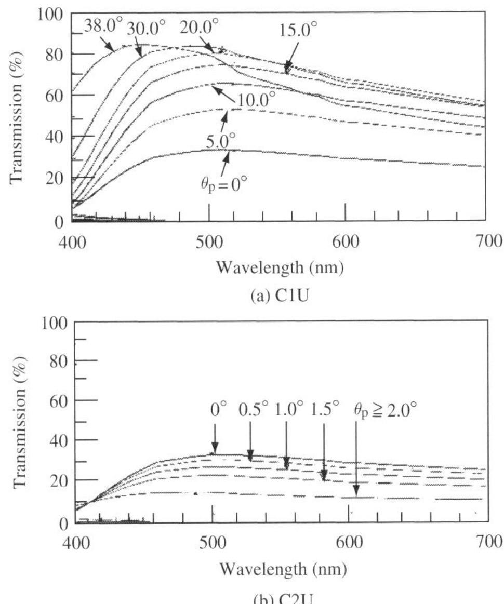

Fig. 5.1.26 Calculated transmission spectra for various surface pretilt angles of (a) C1U and (b) C2U states.

$$\theta_{\mathrm{msat.}} \fallingdotseq (\phi_{\mathrm{IN}} + \theta_{\mathrm{app.}})/2, \tag{5.1.11}$$

On the other hand, the memory angle of the C2U state increases as the surface pretilt angle decreases. However, the memory angle of the C2U state seems from observations to be almost constant, because it depends only slightly on surface pretilt angle.

5.1.6 Summary

This chapter has described the layer structure and molecular orientation of SSFLCs. The orientational states, expressed by c-director orientations and director profiles, which appear in SSFLC cells subjected to parallel rubbing are summarized in Fig. 5.1.27. It is clear that these molecular orientational states produce different optical properties even in an SSFLC cell made of one FLC material. Tsuboyama *et al.* demonstrated a 15 inch diagonal high-resolution mulit-colour FLC display by utilizing the C1U state [25]. Itoh *et al.* reported a 17 inch diagonal video-rate full colour FLC display by utilizing the C2U state [26]. The alignment technology for preparing a suitable state for the purpose of each device is a subject of major industrial importance. The relationship between molecular orientation and display performance, and the alignment methods to prepare preferred molecular orientational states for each display are described in Chapter 6.

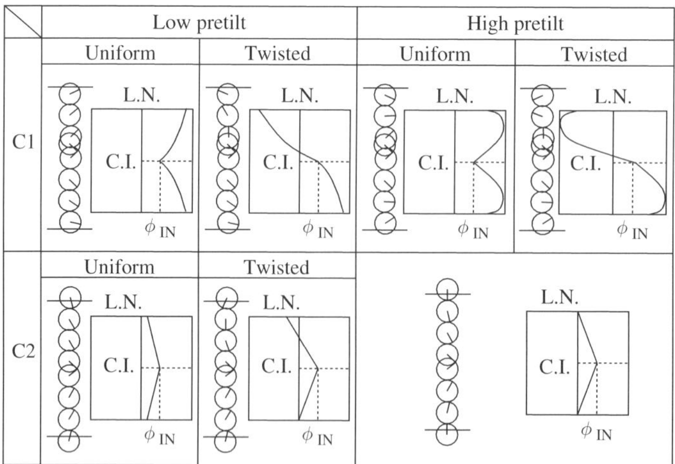

Fig. 5.1.27 Summary of the orientational states in SSFLCs with a chevron layer structure.

5.2 Alignment and Bistability of Ferroelectric Liquid Crystals

Kohki Takatoh

5.2.1 Introduction

In this section, the relationship between the alignment and bistability of ferroelectric liquid crystals will be discussed. From 1985, studies of surface stabilized ferroelectric liquid crystals (SSFLCs) expanded in the quest for large size liquid crystal displays with simple multiplex driving. In those days, achieving active matrix LCD technologies without defects, especially for large size displays of high duty ratio, was considered doubtful. The SSFLCDs attracted a lot of interest because of their memory properties and the realization of mutiplex driving with high duty ratio by a simple matrix without TFTs, and with fast response and wide viewing angle.

For the alignment of SSFLCs, four aspects were considered to be of major importance [1]:

(1) Defect density or the transmission of the dark state
(2) Cone angle between the two relaxed states without fields
(3) Time taken to switch the cell to a stable state
(4) Stability of the director orientation towards small disturbing fields.

Of these four parameters, the second and the fourth are related to the bistability. In this section, the alignment phenomena or the problems related to bistability will be mainly discussed, together with the control of defects.

5.2.2 Driving scheme of SSFLCs

In this chapter, the driving scheme of SSFLC displays will be explained in relation to the bistability, to clarify the reason why the bistability is necessary to demonstrate good quality images for SSFLC displays. The proposed driving schemes of SSFLCs can be categorized into two types. One is a line-sequential scheme, the common multiplex driving scheme for such as TN- or STN-LCDs. Pulse signals are applied to every scanning electrode from the top line to bottom line. Another type of scheme is the "partial access scheme" whereby only the parts of changing images are selectively rewritten. In this scheme, electric voltage is applied to only the pixels in which the pictures change.

Consider the SXGA display for monitor usage (see Fig. 5.2.1). In the case of the SXGA display, the number of scanning electrodes is 1024, or the duty ratio is 1/1024. For the usual line-sequential scheme, the frame frequency of the LCD is 60 Hz, or 60 pictures are formed every second. So, the period of one frame (frame rate) or

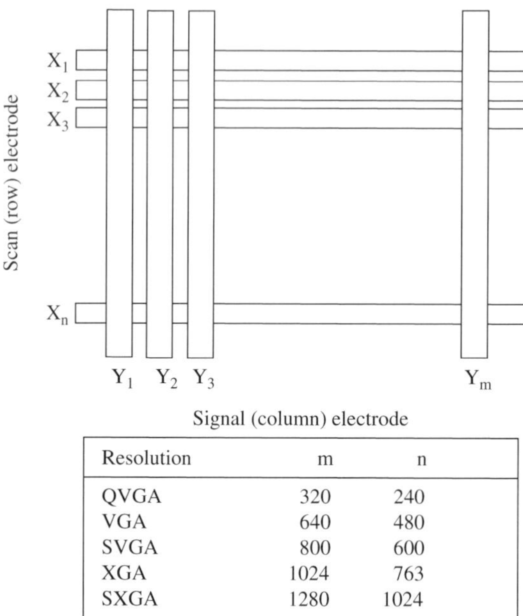

Resolution	m	n
QVGA	320	240
VGA	640	480
SVGA	800	600
XGA	1024	763
SXGA	1280	1024

Fig. 5.2.1 Simple matrix of an SSFLC display.

one picture is 16 ms (1/60 s). Flicker becomes noticeable if the frame frequency becomes too low, for example 1/30 s. The charging time, or time for applying electric voltage to the liquid crystal, becomes 16 μs (16 ms/1024). This means the switching of the liquid crystal must finish within 16 μs. In this case, the memory state should be maintained for 16 ms.

For high duty ratio displays, the partial access scheme is more desirable [5]. Here, the 1/3 bias driving scheme, a typical partial access scheme method will be explained [6]. In Fig. 5.2.2, the scanning pulses and data pulses for the 1/3 bias driving method are illustrated. In this figure, (6) and (7) are the wave forms which induce switching. In the case of other forms, the initial states are maintained. On the lines to be renewed, a voltage, V_{cs} is applied. In this manner, the number of scanning lines can be increased. However, the liquid crystal must maintain the memory state until the next signals (6) or (7) are applied. In fact, in practice, perfect bistability cannot be realized and the liquid crystalline molecules return to their inherent stable positions. To prevent this, a method to rewrite all scanning lines during regular intervals was proposed. For example, a half period is used for the partial access scheme and another half period is used for the rewriting process. To reduce the flicker phenomenon, every n-th scanning line, for example every 16th scanning line is selected. Refreshing is carried out in the order of (1st, 17th, ..., 1 + 16nth, 2nd, ..., 2 + 16nth, ..., 16th, ..., 16 + 16nth) scanning line. In the case of SXGA displays in which 400 μs is

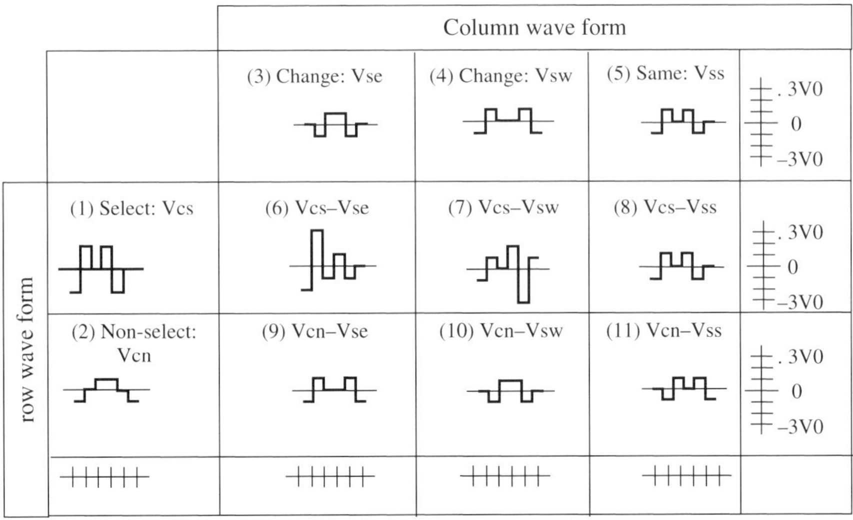

Fig. 5.2.2 1/3 Bias Driving Wave Form Signal (1) is applied to the selected scanning lines. Signals (3), (4) or (5) are applied to the signal lines [6].

necessary for rewriting the FLC materials, one pixel is refreshed once in every $1024 \times 400\,\mu s \times 2 = 819.2\,ms$. This means the memory state must be maintained for 820 ms.

If a perfect memory property could be realized, the refreshing process would not be necessary. Only by a partial access scheme could a high contrast ratio be obtained. On the other hand, in the case of SSFLCs with a poor memory property, the refreshing process must be carried out frequently, and the number of scanning lines would be limited. For SSFLCs, memory property determines both the number of scanning lines and the contrast ratio.

5.2.3 FLC materials for excellent alignment and bistability

From the early stages of SSFLC development, it is known that the phase sequence

$$I - N^* - SmA - SmC^*$$

is preferable for excellent alignment. The best alignment can be obtained where the helical structure of the cholesteric (N*) phase is unwound. Bradshaw *et al.* [7] pointed out that if the pitch of the cholesteric phase p satisfies the relationship $p > 4d$ (d is the cell gap), the helical structure of the cholesteric phase is unwound by alignment layers rubbed in parallel directions. However, the chirality necessary for producing ferroelectricity in the chiral smectic C phase induces a short pitch length (p < 1 μm) in the cholesteric phase. One method of realizing a long helical pitch of the cholesteric phase is a reduction in the amount of chiral dopant. However, by this method, properties such as spontaneous polarization become limited. A more suitable way to realize a long pitch of the cholesteric phase is the use of more than two chiral components with opposite cholesteric twist senses. A left-handed cholesteric component and a right-handed cholesteric component can be mixed to extend the cholesteric pitch. The infinite pitch of the cholesteric phase appears only within a limited temperature range. The composition of the mixture can be adjusted so that the pitch of the cholesteric phase extends to an infinite length at a temperature just above the transition temperature between the cholesteric phase and the smectic A phase.

Concerning the smectic C phase, a large pitch length in the chiral smectic C phase stabilizes the surface stabilized state. Therefore, a large pitch length of the smectic C phase is again preferable. Ferroelectric liquid crystalline materials with opposite twist senses of the chiral smectic C phase, but the same sign of spontaneous polarization, are used for a large helical pitch length and a large spontaneous polarization.

5.2.4 Polymer crystallinity, linearity and molecular structures

5.2.4.1 Criteria for bistability

For the alignment process of SSFLCDs, several methods have been proposed, such as shearing, application of magnetic fields, temperature gradients, and other procedures. However, for practical purposes, only the rubbing method has been adopted. In this chapter, the author focuses on the practical rubbing process, and the effects of the alignment layer surfaces on the bistability of SSFLCs are discussed. However, in many cases it is difficult to separate the factors for improving the bistability from those for improving the alignment quality itself, which can be evaluated by the defect density.

For the evaluation of bistability, the angle between two relaxed states without an electric field is a useful criterion. Moreover, in practical devices with simple matrix driving, small bipolar disturbing pulses are applied during the period between signal pulses. Memory states respond to the pulses and the transmission is modulated, resulting in decrease in contrast. Stability in the pulses is also an important criterion for bistability [8].

5.2.4.2 Polymer linearity

Thermoplastic Polymers and Thermosetting Polymers

In Table 5.2.1, polymers which align well on rubbing and do not align FLC materials are shown [9, 10]. The "good" polymer materials possess linear molecular structures without crosslinking. This rule also applies for the polymers containing

Table 5.2.1 Polymers which align well and which do not align FLC materials

Polymers that align
poly(ethylene)
$-[-CH_2CH_2-]_n-$
poly(vinyl alcohol)
$-[-CH_2-CH(OH)]_n-$
poly(hexamethylene adipamide)
$-[-NH(CH_2)_6-NH-CO(CH_2)_4-CO-]_n-$
poly(hexamethylene nonanediamide)
$-[-NH(CH_2)_6-NH-CO(CH_2)_7-CO-]_n-$
poly(hexamethylene terephthalamide)
$-[-NH(CH_2)_6-NH-CO-C_6H_4-CO-]_n-$
poly(1,4-butylene terephthalate)
$-[-(CH_2)_4-OOC-C_6H_4-COO-]_n-$
poly(1,4-ethylene terephthalate)
$-[-(CH_2)_2-OOC-C_6H_4-COO-]_n-$

aromatic structures. The aromatic polymers with *para*-substitution show better alignment than *meta*- or *ortho*-substituted polymers.

Other aspects are that linear polymers without crosslinking possess thermoplastic properties, which means they can be plastically deformed on heating and can retain the deformed situation on cooling. In the rubbing process, the plastic polymer is sheared and retains the extended state. These processes change the chain orientation and can cause partial crystallization, which can induce epitaxial growth of smectic liquid crystalline alignment.

Nematic liquid crystals are well aligned by thermosetting polymers. However, they are not suitable for smectic liquid crystalline materials. It was also found that smectic liquid crystals were aligned on the surface of a cleaved crystal of sodium chloride. From these observations, it can be expected that the alignment mechanism of smectic liquid crystals is analogous to the epitaxial growth of conventional solid crystals [12, 13]. Thermoplastic polymers are crystallized by the rubbing process, and smectic liquid crystals are aligned by epitaxial growth on the crystallized surfaces.

In Table 5.2.2, thermoplastic polymers with substantial side groups on which good alignment could not be observed, are shown. The deterioration of smectic liquid crystalline alignment due to the lack of linearity of the polymers with substantial side groups can be confirmed.

Table 5.2.2 Thermoplastic polymers with substantial side groups on which good alignment could not be observed

Polymers that do not align
poly(cyclohexyl methacrylate)
$-[CH_2-C(CH_3)(COO-C_6H_{11})-]_n-$
poly(amide resin) crosslinked nylons
poly(vinylmethyl ketone)
$-[-CH_2-CH(CO-CH_3)-]_n-$
poly(vinyl cinnamate)
$-[-CH_2-CH(OOC-CH=CH-C_6H_5)-]_n-$
poly(acetal)
$-[-CH_2-C(OC_2H_5)_2-]_n-$
poly(benzyl methacrylate)
$-[-CH_2-C(CH_3)(COOCH_2-C_6H_5)-]_n-$
poly(brene)
$-[-N^+(CH_3)_2-(CH_2)_6-N^+(CH_3)_2-(CH_2)_3-]_n-\ 2Br^-$

5.2.4.3 Crystallographic class

It has been pointed out that the crystallographic class of alignment layers also affects the bistability [11]. There are two clear groups of polymers, those with monoclinic or triclinic crystal structures, and those with other crystal structures. The alignment layers with monoclinic or triclinic structures show good bistability, and other groups of alignment layers show poor bistabilities.

Monoclinic and triclinic crystals have the lowest number of symmetry elements, and the smectic C phase also has a monoclinic symmetry. Therefore, the symmetry of alignment layers with monoclinic and triclinic crystal structures resembles the symmetry of the smectic C phase. It is plausible to think that the epitaxial growth of a smectic C liquid crystal proceeds most smoothly on alignment layers which resemble their crystallographic structures [12, 13].

5.2.4.4 Kinds of polymers

The relationship between the molecular structures of polymer materials for alignment layers and the bistability has been discussed by Myrvold [14]. Here, the bistabilities resulting from alignment layers made from polyethylene, polyfumarate esters, polyvinyl alcohol (PVA), polyacrylonitrile (PAN) and polyimide (PI) will be overviewed according to his studies.

Low-Density Polyethylene (LDPE) and High-Density Polyethylene (HDPE)
LDPEs give no bistability while HDPEs give bistability. Moreover, LDPEs show only poor alignment with ten times as many zigzag defects as HDPEs. LDPEs are made by free radical polymerization and contain numerous side chains with five or six carbons. These side chains cause deterioration of the crystallinity of the polymers. HDPEs are made by catalytic polymerization and consist of long unbranched chains. Long unbranched chains can realize good crystallinity. Polyethylene shows two different crystal structures of orthorhombic and monoclinic form. The orthorhombic form is stable and is normally obtained. The monoclinic form is obtained by mechanical distortion of the orthorhombic form. The rubbing process is expected to induce a transition from the orthorhombic form to monoclinic form which can induce bistability of SSFLCs.

Polyfumarate Esters
Polyfumarate esters with bulky substituents in the backbone give good alignment [15]. Because of the bulky substituents, the polymer chains form a helical structure which gives the polymer a rigid and rod-like nature, which makes crystal packing easier.

Polyvinyl Alcohol (PVA) and Polyacrylonitrile (PAN)
PVA and PAN are easily obtained commercially and can be used to form high-quality thin films. They are therefore convenient materials for laboratory experiments. The small hydroxy group of PVA allows crystallization of the atactic polymer, and good bistability can be obtained. However, only PVA with straight chains gives good alignment, and cross-linked PVA gives poor alignment. This is because of the difference in the degree of crystallinity of these polymers. PAN also gives high-quality alignment and quite good bistability.

Polyimides (PI)
In practical LCDs, only polyimides are used for alignment layers, and for the realization of practical SSFLCDs, polyimides would be most appropriate. By the

selection of the right diamine and tetracarboxylic acid, the appropriate physical properties, such as curing temperature, adhesion to glass, dielectric constant and chemical resistance, can be realized.

Two factors lead to increase in the crystallinity of polyimides. One is the simple combination of only two monomer units, which facilitates the packing necessary for good crystallinity. Another factor is the flexibility required for the fold-back observed in crystalline regions of polymers.

5.2.4.5 Odd–even effect

It is well known that for liquid crystalline molecules like the cyanobiphenyl compounds, physical properties such as the clearing points change regularly depending on whether the number of carbon atoms of the alkyl chain is odd or even. This phenomenon has been explained by the molecular linearity for odd or even carbon numbers of the alkyl chains. A similar phenomenon is observed in the case of polyimide (1) with the alkyl chain structures shown in Fig. 5.2.3 [16, 17].

Only those polyimides (1) with alkyl chains having an even number of carbon atoms show good bistability. This clear odd–even effect can be explained by using a picture in which the alkyl chains exist in the most outstretched, *all-trans*-conformation. For polyimides with alkyl chains with an odd number of carbon atoms, the polymer chain is bent and good packing is difficult. On the other hand, an even number of carbon atoms gives extended, linear alkyl groups. It can be easily understood that molecular structures of methylene derivative are odd and

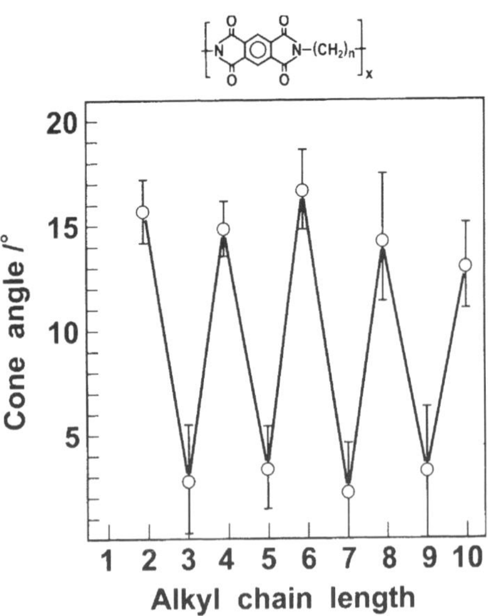

Fig. 5.2.3 Odd–even effect observed for polyimide (1).

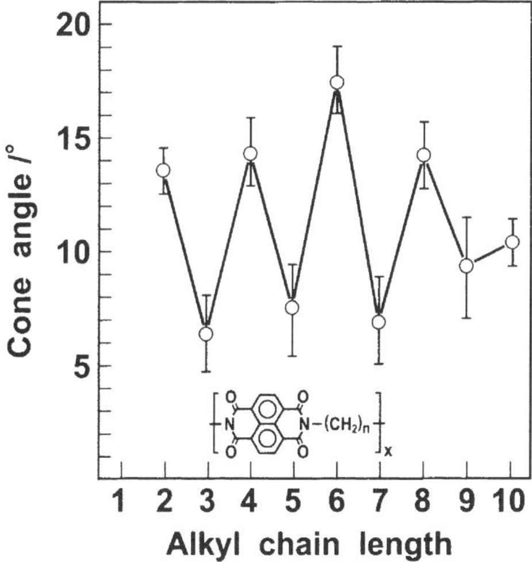

Fig. 5.2.4 Odd–even effect observed for polyimide (2).

of ethylene derivative even. Linear alkyl chains and aromatic groups realize good packing [18].

In the case of the polyimides (2) in which the phenyl group is replaced with a naphthalene group, the crystallinity is expected to increase because of the rigid structure. In fact, as shown in Fig. 5.2.4, in the case of the members with odd number alkyl chains, a larger cone angle can be observed for the polyimides (2) compared with the polyimides(1). However, the change is not clear in the case of the members with even alkyl chains.

In the case of the polyimides (3) based on *cis, cis, cis, cis*-1,2,3,4-cyclopentane teracarboxylic acid anhydride, the core moiety of the polymer is bent. The cyclopentane ring is flat, but the four carboxyl groups are on the same side of the plane. As a result, the molecule is bent as shown in Fig. 5.2.5. In the case of polymers with linear tetracarboxylic acids like polyimide (1), the derivatives with even numbered carbon chains are expected to have a linear structure, and the derivatives with odd numbered carbon chains are expected to have a bent structure. On the contrary, in the case of the polyimide (3) with the bent tetracarboxylic acid moiety, the odd numbered derivatives have linear structures and the even numbered derivatives have bent structures.

In Fig. 5.2.5, the cone angles (bistabilities) of the cyclopentane derivatives are shown. The marked odd–even effect can be observed. However, the tendency in the values of the cone angles is opposite to that observed for linear tetracarboxylic acid derivatives. The even numbered derivatives show a high degree of bistability and the bistability of the odd numbered derivatives is low. This result clearly shows that linearity of the polymers of the alignment layers is a dominant factor for bistability, because the linearity strongly affects the crystallinity of the

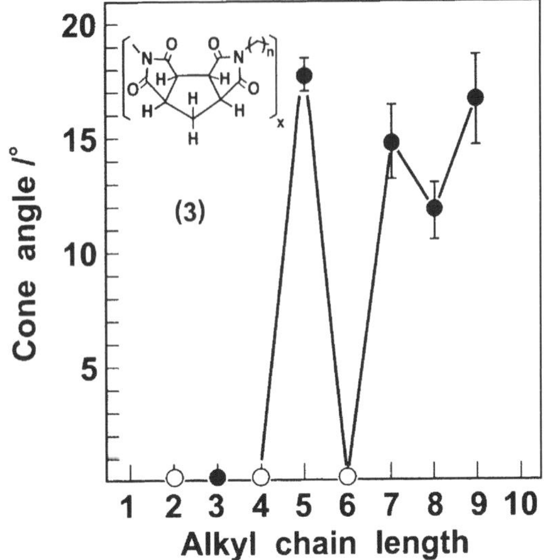

Fig. 5.2.5 Odd–even effect observed for polyimide (3).

polymers. The odd–even effect of the number of alkyl carbons in polyimides was also observed for the magnitudes of the pretilt angles of nematic liquid crystals [19, 20, 21]. This phenomenon is also expected to be based on the crystallinity of the polyimides.

5.2.5 Polarity on polymer surfaces

5.2.5.1 Polarity on polymer surface and alignment of SSFLC

The polarity of the alignment layer surface does not have much influence on alignment phenomena for nematic liquid crystalline materials. However, in the case of FLC materials, the polarity of the alignment layer surface shows an important effect. This is because the interaction between the spontaneous polarization and the polarity of the surface becomes important. This matter has been approached theoretically [27]. The stable director orientation in the SSFLC device was determined by minimizing the total free energy of the surfaces and the bulk elastic distortion as functions of cell thickness, cone angle, helical pitch, elastic constant and surface interaction coefficient. Because of the tendency of the direction of the spontaneous polarization to point either into or out of the substrate surface due to polar surface interaction, the director of the molecules twists from the top to the bottom surface. Therefore, the uniform state can only be stabilized in the case of a small surface interaction coefficient.

In addition to the effect of the surface interaction, smectic C* liquid crystals prefer to form a spontaneous twist and bend form of the director based on the chiral properties. Although the helix is suppressed by the substrate surfaces, the direction can remain in a preferred distortion state, resulting in splay of the polarization. Based on the two factors described above, the choice of the helix handedness, the material constants and the constraint on the small polar coefficient are important for realizing the stable uniform state.

In Figs 5.2.6 and 5.2.7, the calculated critical cell thickness for the various stable states is given as a function of surface interaction γ, the ratio of polar γ_2 to non-polar γ_1 surface interaction coefficient, for $q_0(K_2 - K_3) < 0$ and $q_0(K_2 - K_3) > 0$. q_0 is $2\pi/p$ (p is the pitch length), and the sign of q_0 specifies the handedness of the LC helical structure. For $q_0(K_2 - K_3) < 0$, a left-handed splay state is stabilized and vice-versa. In both cases, γ_2 is positive, which stabilizes the right-handed splay state. For small values of γ, d_c is not sensitive to the variation of γ. Above a critical value of γ, d_c decreases rapidly. The critical value of γ is affected by the tilt angle, the pitch, and the elastic constants of the FLC materials. According to these discussions, *to realize a stable SSFLC device, a small polar surface interaction is necessary with an appropriately large elastic constant of the LC materials.* On the other hand, if the polar surface interaction coefficient is large, the splayed or twisted state becomes more stable.

Dijon [28] has investigated the effect of the surface polarity on the FLC alignment experimentally.

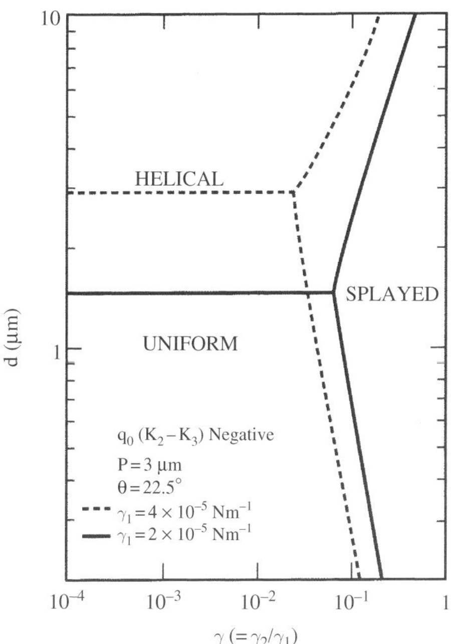

Fig. 5.2.6 Phase diagram as a function of cell thickness and ratio of polar to non-polar surface interaction coefficients for $q_0(K_2 - K_3) < 0$.

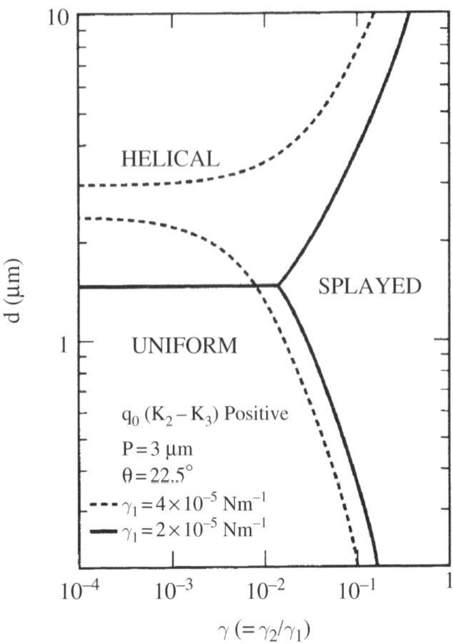

Fig. 5.2.7 Phase diagram as a function of cell thickness and ratio of polar to non-polar surface interaction coefficients for $q_0(K_2 - K_3) > 0$.

The surface tension of a solid surface, γ_s, is composed of two parts: the dispersive part, γ_{sd}, and the polar part, γ_{sp}
where

$$\gamma_s = \gamma_{sd} + \gamma_{sp}.$$

In the case of SSFLCs, asymmetry of the polarities of the top and bottom surfaces determines whether the cell shows bistability, a twisted state or monostability. Because the γ_d value does not change greatly for the surfaces of different polymers, the asymmetry of the cell can be evaluated by the difference, $\Delta\gamma_p$

$$\Delta\gamma_p = \mathrm{St}\cdot\gamma \mathrm{t}_p - \mathrm{Sb}\cdot\gamma \mathrm{b}_p.$$

St, Sb: sign of surface on top plate or bottom plate.

The correlation between $\Delta\gamma_p$ and the cell characteristics was confirmed.

$\Delta\gamma_p \leqq 10$ (dynes/cm)	bistable
$10 \leqq \Delta\gamma_p \leqq 20$	bistable but relaxed to more stable state
$20 \leqq \Delta\gamma_p$	monostable untwisted state

These observations suggest that in the case of SSFLCs, the polarity of the alignment layer is influential in determining the alignment and characteristics of the cell, although the polarity does not have any important influence in the case of TN or STN-LCD.

5.2.5.2 Sign of the surface polarity, S

Polarity must be defined by the magnitude and *the direction*. The direction can be expressed by the sign of the polarity. The sign of the polarity is defined as positive when the dipole on the surface is directed outward from the surface to the liquid crystalline material and vice versa as shown in Fig. 5.2.8 [22]. The sign of the polarity on the alignment layer can be determined by using the disymmetric cell with the two different alignment layers on the two substrates. If a bistable state or a twisted state is observed for the disymmetric cell, the surfaces of the two alignment layers are considered to have the same sign of S. On the other hand, if only the monostable state is observed, the sign S is considered to be different, or the magnitude of the polarity must be greatly different, even with the same sign of S. The direction of tilt angle, θ, is shown by its sign. The sign of θ is defined as positive if the molecular direction is obtained by rotation of the cell in a clockwise direction from the rubbing direction.

For example, from the fact that the sign of θ for the monostable state obtained with the polymer A on the top plate and polymer B on the bottom plate is positive, and that the FLC material has a negative spontaneous polarization, it can be deduced that the polarity for polymer A points OUT of the surface, while the polarity for polymer B points INTO the surface. Polymer materials were categorized into two groups by this method.

The first group was composed of PA6, PVA, MAP, DMOP and ITO in which the dipole points out of the surface, or the polarity on the surface is positive. The second group was composed of polyimides, cyclohexyldimethylchlorosilane and SiO_2 in which the dipole points into the surface, or the polarity on the surface is negative (see Table 5.2.3).

The author would like to propose another method of determining the alignment layer polarity. In this method, the cell structure of the twisted FLC mode described in Ref. 35 is used. Consider the cell structure for which the rubbing directions on the top and bottom plates are set rectangularly, as shown in Fig. 5.2.9. A ferroelectric liquid crystal of 45° tilt angle, showing no smectic

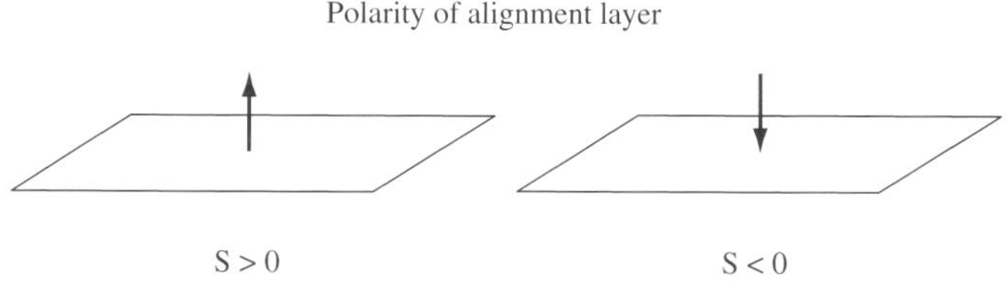

Fig. 5.2.8 Sign of surface polarity and the direction of the dipole moment.

Table 5.2.3 The sign of the polarity on a surface treated by several chemical reagents

Surface showing + polarity	Surface showing − polarity
Surface treated by $(CH_3)_3SiNHSi(CH_3)_3$ HMDS $CH_3NH(CH_2)_3Si(OCH_3)_3$ MAP	Surface treated by CH_3SiCl_3 $CH_3(CH_3O)_3Si$
Polyamide PA6	⬡-$(CH_3)_2SiCl$
polyvinylalchol PVA Surface on ITO layer	$(CH_3)_2ClSi(CH_2)_8SiCl(CH_3)_2$ Surface on polyimide PIX from Hitachi SiO_2 layer

J. Dijon, Ferroelectrics, 85, 47 (1988)

A phase, is injected into the cell. In this cell, the layer direction is determined by the sign of the spontaneous polarization and the polarity direction of the alignment layer surfaces. If the sign of the spontaneous polarization is negative, the layers are formed in the direction shown in Fig. 5.2.9, depending on

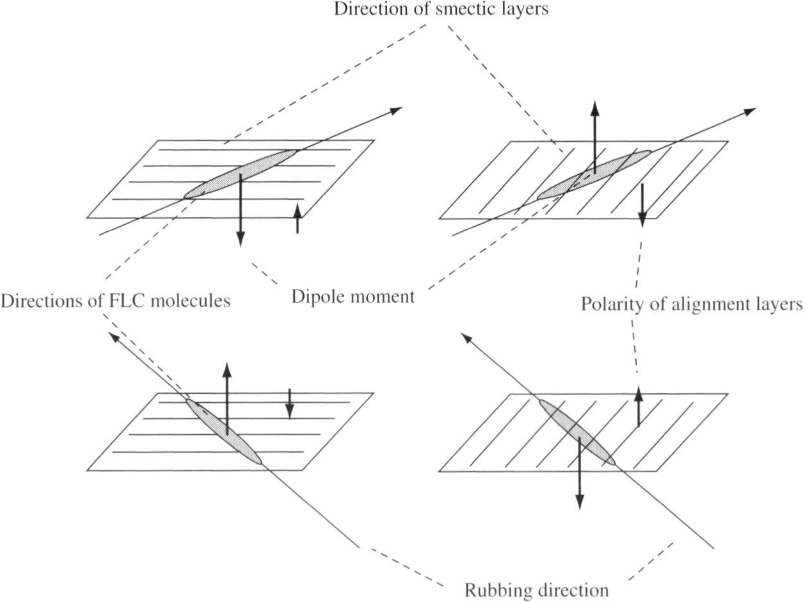

Fig. 5.2.9 The relationship between layer direction and the polarity direction of the alignment layers in the twisted FLC mode when the sign of the spontaneous polarization is negative.

the polarity of the alignment layer. By this method, the polarity of the alignment layers can be determined independently, without other alignment layers.

5.2.5.3 Rubbing process and polarity

The rubbing process is known to enhance the anisotropy of the refractive index of the polymer. To investigate the effect of the rubbing process, an asymmetric cell with rubbed PA6 on the top surface and unrubbed PA6 on the bottom face was produced [22]. Without applying voltage, only one stable state could be observed. θ was positive with negative Ps materials, and negative with positive Ps materials. These observations suggest that the rubbing process enhanced the polarity on the surface.

5.2.6 Rubbing and annealing process

5.2.6.1 Rubbing process [21]

The rubbing process and annealing process are expected to cause a change in crystallinity of the polymer surface. For the rubbing process, the alignment layer is heated and uniaxially stretched. During this process, the crystallinity of the polymer will increase. For SSFLCDs, it is well known that a weak rubbing condition results in better bistability. This phenomenon was explained from the viewpoint of crystallinity. The light rubbing pressure will heat the polymers above the glass transition temperature but below the temperature at which the crystallinity deteriorates. This is a temperature range known to be advantageous for the polymer annealing process. As the result, it was concluded that weak rubbing is effective in creating a highly crystalline surface. Moreover, stretching in the rubbing process is also effective for producing a high degree of crystallization. Crystalline polymers are observed to cause epitaxial growth over a microscopic area, so it is not essential that the whole area is crystallized to obtain a high quality of alignment [23].

5.2.6.2 Annealing process

"Annealing" is quite convenient and practical for a production line process. The annealing process involves only simple procedures, heating the substrate with its alignment layers in an oven or on a hot plate, and cooling to room temperature. These processes are not important for the alignment layers for TN or STN devices. However, for production of SSFLCDs, annealing of alignment layers may affect the electro-optical behaviour of the device. The effect of the annealing process on bistability was investigated by Hartmann *et al.* [24].

The change in alignment layer crystallinity by the annealing process was evaluated by SEM and FTIR spectroscopy, and the bistability was evaluated by the response of the memory state to small disturbing electric fields. Three kinds of layers, S, A and C, were produced. The layer S was produced by the usual process, that is, a methanol/*m*-cresol solution of nylon 6,6 was dried at 160 °C for 1.5 h in a vacuum. By heating to a high temperature, this nylon 6,6 layer melts and becomes an isotropic liquid and, on cooling to room temperature, various degrees of crystallization of the polymer layer occur, depending on the cooling process. Two kinds of layers, A and C were prepared by two typical thermal treatments. Layer A was prepared by heating layer S to 280 °C, creating an isotropic liquid, and by quenching to 20 °C. This process was expected to create an amorphous layer. Layer C was prepared by quenching the polymer layer to 160 °C, and holding it at the same temperature for 4 hours, after heating it at 280 °C. This treatment was expected to make layer C more crystalline.

By observing the surfaces using SEM, a clear difference in the number of spherulites, suggesting a difference in crystallinity, could be observed. In the case of layer A, the number of spherulites was either small or almost no spherulites could be observed showing that layer A was as amorphous. On the contrary, in the case of layer C, many spherulites could be observed, which means that layer C possesses a high degree of crystallinity. The crystallinity of the nylon 6,6 was also evaluated by FTIR spectroscopy [25]. The results showed that the crystallinity increased in the order of A, S and C. The percentages of crystallization were 43.5%, 45.7% and 50.8% for A, S and C, respectively.

The bistability was evaluated by the stability of memory states on applying small disturbing pulses during the period between switching pulses after the rubbing treatment. The amplitudes of the disturbing pulses were 1/2 and 1/6 of the switching pulse amplitude. The contrast decreased for this transmission modulation. A clear difference could be observed between C and A. In the case of layer A of poor crystallinity, a larger modulation could be observed than in the case of layer C of high crystallinity. The contrast of sample C was 8:1 and that of sample A was 3:1. This phenomenon was explained as follows. The high degree of crystallinity of the alignment layer increases the surface anchoring which suppresses the response to the disturbing pulses.

For a thermoplastic resin such as Nylon 6/6, the annealing effect is plausible. However, the annealing effect on thermosetting polymers was also investigated by using polyimide composed of trimellitic acid and aliphatic diols [26]. The stability of the relaxed state towards small pulses was again investigated. For an annealing temperature below 170 °C, the fluctuations caused by 4 V peak to peak disturbing pulses were large enough to give the same transmission as that for both bright and dark relaxed states. So, there is no bistability when subjected to even a low field of 4 V. In the case of annealing temperatures from 200 °C to 235 °C, bistability could be observed when applying 4 V disturbing pulses. The bistability of the cell could be maintained even on the application of 6 V peak to peak pulses, when the cell was annealed at 225 °C to 235 °C for one hour.

This phenomenon is explained as follows. For amorphous polymers, no energy barrier exists between the different memory states. On the other hand, crystalline polymers show good bistability. Stroboscopic micrographs of SSFLC switching show that after the first switched domains, or the cores, are formed, they grow into larger areas. The switched areas become the seeds for the switching of neighbouring areas on the crystalline polymers. Because no barrier exists between the two possible states on an amorphous polymer, switching will occur easily at the amorphous area. The energy barrier from one memory state to another will be lowered by the inclusion of amorphous regions in a crystalline surface. Through the annealing process, the area of the amorphous region decreases, and the energy barrier for switching increases. As the result, the bistability towards a small disturbing field will increase with the annealing process.

5.2.7 Electric double layer influence

In many cases, a decrease in bistability was explained by *the existence of ionic impurities.* Figure 5.2.10 shows the behaviour of an FLC material with ionic impurity. During the writing period, inversion of spontaneous polarization occurs. This period is too short for the ionic impurity to transfer. The holding period is much longer. It is 16 ms in the 60 Hz line-sequential scheme and longer in the partial access scheme. Ionic impurity transfer is then possible during the writing period.

In active matrix addressing of the TN LCD, the applied voltage is maintained during the holding period. In such a situation, ionic impurity transfer can occur because of the electric field maintained between electrodes. The polarity of the

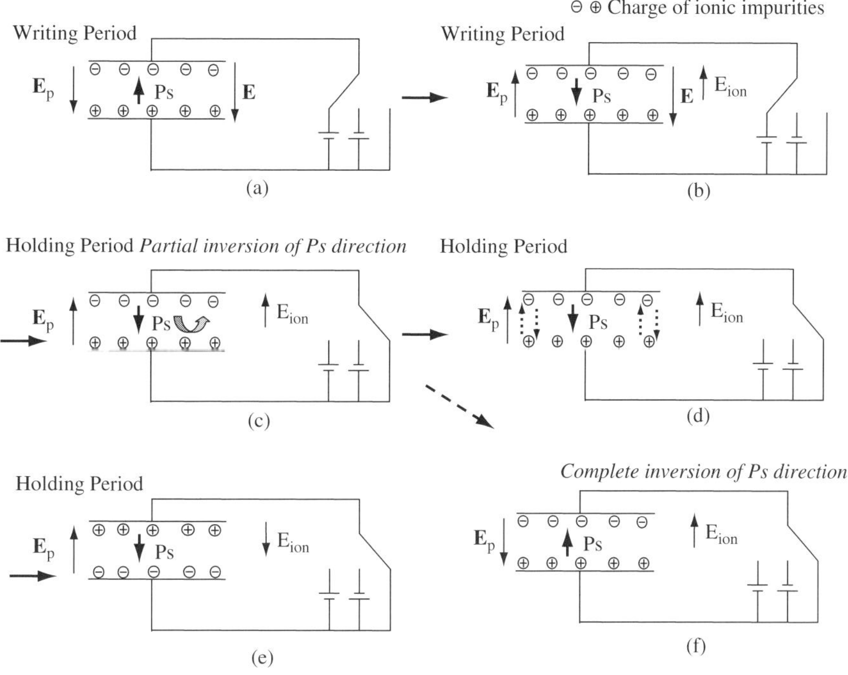

Fig. 5.2.10 Behaviour of a FLC material with ionic impurity in SSFLCD.

applied voltage is changed in every frame to prevent the accumulation of ionic impurities of a specific charge sign.

In the case of the simple matrix driving of SSFLCs, two electrodes are shorted during the holding period.

When ionic impurities exist in an FLC, like (a) and (b), the spontaneous polarization is affected by the electric double layer formed by the ionic impurity (c). As a result, partial inversion of the spontaneous polarization direction to the opposite direction occurs. In extreme cases, the molecules return to the initial position (f). In this case, the result is the same as if no switching had occurred. The spontaneous polarization induces a dc electric field. This electric field causes the transfer of ionic impurities (d), which form a new electric double layer (e).

Inaba *et al.* [29] reported the phenomenon in which the bistability decreased, and moreover that the switchability was lost due to the formation of an electric double layer with ionic impurities. In most cases, the bistable uniform states are formed from the initial twisted state by applying an ac field. This uniform alignment loses its bistability after a few hours. Figure 5.2.11 shows that the threshold voltage from one state to another increases gradually by keeping Ps in one direction. Finally, the liquid crystal becomes unswitchable into the state with Ps in the down direction. Since the spontaneous polarization induces a dc electric field, as shown in Fig. 5.2.10, ionic impurities are attracted in the up or down directions according to their charge sign and accumulate at the boundaries between the LC and the alignment layers (Fig. 5.2.10(d)). After this accumulation process, the electric field due to the impurities increases to make the existing dipole direction more stable and the other direction unstable. Only one state with Ps in the present direction becomes stable, and the FLC becomes unswitchable.

In Fig. 5.2.10(a) [29], the charge on the upper surface due to the spontaneous polarization is in Ps per unit area. This charge can be cancelled by the charge − Ps of the accumulated ionic impurities on the surface. When an external voltage

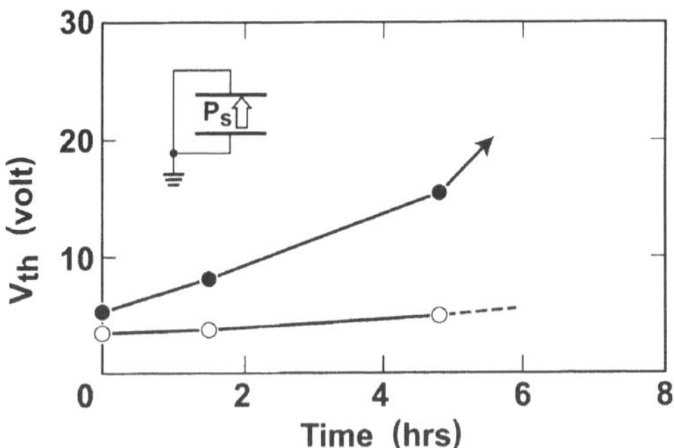

Fig. 5.2.11 Time evolution of threshold voltages after the uniformization at $t = 0$. Dots and open circles indicate V_{th} (up→down) and V_{th} (down→up), respectively [29]. The dipoles are kept in the UP direction during the measurement.

is applied to turn the spontaneous polarization to the downward direction, the charge density due to the spontaneous polarization on the upper surface becomes $-Ps$ (Fig. 5.2.10(b)). When the external pulse is turned off, a built-in voltage V_{rev} appears.

$$V_{rev} = 2Ps/(C_D + C_{LC})$$

C_D and C_{LC} are capacitances per unit area of the upper and lower alignment layers and liquid crystal layer.

The direction of this voltage is upward. Therefore, this voltage destabilizes the existing spontaneous polarization. For $Ps = 30\,nC\,cm^{-2}$, $C_D = 10\,nF\,cm^{-2}$ and $C_{LC} = 2\,nF\,cm^{-2}$, V_{rev} is $5\,V\,\mu m^{-1}$. This value is not large enough for a quick response in the reverse direction. However, the decay time $\tau = (C_D + C_{LC})R$ (R is the resistance of the liquid crystal layer) is of the order of a hundred milliseconds. This is long enough for the molecules to be turned back into their initial position even with $5\,V\mu m^{-1}$. As mentioned above, in the case of the partial access scheme, memory properties for ca. 1 s are required. Therefore, the effect of the electric double layer cannot be neglected in the practical displays. The equation also explains why the high spontaneous polarization and small capacitance of the alignment layer result in an unswitchable SSFLC display. It was also reported that FLC materials with high spontaneous polarization tend to show slow switching due to the existence of ionic impurity [30].

From theoretical analysis [31], Yang *et al.* assumed the following.

(1) Thinner alignment layers with high dielectric permitivity will reduce the depolarization field effect and enhance the bistability.
(2) Conducting alignment layers with a charge injected into the FLC material might also be able to provide a mechanism to neutralize the depolarization field effect so that bistability can be achieved even for FLC materials with high spontaneous polarization.

The application of a Langmuir-Blodgett (LB) film for the alignment layer was found to be efficient for improving the bistability [32]. Very thin films are responsible for realizing excellent bistability, because of the transmissivity of electrical charges accumulated on the surfaces.

In Figs 5.2.12 and 5.2.13, the calculated relationship between the bistability of the FLC and the alignment layer conductivity are shown for two FLC materials of different spontaneous polarizations [33].

The transmission begins to decrease after the external voltage is turned off. The decrease is large when the conductivity of the alignment layer is low, and the spontaneous polarization is large. This decrease is caused by the polarization charges at the interfaces between the FLC material and the alignment layers. The transmission drops from its saturated value until the depolarization effect vanishes. Complete reversal of the transmission can occur if the effect of ion

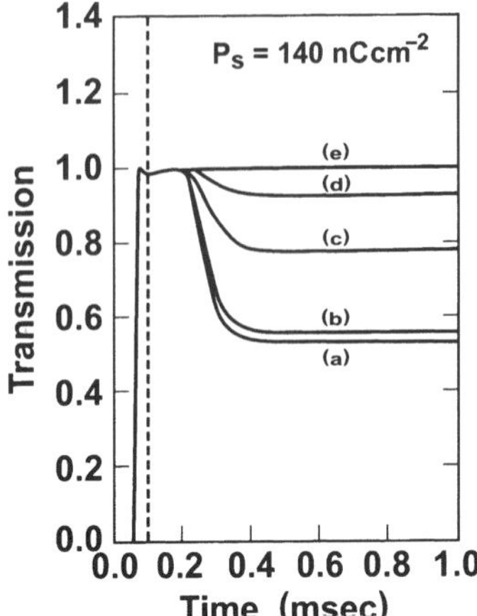

Fig. 5.2.12 Calculated transmissions as a function of time for $Ps = 140\,nC\ cm^{-2}$ using the conductivity of the alignment layers, σ', as a parameter. Curves (a), (b), (c), (d) and (e) correspond to σ' equal to 1×10^{-11}, 1×10^{-10}, 1×10^{-9}, 2×10^{-9} and 5×10^{-9} $(\Omega\,cm)^{-1}$, respectively.

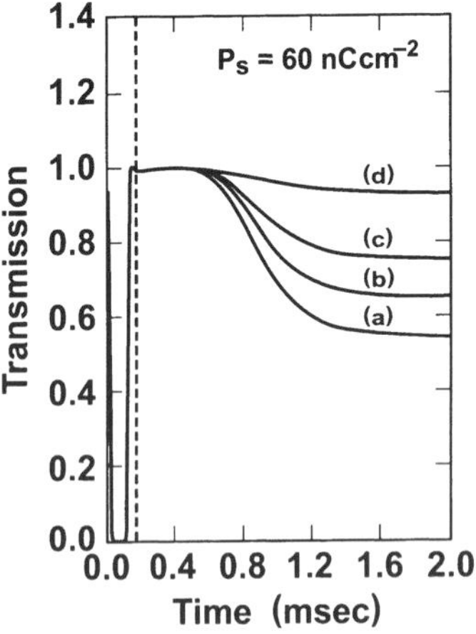

Fig. 5.2.13 Calculated transmissions as a function of time for $Ps = 60\,nC\,cm^{-2}$.

accumulation is included. For conductive alignment layers, free charges can be injected into the interfaces to neutralize the interfacial polarization charges, and the depolarizing effect vanishes before the azimuthal angle reaches its final value and the transmission stays at a higher level.

Enhancement of bistability by doping polyimide alignment layers with a charge transfer complex (CTC) and an appropriate rubbing process has been reported [34].

This enhancement can be attributed to the appropriate conductivity of CTC-doped polyimide films which can neutralize accumulated surface charges. Although there are many candidates for CTCs, most of them show only very restricted solubility to organic solvents for polyimide. Tetramethyltetrathiafulvalene-octadecyltetracyanoquinodimethane (TMTTF-ODTCNQ) was appropriate for doping in amounts of 0.6 wt% to 20 wt% in polyimide. The layer thickness was 50 nm. The resistivities were from $10^7\ \Omega$ cm to $10^9\ \Omega$ cm across the film. In Figs 5.2.14 and 5.2.15, the optical responses of SSFLCDs using CTC-doped and undoped polyimide alignment layers are shown. Excellent bistability was realized by the doping charge transfer complex and no degradation could be observed.

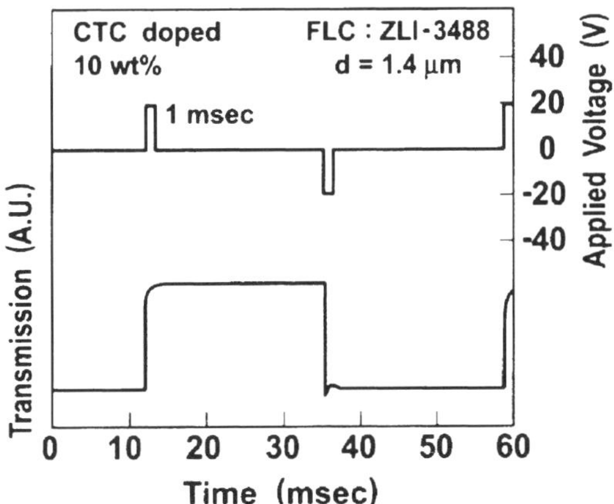

Fig. 5.2.14 Optical response of the SSFLC using CTC-doped polyimide orientation films.

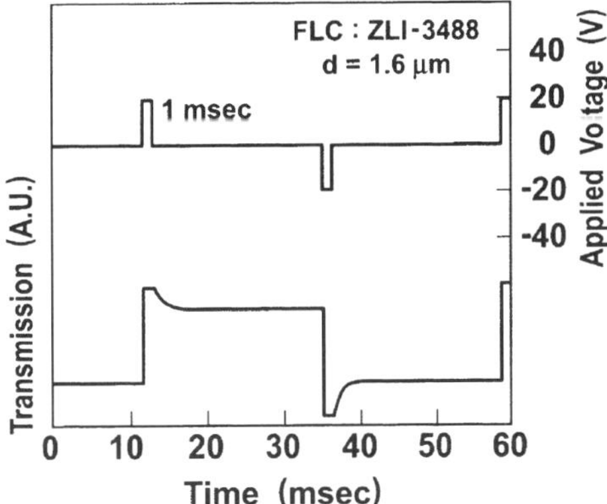

Fig. 5.2.15 Optical response of the SSFLC using CTC-undoped polyimide orientation films.

References

5.1 Layer Structure and Molecular Orientation of Ferroelectric Liquid Crystals

[1] R. B. Meyer, L. Liébert, L. Strzelecki and P. Keller, J. Phys., **36**, L69 (1975).

[2] N. A. Clark and S. T. Lagerwall, Appl. Phys. Lett., **36**, 899 (1980).

[3] G. W. Gray and J. W. Goodby, Smectic Liquid Crystals (Leonard Hill) (1984).

[4] Y. Ouchi, J. Lee, H. Takezoe, A. Fukuda, K. Kondo, T. Kitamura and A. Mukoh, Jpn. J. Appl. Phys., **27**, L1993 (1988).

[5] T. P. Rieker, N. A. Clark, G. S. Smith, D. S. Parmar, E. B. Sirota and C. R. Safinya, Phys. Rev. Lett., **59**, 2658 (1987).

[6] A. Mochizuki, T. Yoshihara, M. Iwasaki, M. Nakatsuka, Y. Takanishi, Y. Ouchi, H. Takezoe and A. Fukuda, SID '90, **31**, 123 (1990).

[7] Y. Takanishi, Y. Ouchi, H. Takezoe, A. Fukuda, A. Mochizuki and M. Nakatsuka, Jpn. J. Appl. Phys., **29**, L984 (1990).

[8] Y. Ouchi, J. Lee, H. Takezoe, A. Fukuda, K. Kondo, T. Kitamura and A. Mukoh, Jpn. J. Appl. Phys., **27**, L725 (1988).

[9] J. W. Goodby and E.Chin, J. Am. Chem. Soc., **108**, 4736 (1986).

[10] E. N. Keller, E. Nachaliel and D. Davidov, Phys. Rev. A, **34**, 4363 (1986).

[11] K. Suzuki, K. Toriyama and A. Fukuhara, Appl. Phys. Lett., **33**, 561(1978).

[12] M. A. Handschy, N. A. Clark and S. T. Lagerwall, Phys. Rev. Lett., **51**, 471 (1983).

[13] Y. Ouchi, H. Takezoe and A. Fukuda, Jpn. J. Appl. Phys., **26**, 1 (1987).

[14] J. Kanbe, H. Inoue, A. Mizutome, Y. Hanyuu, K. Katagiri and S. Yoshihara, Ferroelectrics, **114**, 3 (1991).

[15] Y. Ouchi, H. Takano, H. Takezoe and A. Fukuda, Jpn. J. Appl. Phys., **27**, 1 (1988).

[16] N. A. Clark and T. P. Rieker, Phys. Rev. **A37**, 1053 (1988).

[17] K. Okano and S. Kobayashi, Ekisyo Oyohen (Application of liquid crystals)[in Japanese](Baihukan Press) (1985).

[18] M. Koden, H. Katsuse, A. Tagawa, K. Tamai, N. Itoh, S. Miyoshi and T. Wada, Jpn. J. Appl. Phys., **31**, 3632 (1992).

[19] A. Tagawa, H. Katsuse, K. Tamai, N. Itoh, M. Koden, S. Miyoshi and T. Wada, Japan Display '92, 519 (1992).

[20] J. E. Maclennan, N. A. Clark, M. A. Handschy and M. R. Meadows, Liquid Crystals, **7**, 753 (1990).

[21] S. J. Elston and J. R. Sambles, Jpn. J. Appl. Phys., **29**, L641 (1990).

[22] D.W. Berreman, J. Opt. Soc. Am., **62**, 502 (1972).

[23] M. Nakagawa and T. Akahane, J. Phys. Soc. Jpn., **55**, 1516 (1986).

[24] M. Kawaida, T. Yamaguchi and T. Akahane, Jpn. J. Appl. Phys., **28**, L1602 (1989).

[25] A. Tsuboyama, Y. Hanyu, S. Yoshihara and J. Kanbe, Japan Display '92, 53 (1992).

[26] N. Itoh, H. Akiyama, Y. Kawabata, M. Koden, S. Miyoshi, T. Numao, M. Shigeta, M. Sugino, M. J. Bradshaw, C.V. Brown, A. Graham, S. D. Haslam, J. R. Hughes, J. C. Jones, D. G. McDonnell, A. J. Slaney, P. Bonnett, P. A. Gass, E. P. Raynes and D. Ulrich, IDW '98, 205 (1998), The Journal of the Institute of Image Information and Television Engineers, **53**, No. 8, 1136 (1999).

5.2 Alignment and Bistability of Ferroelectric Liquid Crystals

[1] B. O. Myrvold, Mol. Cryst. Liq. Cryst., **202**, 123 (1991).

[2] A. D. L. Chandani, T. Hagiwara, Y. Suzuki, Y. Ouchi, H. Takezoe and A. Fukuda, Jpn. J. Appl. Phys., **27**, L279 (1988).

[3] Y. Yamada, N. Yamamoto, K. Mori, N. Koshoubu, K. Nakamura, I. Kawamura and Y. Suzuki. J. of SID, 1/3 (1993).

[4] K. Takatoh, H. Yamaguchi, R. Hasegawa, T. Saishu and R. Fukushima, Polym. Adv. Technol., **11**, 413 (2000).

[5] J. Kanbe, H. Inoue, A. Mizutome, Y. Hanyuu, K. Katagiri and S. Yoshihara, Ferroelectrics, **114**, 3 (1991).

[6] K. Tamai, M. Shiomi, N. Itoh, T. Numao, M. Koden, S. Miyoshi and T. Wada, Ouyou-Butsuri, **63**, (6) 548 (1994).

[7] M. J. Bradshaw, V. Brimmell and E. P. Raynes, Liq. Cryst., **2**, (1) 107 (1987).

[8] B. O. Myrvold, Mol. Cryst. Liq. Cryst., **202**, 123 (1991).

[9] J. S. Patel, T. M. Leslie and J. W. Goodby, Ferroelectrics **59**, 137 (1984).

[10] J. W. Goodby *et al.*, "Ferroelectric Liquid Crystals: Principles, Properties and Applications", 235–239, Gordon and Breach Science Publisher (1991).

[11] B. O. Myrvold, Mol. Cryst. Liq. Cryst., **202**, 123 (1991).

[12] J. M. Geary, J. Appl. Phys., **62**, 4100 (1987).

[13] K. Kondo, H. Takezoe, A. Fukuda and E. Kuze, Jpn. J. Appl. Phys., **22**, L85 (1983).

[14] B. O. Myrvold, Liq. Cryst., **3** (9), 1255 (1988).

[15] M. Taguchi, K. Iwasa, H. Suenaga and N. Ohwaki, 2nd International Conference on Ferroelectric Liquid Crystals, Gothenburg (1989).

[16] B. O. Myrvold, Liq. Cryst., **4** (6), 637 (1989).

[17] B. O. Myrvold, Liq. Cryst., **7** (2), 261 (1990).

[18] V. V. Korshak et al., J. Polymer Soc. Polymer Phys., **18**, 247 (1980).

[19] H. Yokokura, M. Oh-e, K. Kondo and S. Oh-hara, Mol. Cryst. Liq. Cryst., **225**, 253 (1993).

[20] B. O. Myrvold, K. Kondo and S. Oh-hara, Liq. Cryst., **15**, 429 (1993).

[21] B. O. Myrvold, K. Kondo and S. Oh-hara, Mol. Cryst. Liq. Cryst., **269**, 99 (1995).

[22] B. O. Myrvold, Liq. Cryst., **3** (9), 1255 (1988).

[23] K. Ishikawa, Jpn. J. Appl. Phys., **23**, L211 (1984).

[24] W. J. A. M. Hartmann, A. M. M. Luyckx-Smolders and R. P. V. Kessel, Appl. Phys. Lett., **55** (12), 1191 (1989).

[25] H. W. Starkweather and R. E. Moynelan, J. Polym. Sci., **22**, 363 (1956).

[26] B. O. Myrvold, Liq. Cryst., **7** (6), 863 (1990).

[27] T. C. Chieu, J. Appl. Phys., **64**, 6234 (1988).

[28] J. Dijon, Ferroelectrics, **85**, 47 (1988).

[29] Y. Inaba, K. Katagiri, H. Inoue, J. Kanbe, S. Yoshihara and S. Iijima, Ferroelectrics, **85**, 255 (1988).

[30] J. Dijon, C. Ebel, C. Vaucher, F. Baume, J.-F. Clere, M. Estor, T. Leroux, P. Maltese and L. Mulatier, Digest of SID '88, 246 (1988).

[31] K. H. Yang, T. C. Chieu and S. Osofsky, Appl. Phys. Lett., **55** (2), 125 (1989).

[32] H. Ikeno, A. Oh-saki, N. Ozaki, M. Nitta, K. Nakaya and S. Kobayashi, Proceedings of SID, 45 (1988).

[33] T. C. Chieu and K. H. Yang, Appl. Phys. Lett., **56** (14), 1326 (1990).

[34] K. Nakaya, B. Y. Zhang, M. Yoshida, I. Isa, S. Shindoh and S. Kobayashi, Jpn. J. Appl. Phys., **28** (1), L116 (1989).

[35] K. Takatoh, H. Nagata and T. Saishu, Ferroelectrics, **179**, 173 (1996).

Chapter 6

Applications of Ferroelectric and Antiferroelectric Liquid Crystals

6.1 Molecular Orientations and Display Performance in FLC Displays

Mitsuhiro Koden

6.1.1 Introduction

In typical FLC (Ferroelectric Liquid Crystal) displays, the FLC material is sandwiched between two substrates separated by around 2 μm. A number of molecular orientations have been found to be suitable for use in FLC displays and investigated. Some of these are summarized in Fig. 6.1.1.

The orientations of the molecules of the FLC materials are classified by the presence or absence of a helical structure. The most famous FLC device is the SSFLC (Surface Stabilized FLC) [1], in which the helical structure of the FLC material is unwound. While a variety of molecular orientations have been applied in SSFLC devices, three molecular orientations appear to be the most useful in practical FLC displays. These are the bookshelf-layered structure and the C1-uniform (C1U) and the C2-uniform (C2U) orientations [2]. Each of these structures shows monostability or bistability, depending on the material and its alignment properties. The monostable orientations are applicable to active matrix FLC displays while the bistable orientations are applicable to passive matrix FLC displays. FLC displays with a helical orientation have also been investigated. One useful FLC mode with the helical orientation is the DHF (deformed-helix ferroelectric) mode [3]. This mode is monostable and is thus suitable for an active matrix drive method.

Molecular orientation is a very important factor in determining the performance of a display and thus in the development of practical FLC displays. The situation is more complex than with nematic liquid crystal displays because FLC displays require consideration of additional properties. These are the layered

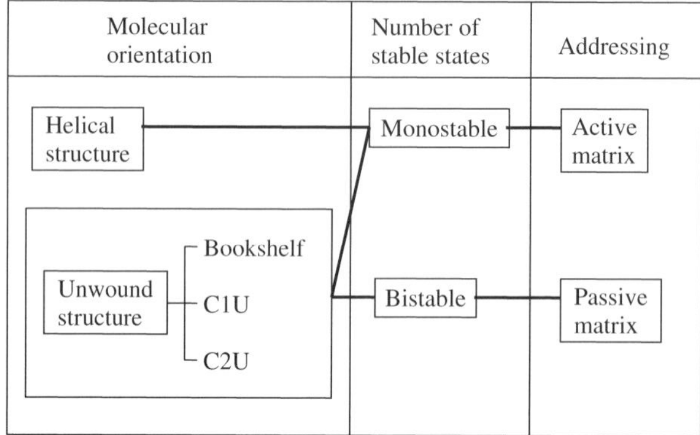

Fig. 6.1.1 Molecular orientations and types of FLC display.

structure, the tilt of the molecular long axis relative to the direction normal to the layer of FLC material, the spontaneous polarization (P_s), the phase sequence of the materials, etc. Due to this complex set of parameters, it is not easy to realize a suitable molecular orientation in FLC displays.

Several typical molecular orientations used in FLC displays and the relationship between molecular orientation and display performance will be described in this chapter. This chapter will also describe how to realize a desired molecular orientation by optimizing material parameters, alignment parameters, etc.

Section 2 covers the DHF mode with a helical structure. The bookshelf structure, the C1-uniform (C1U) orientation, and the C2-uniform (C2U) orientation are described in Sections 3, 4, and 5, respectively. Section 6 covers the stability of the molecular orientations in FLC devices exposed to shock. Weakness in this respect is one of the intrinsic problems of FLC displays. The layer rotation behaviour, an intrinsic behaviour of FLC displays, is described in Section 7.

6.1.2 The DHF mode

A helical molecular orientation is used in the DHF (Deformed Helix Ferroelectric) liquid crystal mode [3]. This is shown in Fig. 6.1.2. The helical pitch of the FLC material must be shorter than wavelength of visible light. When no voltage is applied, a dark state is obtained with crossed polarizers due to the helical structure of the FLC material. When a voltage is applied, the orientation of the FLC molecules is switched by the interaction between the electric field and the spontaneous polarization (P_s) as shown in Fig. 6.1.2. The characteristics of operation in the DHF mode are fast response times, a wide viewing angle, monostable properties, and an analogue grey scale. Active matrix drive methods apply to this mode. The specifications of some prototype DHF displays [4, 5] are given in Table 6.1.1.

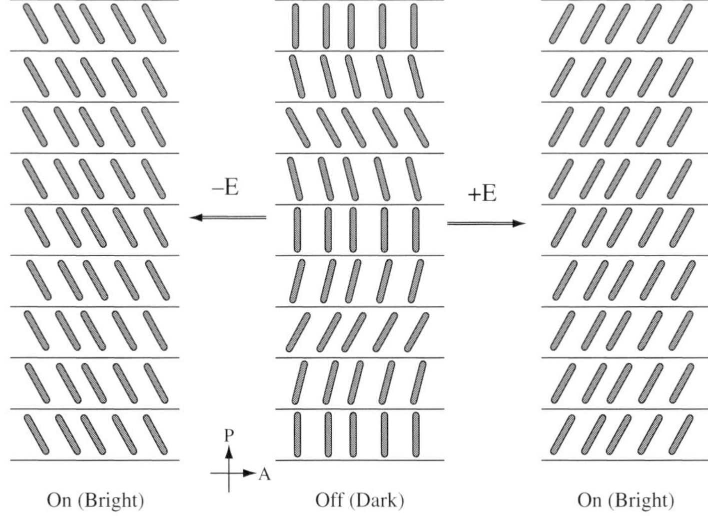

Fig. 6.1.2 The DHF (Deformed Helix Ferroelectric) liquid crystal mode.

Table 6.1.1 The specifications of active matrix prototype DHF displays reported by Tanaka *et al.* [4] and Verhulst *et al.* [5]

Display size (diagonal)	1.8″	0.75″
Pixel number	220 × 280 (×RGB)	140 × 170
Addressing method	Active matrix (s-Si-TFT)	Active matrix (4 diodes)
Molecular orientation	Helical structure with short pitch	
Drive voltage	17 V	Vrow = 16.5 V Vcolumn = 3 V ± 4.5 V
Drive frequency	20 Hz	50 Hz
Drive speed	200 μs/line	64 μs/line
Contrast ratio	20:1	25:1
Grey scale	Analogue grey scale	Analogue grey scale
Monochrome/Colour	Full colour	Monochrome
Reference	[4]	[5]

6.1.3 The bookshelf orientation

In a sense, the bookshelf layer structure (shown in Fig. 6.1.3) is the ideal molecular orientation for an SSFLC. One of the greatest merits of the bookshelf layer structure is its wide memory angle. The ideal memory angle θ_m in an SSFLC is 22.5° because the transmission of light of a wavelength λ is given by the following equation.

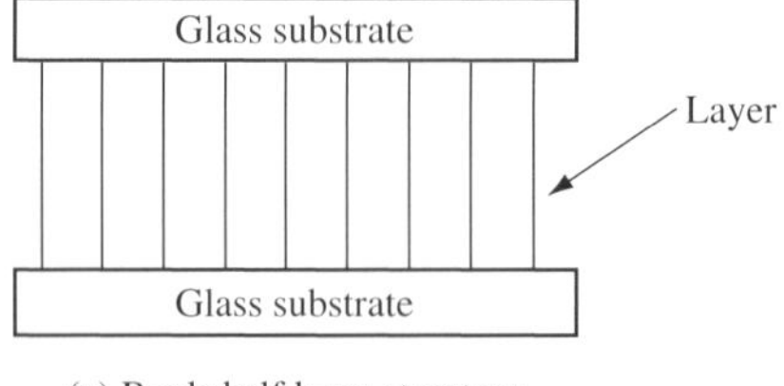

(a) Bookshelf layer structure

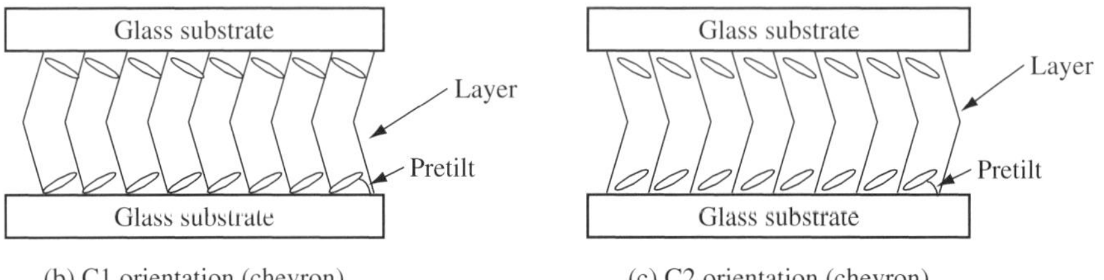

(b) C1 orientation (chevron)

(c) C2 orientation (chevron)

Fig. 6.1.3 Bookshelf and chevron layer structures. (a) Bookshelf layer structure, (b) C1 orientation of a chevron layer structure, (c) C2 orientation of a chevron layer structure.

$$I = I_0 \sin^2(4\theta_m)(\pi \cdot \Delta n \cdot d)/\lambda$$

where θ_m is the memory angle, which is the angle between the molecular long axis of the FLC molecules in their memory state and the direction normal to the layer, Δn is the birefringence of the FLC material, and d is the cell thickness. Although the tilt angles of many FLC materials are around 22.5°, the memory angles in chevron layer structures tend to be smaller than 22.5° because of the tilted layer structure. In the bookshelf layer structure, on the other hand, the theoretical model indicates that the memory angle will be equal to the tilt angle. It is thus rather easy to obtain the memory angle of 22.5° from the bookshelf layer structure. The wide memory angles that are typically obtained with the bookshelf layer structure are suitable for display applications that require high contrast ratios and high levels of brightness.

While the bookshelf layer structure has strong merits for display applications, it is not easy to obtain the bookshelf structure due to the intrinsic temperature dependence of the tilt angles of FLC materials. Three major approaches to obtaining a bookshelf layer structure have been investigated.

The first is AC-field treatment. In this method, an initial chevron layer structure is turned into a bookshelf layer structure by applying a strong low frequency AC field. The FLC material used must possess a high P_s because the interaction between the electric field and P_s induces a torque that changes the layer structure from chevron to bookshelf.

As the strength of the AC electric field is increased, the observed texture changes from the initial virgin texture (texture I), via a rather inhomogeneous rooftop texture (texture II), to a more uniform striped texture (texture III) [6]. These textures are shown in Fig. 6.1.4. This method is also called the "Texture Change Method" because of these changes in texture which reflect changes in the layer structure. Sato *et al.* have reported that zigzag defects in the initial texture vanished and that slender domains perpendicular to the smectic layers appeared after AC-field treatment with a square waveform (±25 V, 15 Hz). They showed, by X-ray measurements, that the layer structure of these slender domains was quasi-bookshelf [7]. They also reported that the quasi-bookshelf orientation

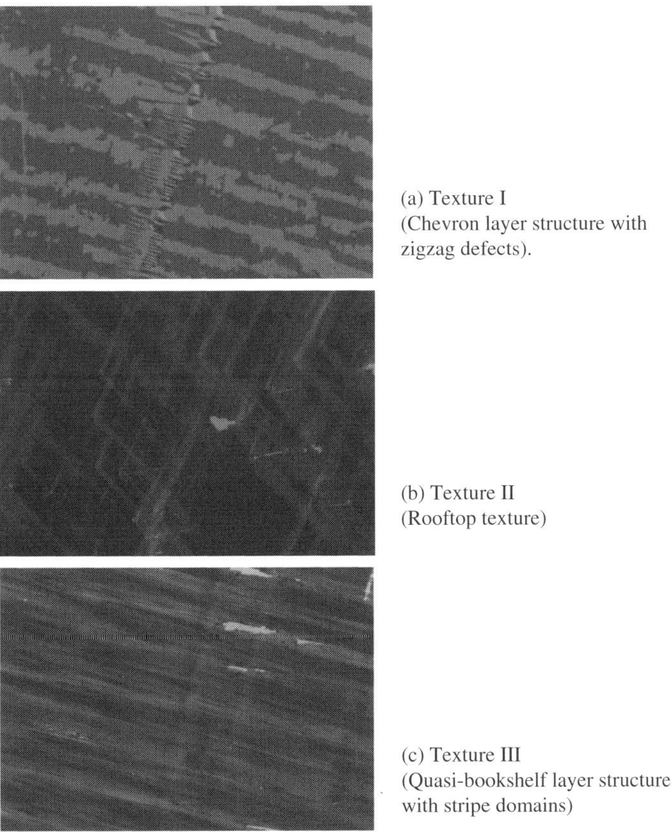

Fig. 6.1.4 Changes in texture induced by AC-field treatment. (a) Texture I. Initial virgin texture with chevron layer structure and zigzag defects. (b) Textures II. A rooftop texture is observed after the application of an AC field of medium strength. (c) Texture III. A quasi-bookshelf layer structure with stripe domains is observed after the application of a strong AC field.

produced a good memory characteristic and good multiplexing behaviour with little perturbation under non-selecting bias waveforms.

The quasi-bookshelf orientations produced by AC-field treatment have been applied to passive matrix FLC displays. Rieger *et al.* reported that a quasi-bookshelf orientation induced by AC-field treatment showed a contrast ratio of 60:1 with a line-address time of 90 μsec under the control of a 4-slot multiplexing addressing waveform with a bias of 4:1. They presented a 1.9″ passive matrix FLC panel with 96 × 128 pixels that they had fabricated by using this method [8]. The specifications of their prototype FLC display are given in Table 6.1.2. T. Kitamura *et al.* also reported use of the AC-field treatment method to fabricate a passive matrix FLC display with 2000 × 2000 pixels [9]. The specifications of this display are summarized in Table 6.1.3.

Hartmann investigated the texture-change phenomena and reported that texture III can be used to implement analogue grey scale switching, in which the ratio of black and white domains is controlled by the addressing waveform [6]. Hartmann then applied the texture III state to a passive matrix FLC display (Table 6.1.4).

The second approach to obtaining a bookshelf layer structure is the utilization of a unique class of materials with tilt angles that are weakly dependent on temperature. As was described in the previous chapter, the predominant behaviour that gives rise to the chevron layer structure is the reduction of layer spacing brought about by cooling. The reduction of layer spacing is usually observed in FLC materials with an INAC (isotropic – nematic – smectic A – smectic C) phase sequence, because the tilt angles of such FLC materials are usually strongly dependent on temperature. However, there is a unique class of FLC materials in

Table 6.1.2 The specifications of a passive matrix FLC display developed using the AC field treatment method by Rieger *et al.* [8]

Display size	29 × 38 mm
Pixel number	96 × 128
Addressing method	Passive matrix
Molecular orientation	Quasi-bookshelf (AC field treatment method)
Cell thickness	1.6 μm
Spontaneous polarization (P_s)	31 nCcm^{-2}
Memory angle (2θ)	46°
Drive waveform	4-slot multiplexing waveform with 4:1 bias
Drive voltage	Vs = 38 V
Line address time	90 μs/line
Contrast ratio	60:1
Grey scale	0/1
Colour	Monochrome

Table 6.1.3 The specifications of a passive matrix FLC display developed using the AC field treatment method by Kitamura *et al.* [9]

Addressing method	Passive matrix drive
Pixel number	2000 × 2000
Addressing method	Passive matrix
Molecular orientation	Quasi-bookshelf (AC field treatment method)
Cell thickness	2 μm
Spontaneous polarization (P_s)	73 nC/cm^2
Response time	50 μs
Contrast ratio	> 30:1
Grey scale	0/1
Colour	Monochrome

which the tilt angle is not strongly dependent on temperature. It is possible to realize a bookshelf layer structure for such materials without electrical manipulation. Mochizuki *et al.* reported that certain FLC mixtures that include an FLC compound with a naphthalene ring will take on a quasi-bookshelf layer structure [10]. They prepared mixtures consisting of ZLI-4139 (E. Merck) having an INC phase sequence (isotropic – nematic – smectic C) and the FLC compound with a naphthalene ring shown in Fig. 6.1.5. They reported that the temperature range over which material was in the smectic A phase was increased and that zigzag defects tended to disappear as the concentration of the naphthalene compound was increased. An FLC mixture with 40 wt% of the naphthalene compound showed a uniform orientation with no zigzag defects. They presented a 3.4″ monochrome passive matrix FLC display with a contrast ratio of 35:1 and a multiplexing drive

Table 6.1.4 The specifications of a FLC display developed using the AC field treatment by Hartmann [6]

Display size (diagonal)	2.5″
Pixel number	140 × 170
Addressing method	Passive matrix
Molecular orientation	Texture III (Quasi-bookshelf by AC field treatment method)
Cell thickness	1.4 μm
Spontaneous polarization (P_s)	46 nC/cm^2
Drive frequency	50 Hz
Response time	128 μs/line (half-reset)
Contrast ratio	30:1
Grey scale	100 grey levels
Colour	Monochrome

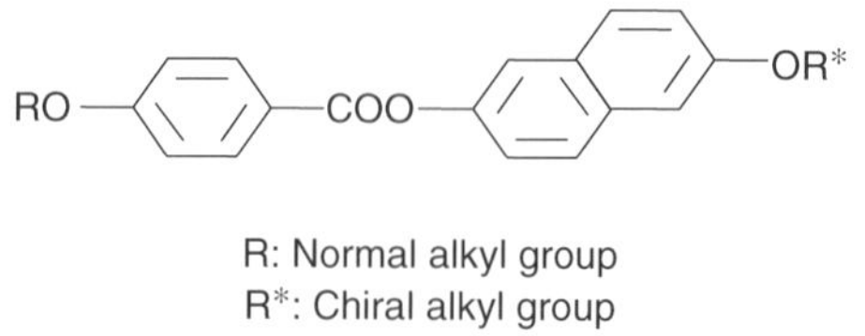

Fig. 6.1.5 An example of the naphthalene compounds that are used in the quasi-bookshelf mode [10].

waveform that was based on an FLC mixture containing compounds with a naphthalene ring [10]. The specifications of this FLC display are summarized in Table 6.1.5.

The third approach is the utilization of FLC materials with an INC phase sequence. A quasi-bookshelf orientation is obtained by applying a DC voltage as the material cools and changes from the nematic to the smectic C phase. This is because the tilt angles of FLC materials with an INC phase sequence are weakly dependent on temperature [11]. The quasi-bookshelf orientation exhibits monostability and has a smooth voltage–transmission characteristic and wide switching angle. An active matrix drive method is thus suitable for a material in this orientation. This mode is called CDR (Continuous Director Rotation) [11] or Half-V shaped FLC [12, 13]. A schematic view of switching in this mode and an example of the V–T characteristic of one such material are shown in Fig. 6.1.6. Asao *et al.* [12] and Furukawa *et al.* [13] have independently used this mode to produce TFT-FLC displays. The specifications of the prototype TFT-FLC display reported by Furukawa *et al.* are given in Table 6.1.6.

Table 6.1.5 The specifications of a FLC display developed using naphthalene compounds by Mochizuki *et al.* [10]

Display size	50 × 70 mm
Pixel number	128 × 196
Addressing method	Passive matrix
Molecular orientation	Quasi-bookshelf (by using naphthalene compounds)
Drive waveform	4-slot multiplexing waveform with 4:1 bias
Drive voltage	20 V
Line address time	160 μs/line
Contrast ratio	35:1
Grey scale	0/1
Colour	Monochrome

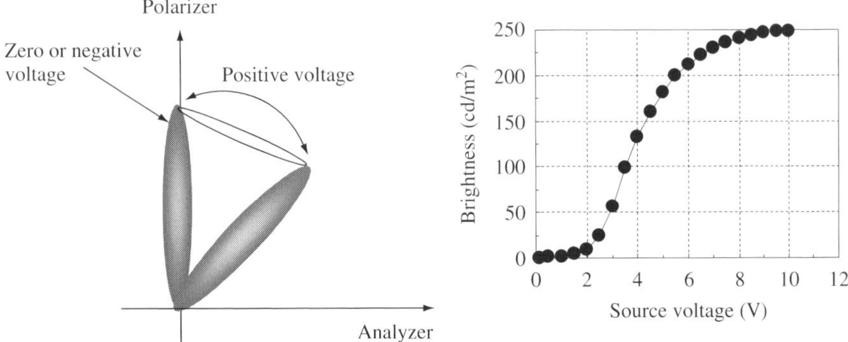

Fig. 6.1.6 A schematic view of switching and an example of the V–T characteristic of the half-V shaped FLC mode [13].

Table 6.1.6 The specifications of a TFT-FLC display developed using the "Half-V shaped FLC" reported by Furukawa *et al.* [13]

Display size (diagonal)	10.4″
Pixel number	480 × 640 (×RGB)
Addressing method	Active matrix (a-Si-TFT)
Molecular orientation	Bookshelf (Half-V shaped mode)
Cell thickness	1.5 μm
Phase sequence of FLC material	INC phase sequence
Spontaneous polarization (P_s)	3.2 nCcm^{-2} (at 25 °C)
Switching angle	45°
Δn	0.186 (at 25 °C)
Response time	Rise 0.5 ms, Decay 0.2 ms
Contrast ratio	350:1
Brightness	250 cdm^{-2}
Grey scale	256 grey levels
Colour	Full colour
Shock stability	20 kgcm^{-2}

6.1.4 C1-uniform (C1U) orientation

The C1-unform (C1U) orientation (shown in Fig. 6.1.7(a)) is also of practical value for use in FLC displays. The C1 and C2 orientations were first reported by Kanbe *et al.* [14]. The C1 orientation they reported on was not the C1-uniform (C1U), but the C1-twisted (C1T) orientation. This was not clearly described in the team's paper, but is suggested by the photographs of the material showing the bluish tinge that characterizes the C1T state and their description of the state as being characterized by a light blue colour.

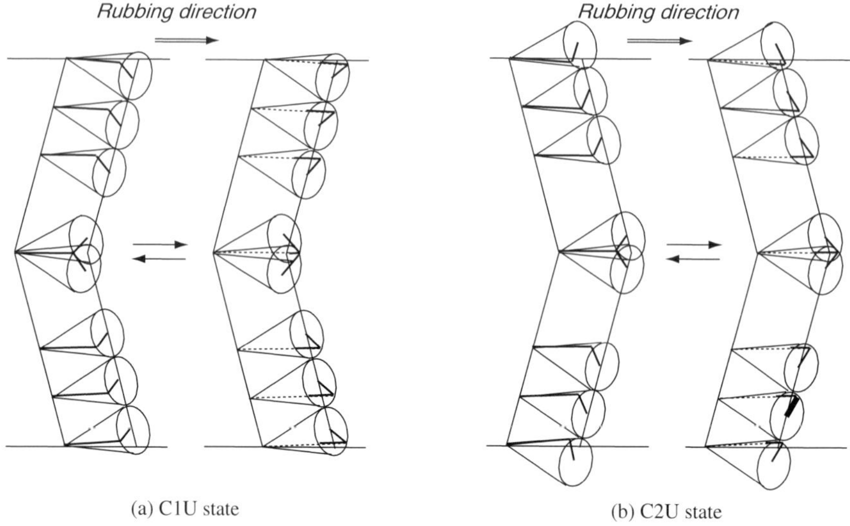

Fig. 6.1.7 Model of molecular orientation in the C1U and C2U orientations.

The C1-uniform (C1U) orientation was first reported by Koden *et al.* [15]. They reported that C1 states of two types were observed in FLC cells with high pretilt aligning film. One state showed extinction positions between crossed polarizers; this state was determined to be the C1U state. The other state showed no extinction positions with crossed polarizers; this state was determined to be the C1T state. The team reported that material in the C1U state also has a larger memory angle than material in the C1T or C2 states. Their report implied that the C1U orientation prepared by using a high pretilt aligning film is applicable to practical FLC displays.

The need for a high pretilt aligning film to obtain the C1U orientation is explained by the report of Kanbe *et al.* [14]. Their proposed conditions for the existence of the C1 and the C2 states is shown in Fig. 6.1.8. According to this team, in the C1 state, it is only possible for liquid crystal molecules on the boundary surface to exist on the cone shown if $\theta + \delta > \theta p$, where θ is the tilt angle, δ is the layer tilt angle and θp is the pretilt angle. In the C2 state, on the other hand, the tilt angle is required to satisfy $\theta - \delta > \theta p$. Therefore, in the C1U mode, a combination of the high pretilt aligning film and a low tilt angle for the FLC material is required to obtain the C1U state and prevent the appearance of the C2 state.

Koden *et al.* [16] and Tsuboyama *et al.* [17] have independently reported that material in the C1U orientation showed good display performance. Both teams prepared materials with a uniform C1U orientation by combining a high pretilt aligning film with a low tilt FLC material. Koden *et al.* reported that the C1U state showed the high contrast ratio of 30:1 for a multiplexing waveform with a 1/3 bias waveform, and that the degree of flicker over a bias period was lower in

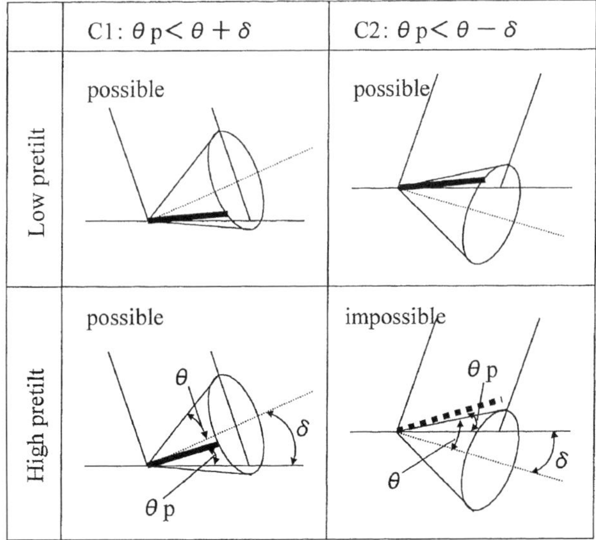

(θ_p: pretilt on aligning film, θ: tilt angle of FLC material , δ: layer leaning angle)

Fig. 6.1.8 The conditions for the appearance of the C1 and C2 states [14].

the C1U state than in the C2. These differences are shown in Fig. 6.1.9. Tsuboyama *et al.* used the C1U orientation to develop a 15″ passive matrix SXGA FLC display with a 4 grey level monochrome or 16 colours. The specifications of the FLC displays are shown in Table 6.1.7.

Terada *et al.* [18] and Hanyu *et al.* [19] have also reported on some methods for stabilizing the C1U orientation of material. Terada *et al.* reported that a particular dependence of tilt angle on temperature is preferable to keep the C1U orientation over a wide range of temperatures [18]. Since the stability of the C1U state depends on the relationship between the tilt angle of the material and the pretilt angle of the alignment films, FLC materials are required to show a relatively small tilt angle and small dependence of tilt angle on temperature. Indeed, Terada *et al.* have reported that an FLC material with a weak dependence of tilt angle on temperature showed the C1U orientation between 10 °C and 50 °C, while other materials with rather strong dependence of tilt angle on temperature only exhibited the C1U orientation over a narrow range of temperatures. Hanyu *et al.* have reported that the cross-rubbing technique is useful for producing a uniform C1U state by preventing the occurrence of C1T [19].

6.1.5 C2-uniform (C2U) orientation

The C2 orientation was first reported by Kanbe *et al.* [14] and the C2-uniform (C2U) orientation (shown in Fig. 6.1.7(b)) was first reported by Koden *et al.* [20];

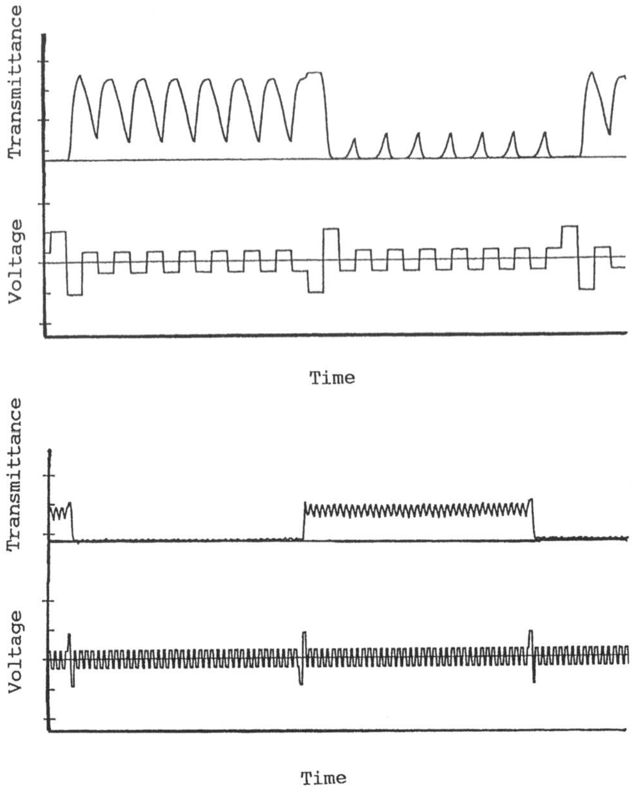

Fig. 6.1.9 A comparison of transmission under a 1/3 bias waveform by materials in the C1U and C2 state [16].

the C2 orientation reported by Kanbe *et al.* is estimated to be the C2U orientation. Kanbe *et al.* reported that, while the C1 state does not exhibit an extinction position, the C2 state does exhibit extinction positions. They used a material in the C2 orientation to develop a 14″ passive matrix SXGA monochrome FLC display, but its contrast ratio was only 5:1. The low contrast ratio of materials in the C2 orientation has also been confirmed by a comparison of molecular fluctuations (light leakage) between materials in the C1U and C2 states under a bias period with a multiplexing drive waveform [16]. As is shown in Fig. 6.1.9, the material in the C2 state exhibited a large leakage of light during the bias period due to a large degree of fluctuation of the states, giving rise to the low contrast ratio.

It was, however, found to be possible to realize a high contrast ratio from a material in the C2 state when it was combined with the τ-Vmin mode [21] by using a FLC material with a negative dielectric anisotropy and/or a large positive dielectric biaxiality [22]. In addition, the C2U state has several advantages over the C1U state and bookshelf state. The C2U state provides faster response times

Table 6.1.7 The specifications of a FLC display developed using the C1U mode by Tsuboyama *et al.* [17]

	Monochrome type	Colour type
Display size (diagonal)	15″	
Pixel number	1024 × 1280	
Pixel construction	3:2 divided	RGBW
Addressing method	Passive matrix	
Molecular orientation	C1U orientation	
Cell thickness	1.5 μm	
Spontaneous polarization (P_s)	5.8 nCcm^{-2}	
Tilt angle	15°	
Drive voltage	20 V	
Line address time	70 μs/line (14 Hz/frame)	
Contrast ratio	40:1	
Grey scale	4 levels	0/1
Colour	Monochrome	16 colours

and wider operating temperature ranges than the C1U state. This is probably due to the lack of significant molecular switching on the surface [23].

The τ-Vmin behaviour is an unusual behaviour that shows a minimum in the response (τ)–voltage (V) curve. The behaviour is exhibited by FLC materials that have modest values of spontaneous polarization to balance the dielectric restoring torque [22]. Therefore, to operate in the τ-Vmin mode, FLC materials must have low P_s values and large positive dielectric biaxiality ($\delta\varepsilon$). A typical τ-Vmin characteristic is shown in Fig. 6.1.10. The C2U orientation in combination with the

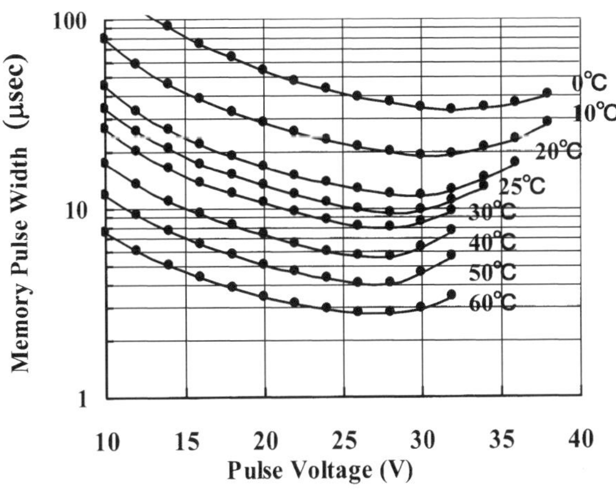

Fig. 6.1.10 The τ-V characteristic of the SF-2692 FLC material [27].

τ-Vmin mode is able to provide a high contrast ratio because the AC-stabilization effect restricts molecular fluctuation during the bias period of practical multiplexing drive waveforms [23].

Two useful approaches to realizing the selective formation of a C2U orientation have been reported. One approach involves an investigation of the method of alignment. An intermediate pretilt angle is preferable [23]; this is indicated by the conditions of the C1 and C2 states as shown in Fig. 6.1.8. Strong rubbing has also been reported as being effective in obtaining a material in the C2 orientation [24]. Figure 6.1.11 shows the relationship between the areal proportion of a material in the C2 state and the retardation of the aligning film, which is changed by the rubbing temperature. Since the rubbing temperature affects the rubbing strength [25], Fig. 6.1.11 clearly shows that strong rubbing is favourable to producing the C2 orientation. This is because the direction of molecules on a surface in the C2 orientation become almost the same as the direction of rubbing and so is insensitive to variations in alignment, surface polarity, memory effects, and surface switching.

The other approach is to investigate the relationship between the properties of materials and the formation of the C2 orientation. Figure 6.1.12 shows the effect of the temperature range of the smectic A phase for a material on the areal proportion of the material in the C2 state. The figure indicates that, to obtain the C2 orientation, a material must be in the smectic A phase over a wide range of temperatures [24]. This phenomenon can be attributed to the fact that FLC materials that are in the smectic A phase over a wide range of temperatures tend to have small tilt angles near the SmC–SmA transition point. This small tilt angle is expected to reduce the energy barrier for transition from the C1 to the C2 state near the SmC–SmA transition point because the small tilt angle leads to chevron layers leaning at a small angle.

It is possible to obtain high contrast ratios by applying adequate addressing waveforms to FLC devices with materials that exhibit the τ-Vmin characteristic

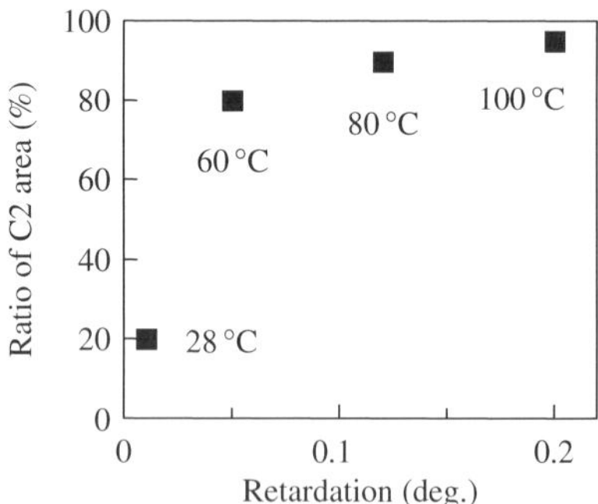

Fig. 6.1.11 The relationship between the retardation of aligning film rubbed at different temperatures and the areal proportion of material in the C2 orientation [24].

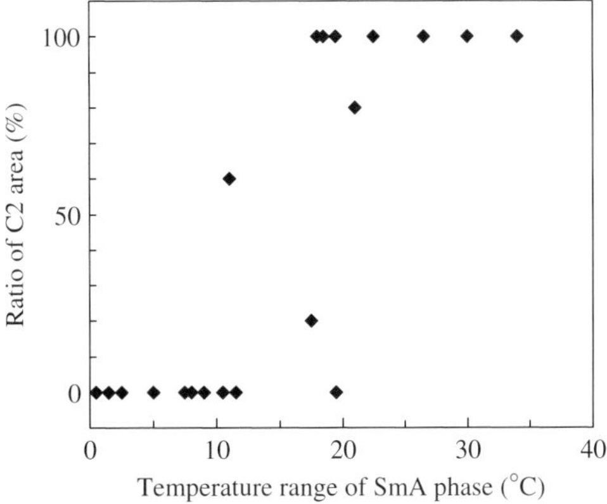

Fig. 6.1.12 The relationship between the temperature range of SmA and the areal proportion of material in the C2U state [24].

and are in the C2U orientation. This is because bias pulses in the passive matrix addressing are able to restrict fluctuations in the FLC molecules that arise from the AC-stabilization effect. Koden *et al.* [24] stated that contrast ratios higher than 100:1 were obtained by combining the FDS-71 FLC material they developed, the C2U orientation, and the DRAMA-3 drive waveforms [26] shown in Fig. 6.1.13. Koden *et al.* [27] achieved an even higher contrast ratio, greater than 200:1, by using an optimized device structure with an improved FLC material SF-2692.

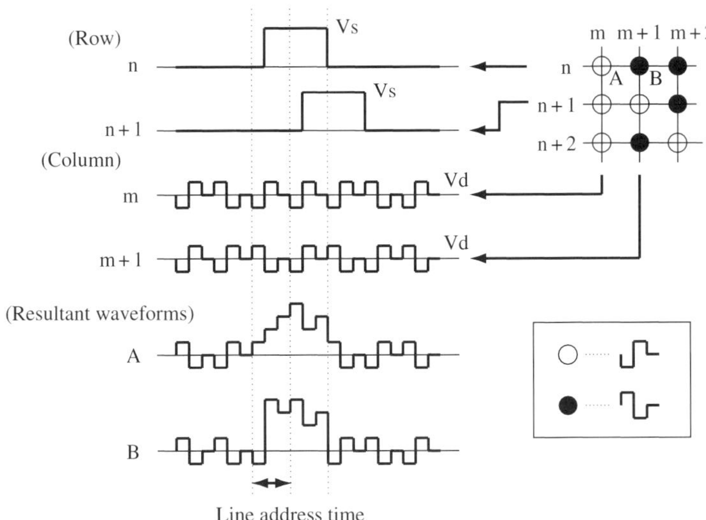

Fig. 6.1.13 The Drama-3 drive waveforms [26].

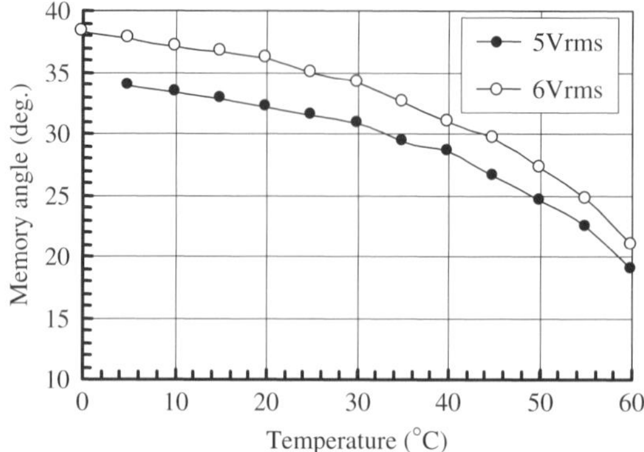

Fig. 6.1.14 Memory angles of the SF-2692 FLC material; the τ-Vmin mode under bias voltages [27].

The τ-Vmin characteristics are shown in Fig. 6.1.10. The material showed a wide memory angle under the bias voltages shown in Fig. 6.1.14, giving it a high brightness.

The key technologies outlined above and the digital grey scale technology were applied in a 15″ passive matrix full colour prototype FLC display [27, 28]. This prototype 15″ FLC display, with its black matrix of metal, and its optimized device structure and material parameters, has a contrast ratio of 150:1, a resolution of 920 × 540 dots. The display is capable of displaying full colour pictures at video frame rate with 256 grey levels and is stable under exposure to a shock of 20 kgcm^{-2}. The specifications of the prototype FLC display are summarized in Table 6.1.8. Figure 6.1.15 is an example of the pictures on this prototype display.

Adoption of the C2U orientation in monostable modes for active matrix FLC displays is also possible. Furue *et al.* have reported that a C2U orientation obtained by a polymer-stabilized method was in a monostable state, was defect-free, and had a smooth voltage-transmission characteristic [29].

6.1.6 Shock stability of FLC displays

One of the intrinsic problems for the practical application of FLC displays is the low mechanical stability of the molecular orientations. The initial molecular orientations in an FLC material are easily destroyed by the application of mechanical pressure and/or mechanical shock. The materials do not return to their initial states, unlike molecular orientations in nematic liquid crystals, which are also disturbed by mechanical pressure and/or shock but usually return to their initial states. The low stability of FLC devices under exposure to shock is attributed to the presence of the smectic layer structure. The molecular orientation of the smectic phases is highly ordered in comparison with that of the nematic phase.

Table 6.1.8 The specifications of a passive matrix full colour FLC display developed using the C2U orientation and the τ-Vmin mode by Koden [27]

Display size (diagonal)	15″
Pixel number	540 × 920 (×RGB)
Addressing method	Passive matrix
Molecular orientation	C2U state
Cell thickness	1.4 μm
FLC material	SF-2692 ($P_s = 10\,nCcm^{-2}$ at 25 °C)
Memory angle 2 θm	35° (under $V_d = 7$ V at 25 °C)
Drive frequency	60 Hz (Interlaced)
Addressing scheme	DRAMA-3
Drive voltage	Vs = 30 V, $V_d = 7$ V
Line address time	15 μs/line
Contrast ratio	150:1
Brightness	200 cdm^{-2}
Grey scale	256 grey levels
Colours	16,700,000 colours
Operating temperature	0~60 °C
Shock stability	20 $kgcm^{-2}$

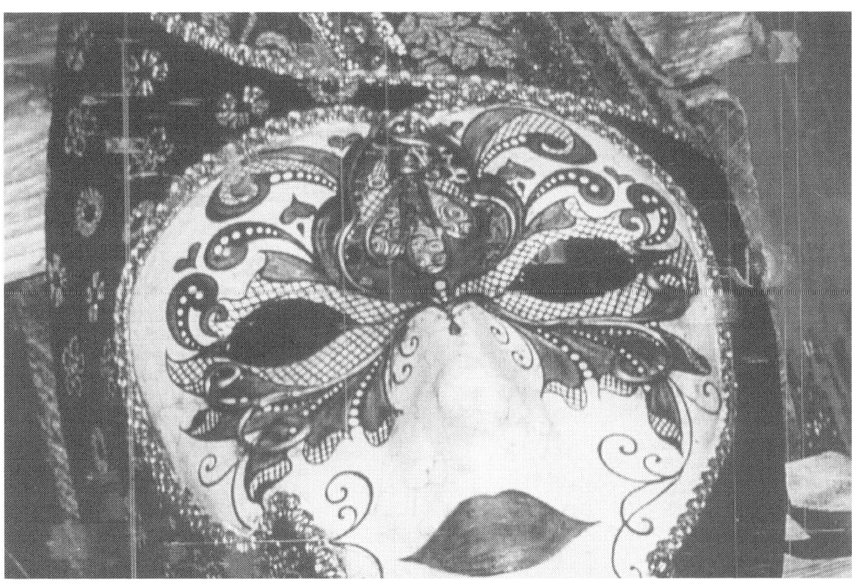

Fig. 6.1.15 An example of pictures of the 15″ passive matrix, full colour FLC display reported by Koden [27].

The processes of damage under exposure to mechanical shock differ according to whether the shock consists of a continuous pressure or a hammer-style shock [30, 31].

The typical process of damage resulting from a continuous pressure is illustrated in Fig. 6.1.16 [31]. When relatively little pressure is applied, zigzag defects appear, but the material recovers to its initial state after the pressure is removed. However, with a higher pressure, the initial state is transformed to a damaged texture (Fig. 6.1.16(d)) and the material does not recover. This phenomenon is caused by a reduction in the cell gap and the resultant flow of material from the compressed region.

The typical process of damage by hammer-style shock is different from the process of damage by the continuous application of pressure. The typical process of damage induced by a continuous pressure has been illustrated above in Fig. 6.1.17 [31]. When, however, a relatively small hammer-style shock is applied, a patch of needle defects forms, in the direction opposite the direction of rubbing from the point of impact (Fig. 6.1.17(b)). When the amount of hammer damage is relatively great, a symmetrical pattern of damage pattern forms, with approximately circular patches of damaged texture and surrounding needle defects (Fig. 6.1.17(c)). When the hammer-style shock is very strong, the patches of damaged texture produce extinction parallel to the rubbing direction (Fig. 6.1.17(d)). This "butterfly" pattern of damage, where the damage occurs in the direction parallel to the direction of rubbing, is attributed to shearing of the FLC layers [31, 32].

Various approaches to realizing a high degree of stability under exposure to shock have been investigated. Adhesive spacers have been proposed as a way of

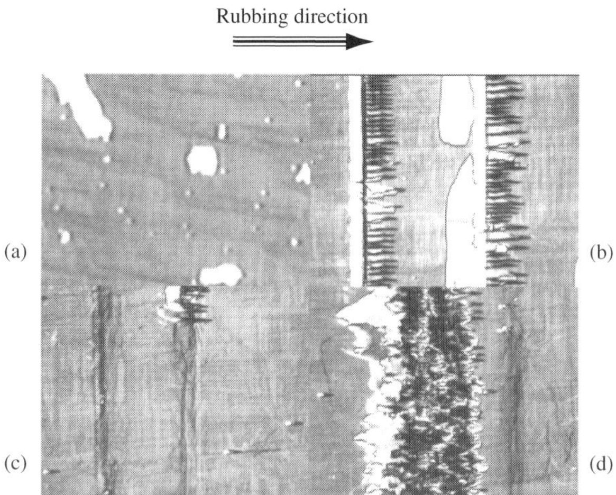

Fig. 6.1.16 Typical damage processes in pressure tests [31]. (a) Initial texture (C2 state), (b) nucleation of zigzag defects due to flow past obstacles, (c) shrinking of zigzag defects leaving perturbed smectic layering, and (d) further flow causes extensive damaged textures.

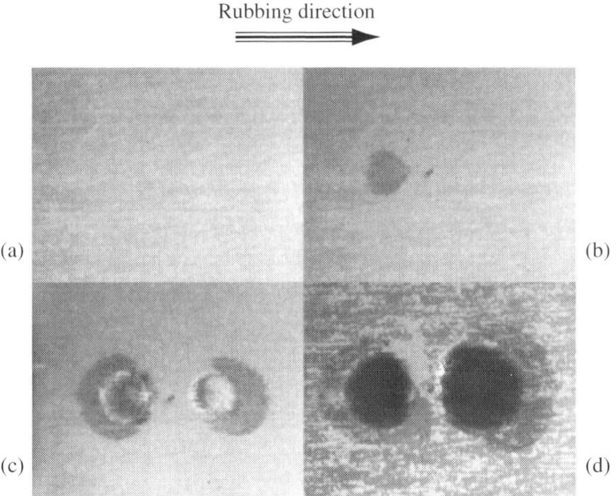

Fig. 6.1.17 Typical damage processes in hammer tests [31]. (a) Initial texture (C2 state), (b) 0.028 Ns blow, needle defects form in the direction opposite to that of rubbing, (c) 0.083 Ns blow, symmetrical damage pattern forms, (d) 0.121 Ns blow, damage texture produces extinction parallel to the direction of rubbing.

increasing shock stability [33, 34], but the mechanical stability reported for this technique is only of the order of $10\,\mathrm{Ncm}^{-2}$. To realize sufficient shock stability, flow, as well as shear, must be prevented in the FLC panel.

Sufficient shock stability can be achieved by introducing polymer walls, running parallel to the direction of rubbing, within the FLC panel [30, 35, 36]. Figure 6.1.18 illustrates an example of such a device structure. Gass *et al.* have

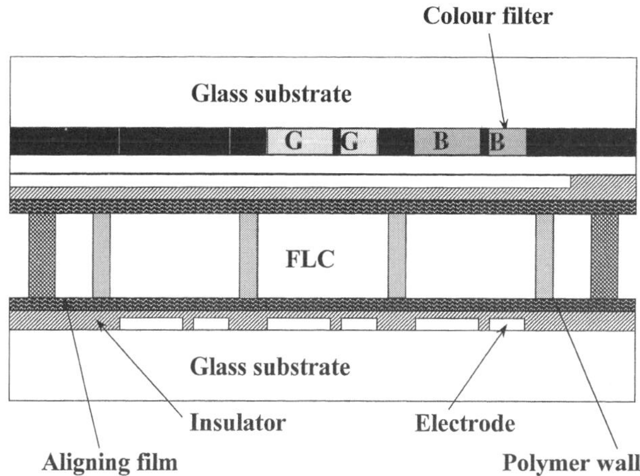

Fig. 6.1.18 An example of the structure of an FLC device with polymer walls [27].

Table 6.1.9 Damage thresholds of STN and FLC panels with and without polymer walls [30]

Test type	Standard STN panel	Standard FLC panel	FLC panel with polymer wall
Hammer impulse with no bending	0.12 Ns	> 0.03 Ns	0.06 N
Hammer impulse with bending	0.38 Ns	> 0.03 Ns	0.11 Ns
Pressure with no bending	225 Ncm^{-2}	20 Ncm^{-2}	200 Ncm^{-2}
Pressure with bending	325 Ncm^{-2}	5 Ncm^{-2}	200 Ncm^{-2}

reported that 20 $kgcm^{-2}$ was achieved by this method. This performance is compared, in Table 6.1.9, with those of STN and FLC panels without the polymer-wall structure [30]. Koden *et al.* have presented prototype FLC displays in which this polymer-wall structure is applied [27, 36].

Another interesting technology involves a novel method of FLC and AFLC orientation that also applies a polymer-wall structure. Minato *et al.* have reported that they were able to obtain uniform orientations by applying a temperature graduation method to FLC panels with polymer walls [37]. This method utilizes the reduction of volume on partial cooling and is able to realize a uniform C2 orientation and a uniform alignment with antiparallel rubbing, etc.

6.1.7 The layer-rotation phenomenon

The layer-rotation phenomenon is one of the unique behaviours of FLC and AFLC devices. It is caused by the presence of spontaneous polarization.

Layer rotation was first reported by Nakagawa *et al.* [38]. They measured the angles between the rubbing direction and the layer normal in FLC cells in which only one substrate had been rubbed and the FLC material had an IAC (isotropic – smectic A – smectic C) phase sequence. The definition of the angle of rotation is shown in Fig. 6.1.19. The cells used in the measurements were cooled from the isotropic phase and their angles of rotation were measured at room temperature. The results are given in Table 6.1.10. The value of the layer rotation δ is not large, but is obvious. In addition, it is obvious that the direction of rotation is correlated with the sign of spontaneous polarization.

This phenomenon can be explained by the surface electroclinic effect of the smectic A phase as shown in Fig. 6.1.20. Nakagawa *et al.* have suggested the existence of a local electric field at the boundary or surface field, which in turn is explained by the contact between the two materials, liquid crystal and aligning film. The induction of a molecular tilt by the electroclinic effect [39] can be expected under exposure to such a surface field. Since FLC molecules are aligned with the direction of rubbing, the layer is formed with a layer deviation angle δ.

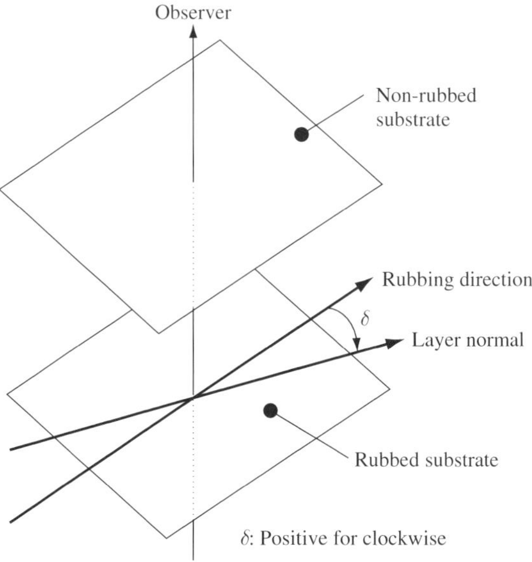

Fig. 6.1.19 The definition of layer-rotation angle δ [38].

The direction of this surface field seems to depend on the materials at the boundary. In the case shown in Table 6.1.10, the relationship between the sign of δ and the sign of P_s indicates that the direction of the surface field is from the liquid crystal layer to the surface.

This layer-rotation phenomenon is clearly visible in FLC cells in which only one substrate has been rubbed and in which the FLC material has an IAC phase sequence. This layer rotation is also observed when FLC materials show the nematic phase and a single substrate is rubbed. This layer rotation is attributed to the helical structure of a material in the nematic phase [38]. When, on the other hand, the same aligning film is used under both substrates and the substrates

Table 6.1.10 The relationship between the sign of layer rotation and the sign of spontaneous polarization in FLC cells in which only one substrate is rubbed and the FLC material shows an IAC phase sequence [38]

Liquid crystal	Layer rotation (δ)	Spontaneous polarization (P_s)
1	+3.6°	−
2	−1.0°	+
3	+0.8°	−
4	+0.6°	−

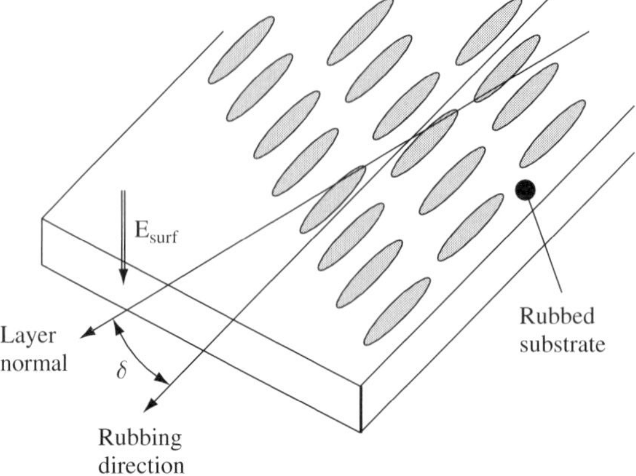

Fig. 6.1.20 The mechanism for the deviation of layer normal from the rubbing direction due to the electroclinic effect [38].

are rubbed, the layer-rotation phenomenon is not observed. This is because rotation in the layers under the two substrates is in the opposite directions.

Investigation of the layer-rotation phenomenon has suggested that the cross-rubbing method is useful in alignment when FLC or AFLC materials with no nematic phase are used. In the cross-rubbing method, the directions of rubbing for the two substrates are not consistent with each other. The angle between the rubbing directions for the two substrates is set to be close to the layer-rotation angle δ. Indeed, Hattori *et al.* have applied the cross-rubbing technique to the alignment of an AFLC material [40]. They reported that an AFLC cell with a cross-rubbing angle of 165° showed a uniform orientation with good extinction, but an AFLC cell with antiparallel rubbing (180°) showed an orientation with many line defects.

6.1.8 Summary

This chapter has described the relationship, for FLC displays, between molecular orientation and performance. Much research and development has been carried out to give us a scientific understanding of the roles of molecular orientation and alignment technologies in FLC displays. However, knowledge and technologies for obtaining molecular orientations in FLC displays are not enough.

One reason for this situation is that the molecular orientations have not been sufficiently analyzed and described. In some papers to do with molecular orientation, the obtained or measured orientations have not been described. For example, some papers have described the obtaining of a uniform orientation by using a certain method, material, method of alignment, etc., but have not described the kind of molecular orientation that was obtained.

Another reason for this is the complexity of molecular orientation in FLC devices. FLC devices have layered structures, three surfaces (that is, two substrate surfaces and a chevron interface), tilt angles, spontaneous polarization, etc. These features lead to more complex phenomena such as layer leaning, twisted orientations, low shock stability, layer rotation, etc.

However, the FLC still has features that make it attractive for use in the flat-panel displays of the future. FLCs are able to realize fast response times and wide viewing angles. These advantages allow the application of bistable FLC devices to passive matrix full colour FLC displays with digital grey scales [27], passive matrix high-resolution FLC displays that apply the memory effect [41], etc. Monostable FLC devices are applicable to high contrast, fast response, full colour TFT-FLC displays [12, 13], FLC on LSI micro-displays driven by the digital grey scale method [42], etc. In particular, the fast response times of TFT-FLC displays have recently drawn attention. Recent studies on the picture quality of LCDs have revealed that the slow response times of nematic liquid crystal are inadequate because they induce blur in dynamic images. This is a severe problem in TV applications and multi-media displays that involve dynamic images. One solution is the application of the FLC to TFT-LCDs because it is possible to realize very fast response times, of the order of microseconds, with FLCs. Indeed, Furukawa *et al.* have presented a 10.4″, high contrast, full colour TFT-FLC display, in which an in-pulse type of drive instead of a normal hold-type drive was realized by applying the fast response times of FLCs [13].

It should be noted that there are still many problems with FLC displays and that they are still under development. It is hoped that the close relationship between fundamental scientific study and the development of practical technologies will lead us to commercial successes with FLC displays.

Finally, although they are not described in this chapter, we must note some important and interesting studies such as those of the oblique evaporation method [43], the twisted FLC [44], an oblique layered structure that exhibits an analogue grey scale [45], etc.

6.2 Alignment and Performance of AFLCD

Kohki Takatoh

6.2.1 AFLC materials and Devices

In 1988, Fukuda and co-workers discovered the existence of liquid crystalline materials showing antiferroelectric properties [1], exhibiting tristable switching, as shown in Fig. 6.2.1. The alignment structure of these materials was concluded to be the one depicted in Fig. 6.2.2.

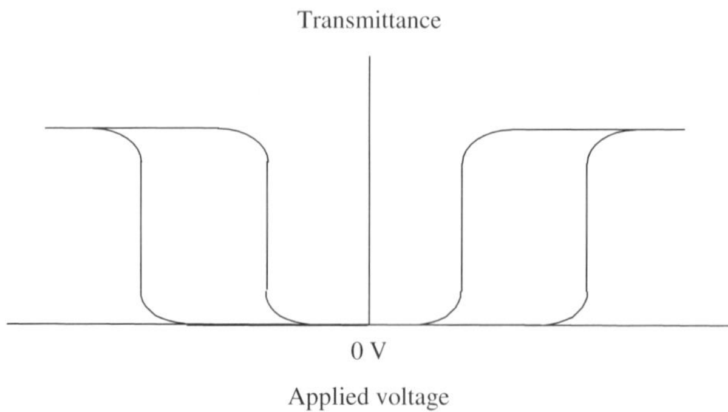

Fig. 6.2.1 The relationship between applied voltage and transmittance.

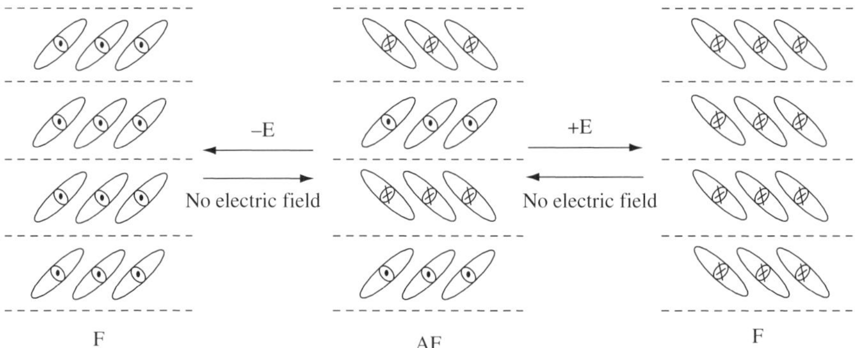

Fig. 6.2.2 The alignment structure of an AFLC with and without applied voltage.

Without applied voltage, the dipole moment of the liquid crystalline molecules in each adjacent layer is directed in the opposite direction, due to the large dipole moment. Generally, the spontaneous polarizations for AFLC materials are larger than 50–60 $nCcm^{-2}$, whilst FLC materials of spontaneous polarization of ca. 1–10 $nCcm^{-2}$ are normal. AFLC materials have also strongly attracted the attention of industry, because grey scale expression is possible by simple matrix driving or without an active matrix. Moreover, the deterioration caused by applying stress in the case of SSFLC can be repaired by applying voltage [2].

A driving method using tristable V–T properties has been proposed to realize the grey scale expression by simple matrix driving [3–9]. Figures 6.2.3 [4] and 6.2.4 show the concept and the scheme for the driving method to realize grey scale expression using AFLC tristable switching.

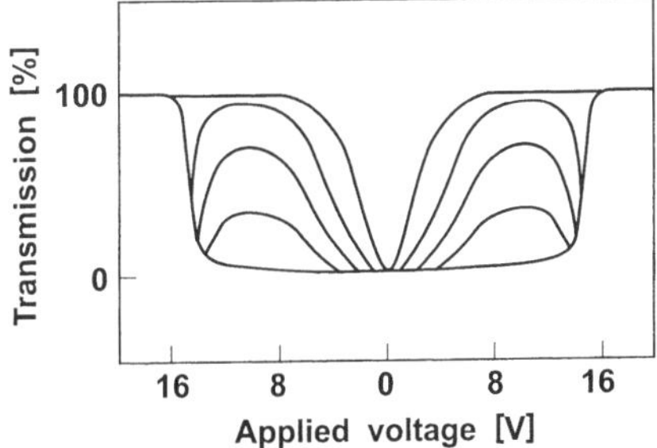

Fig. 6.2.3 Concept for the driving method to realize grey scale expression using AFLC tristable switching.

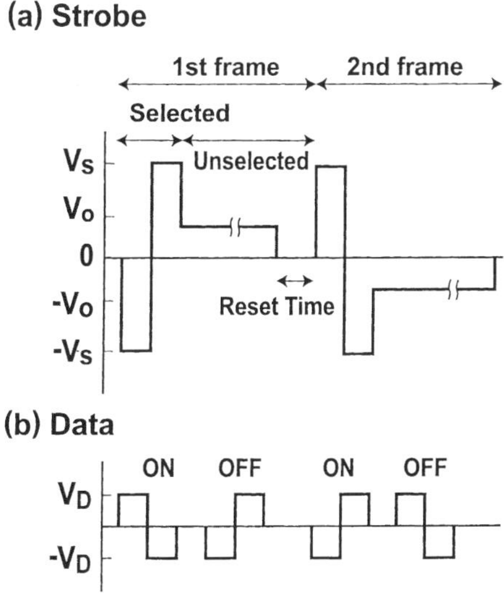

Fig. 6.2.4 Scheme for the driving method to realize grey scale expression.

During an unselected period, the offset voltage, V_0, is applied and maintains the transmission corresponding to the voltage applied during the selected period. Figure 6.2.3 shows the relationship between the applied voltage ($V_S + V_D$) and the transmission, and Fig. 6.2.3 suggests that simple matrix driving for AFLC materials can realize grey scale expression by selecting the appropriate offset voltage, V_0.

The switching between the AF-state and F-state proceeds through the nuclear generation and growth of AF- or F-state domains in stripe form parallel to the layer direction. This switching is sometimes called "multidomain switching" or . "domain switching".

The grey scale can be realized by change in the ratio of the area showing AF and F states in one pixel. The switching from AF to F states proceeds through two step processes [10]. The first step is the change from the initial AF state to an AF intermediate state through the continuous shift of the extinction position. In the second step, the intermediate state switches to the F state with the formation of a stripe-shaped domains parallel to the smectic layers characteristic of AFLC materials. For FLC materials, the switching proceeds through the formation of boat-shaped domains.

6.2.2 Alignment of AFLC

To obtain high contrast, AFLC alignment without defects must be realized. Moreover, to realize the stable grey scale expression of the high duty ratio by simple matrix driving, large hysteresis and sharp threshold characteristics are indispensible. These characteristics are determined not only by the AFLC material properties, but also by the quality of the alignment states.

6.2.2.1 Polymer molecular structure for alignment layers

To obtain good initial alignment, varieties of polyimide materials for the alignment layer were investigated. The symmetry of the polymer molecules is critically important. In Table 6.2.1 [11], the molecular structures of polyimide materials, the evaluation of the alignment state and the contrast ratios of AFLCDs using the polyimides are shown.

In Table 6.2.1, the contrast ratios depend largely on the polymer molecular structures. The higher the polyimide molecular symmetry is, the higher the contrast ratio becomes. The polyimide of highest molecular symmetry, PI-1, shows the highest contrast ratio. A highly symmetric polymer structure induces rigid linear rods on the surface, which induces the high degree of alignment. The polyimide material with the lowest symmetry, PI-6, shows random alignment. The degree of molecular symmetry of polyimide, polyimide-amide, and polyamide, and the values of the contrast ratio of the AFLCDs using these polymers, decrease in that order. Among similar categories of polymer materials of polyimide, polyimide-amide and polyamide type, the order of the contrast ratio on the basis of the acid or diamine structure increases in the following order.

Molecular Structure 1

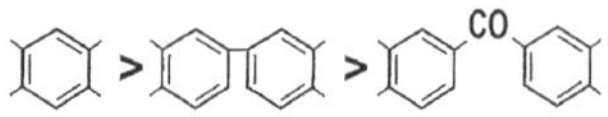

acid part

Molecular Structure 2

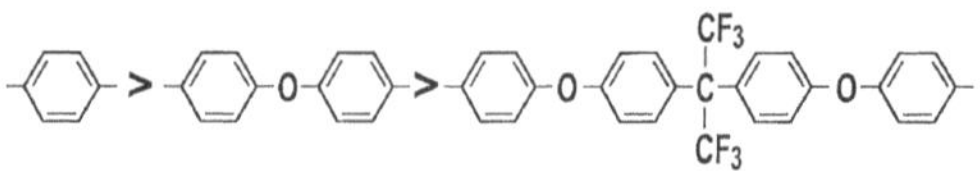

diamine part

Table 6.2.1 Molecular structures of polyimide materials, the evaluation of the alignment state and the contrast ratios of AFLCDs using the polyimides

Polymer code	Polymer structure	Alignment		Contrast (AFLC)
		FLC	AFLC	
Effect of polyimide on alignment properties				
PI-1	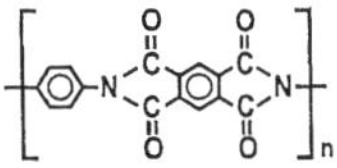	Good zigzag defect	Good	61
PI-2	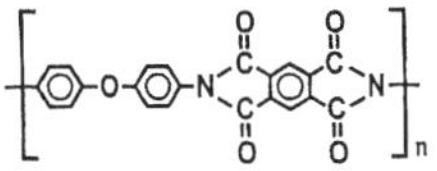	Good zigzag defect	Good	35
PI-3	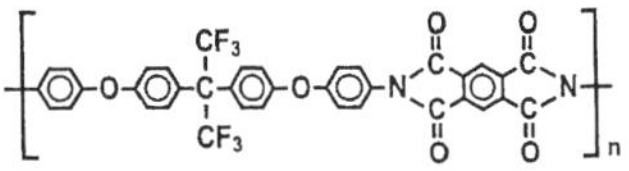	Good zigzag defect free	Good	13
PI-4	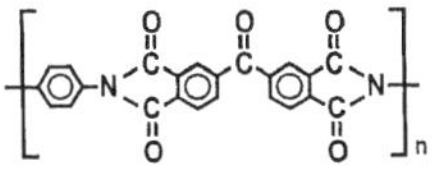	Good zigzag defect	Good	32
PI-5	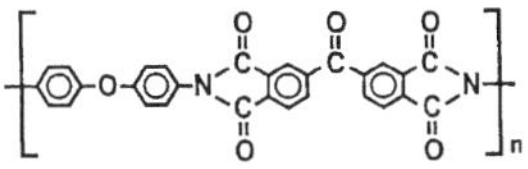	Good zigzag defect	Good	28
PI-6	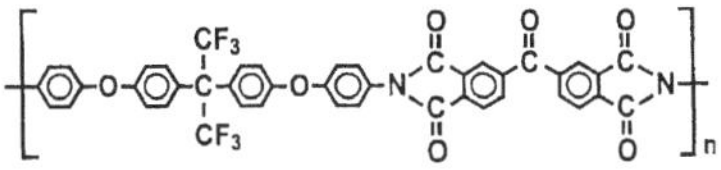	Random	Random	<10
PI-7	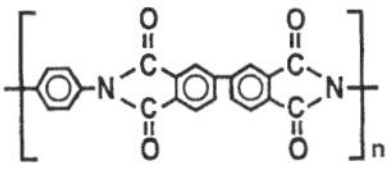	Good zigzag defect	Good	41
PI-8	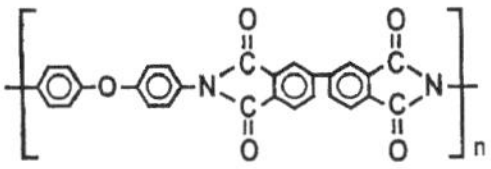	Good zigzag defect	Good	35
PI-9	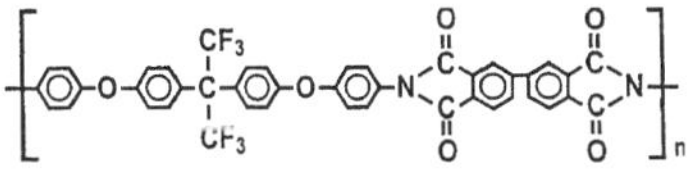	Random	Random	<10
Effect of polyamide-imide structure on alignment properties				
PAI-1	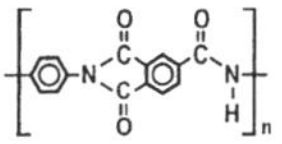	Good zigzag defect	Good	40
PAI-2	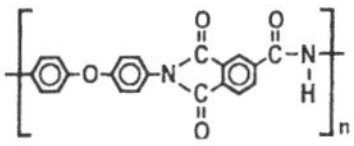	Good zigzag defect	Good	22
PAI-3	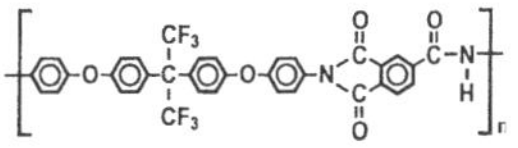	Good zigzag defect	Good	10

Table 6.2.1 (Continued)

Polymer code	Polymer structure	Alignment		Contrast (AFLC)
		FLC	AFLC	
PA-1	$-[\!\bigcirc\!-N(H)-C(=O)-\bigcirc-C(=O)-N(H)-]_n$	Good zigzag defect	Good	38
PA-2	$-[\bigcirc-O-\bigcirc-N(H)-C(=O)-\bigcirc-C(=O)-N(H)-]_n$	Random	Random	<10
PA-3	$-[\bigcirc-O-\bigcirc-C(CF_3)_2-\bigcirc-O-\bigcirc-N(H)-C(=O)-\bigcirc-C(=O)-N(H)-]_n$	Good zigzag defect free	Good	15

Furthermore, aromatic polyimides can produce better alignment than aliphatic polyimides [6].

Other aspects described below were pointed out.

(1) A significant variation in contrast properties was observed, which might be related to the specific interaction between the polar group of liquid crystals and the polymer molecular structure of high polarity.
(2) In spite of the presence of a side group, a hexafluoropropyl moiety in the polymer molecular structure can realize good alignment of the FLC and AFLC, similar to the non-substituted polymer structure. For example, it is reported that polyimide (PI-3) and polyamide (PA-3) gave a zigzag defect-free alignment in the FLC. The hexafluoropropyl moiety might have a significant influence on the defect-free alignment of FLC cells.
(3) The degree of alignment also depends on the degree of imidization. In the case of polyimide baked at high temperature with a resultant higher degree of imidization, a better initial texture is obtained.

6.2.2.2 Surface energy

The influence of surface energy on the alignment layer was also investigated, and it was confirmed that the resultant AFLC alignment also depends on the surface energy. It was reported that the best AFLC alignment was obtained when polar and non-polar components give 2–6 dynescm^{-1} and 42–46 dynescm^{-1}, respectively [6].

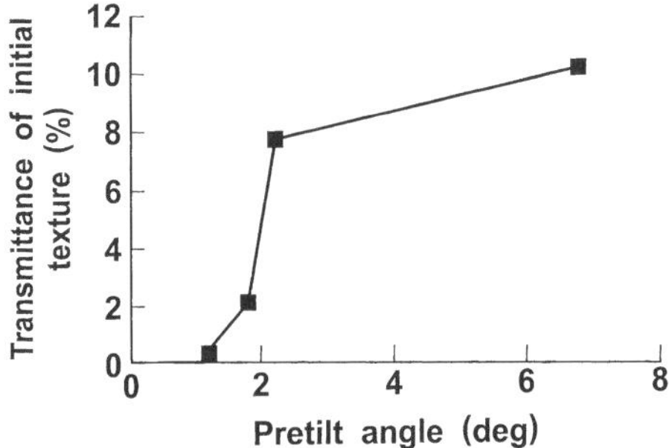

Fig. 6.2.5 The relationship between pretilt angle and transmittance of initial texture [6].

6.2.2.3 The thickness of the alignment layers

The thickness of the alignment layer is an important factor for AFLCD performance as well as for SSFLC performance. The best performance is obtained with a 20–30 nm layer thickness. The thicker layers increase threshold voltage and decrease the sharpness of the threshold. The thinner layers cause deterioration of the liquid crystal alignment. One of the reasons appears to be the direct influence of the substrates [6].

6.2.2.4 Pretilt angle

For SSFLCDs, a high pretilt angle, 18° for example, is recommended to obtain uniform C1 alignment, and a low pretilt angle is recommended to obtain uniform C2 alignment and bookshelf alignment. For AFLCDs, low pretilt angles result in good alignment without defects. Pretilt angles lower than 1.5° are recommended (see in Fig. 6.2.5). The best pretilt angles were 0.8–1.2° in the case of the commercial AFLC material, Merck ZLI 4792 [6]. Alignment layers of higher pretilt angles resulted in a poor initial texture with lots of defects.

6.2.2.5 Rubbing conditions

Table 6.2.2 shows examples of the conditions of the rubbing process and the initial alignment obtained for AFLCs. The best alignment was obtained with a soft rubbing cloth and strong rubbing conditions. Rubbing cloths of natural fibres such as cotton have soft characteristics. On the other hand, cloths of synthetic fibres such as rayon are hard. Cloths made from tough fibres always result in poor alignment. The rubbing process can also be classified into two categories, strong rubbing and weak rubbing. The best results were obtained with 2–5 gcm^{-2} rubbing pressure. For example, rubbing conditions involving 0.6 mm contact length between the rubbing cloths and the alignment layer surface, and 2.8 gcm^{-2} rubbing pressure are recommended.

Table 6.2.2 Relationships between rubbing conditions and initial alignment [6]

		Contact length of rubbing cloth	Rubbing pressure	Initial texture
Strong rubbing	Cotton	0.6 mm	2.8 gcm^{-2}	Good
	Rayon	0.6	16.7	Bad
Weak rubbing	Cotton	0.2	0.5	Bad
	Rayon	0.2	12.7	Bad

6.2.3 Driving margin hysteresis and threshold properties [10]

6.2.3.1 Symmetry of alignment layers

To obtain a high contrast, not only excellent initial alignment, but also a low transmission to bias voltage V_0 is indispensable. Moreover, the magnitude of the hysteresis and the sharpness of the threshold are critical factors for realizing high contrast and high duty ratio by the simple matrix described above. To evaluate these two factors, the Driving Margin (M) was defined.

$$M = \frac{V_{th}(10) - V_{th}(90)}{V_{sat}(90) - V_{th}(10)}$$

The values of $V_{th}(10)$, $V_{sat}(90)$, and $V_{th}(90)$ are defined as shown in Fig. 6.2.6 [5].

The sharper the threshold and the larger the hysteresis, the larger is the driving margin M obtained. To obtain a high contrast ratio, M > 1 or 2 is necessary. The driving margin M depends on the molecular structure of the alignment polymers. The symmetries of polyimide molecular structures are high compared with

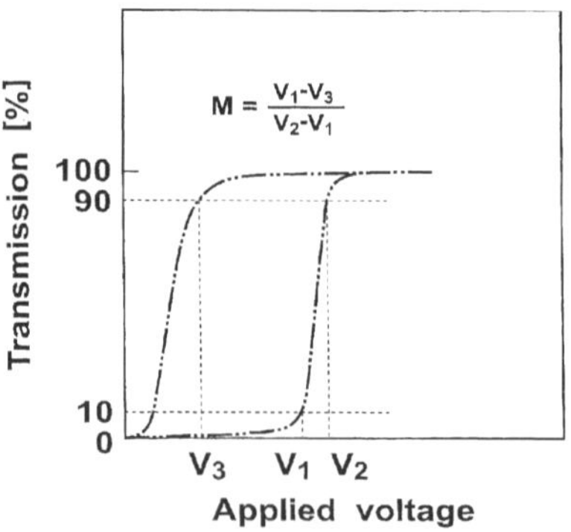

Fig. 6.2.6 The definition of the values of $V_{th}(10)(V_1)$, $V_{sat}(90)(V_2)$ and $V_{th}(90)(V_3)$.

other polymer materials. Polyimides show a large value of the driving margin, M, compared with other kinds of polymer. However, the polyimides with large substituent groups like hexafluoropropyl or propyl groups have low structural symmetry and show low values of the driving margin, M, or even zero. The hexafluoropropyl group greatly reduces the driving margin, M, compared to the propyl group. Polyamides with low symmetry of molecular structures exhibit only random alignment, or a quite low value of the driving margin. Table 6.2.3 shows the relationship between polymer molecular structures of the alignment layer and the quality of the alignment and the magnitude of the driving margin. It is clear that polymer materials of high symmetry show a high driving margin, even though the alignment of the AFLC on the alignment layers shows the same quality.

Figure 6.2.7 shows the relationship between the structure of diamine units and the driving margin. The driving margin decreases according to the decrease in the symmetry of the unit structures.

Table 6.2.3 Effects of polyimide, polyamide and polyamide-imide alignment layer on hysteresis and driving margin using an AFLC mixture

Polymer code	Polymer structure	Driving margin	Alignment with AFLC
LQ-1800 for comparison	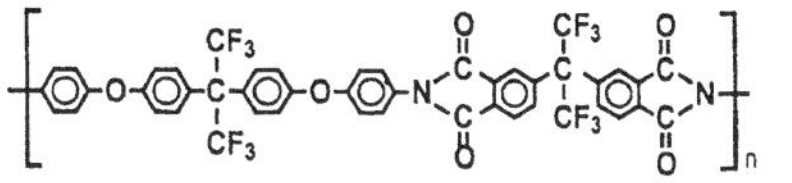 $\left[-C_6H_4-O-C_6H_4-C(CF_3)_2-C_6H_4-O-C_6H_4-N(CO)_2C_6H_3-C(CF_3)_2-C_6H_3(CO)_2N-\right]_n$	0.0	GOOD
PI-1	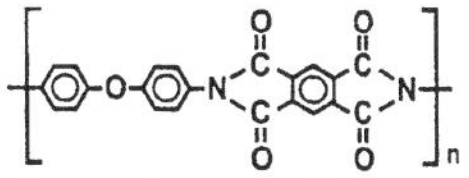 $\left[-C_6H_4-O-C_6H_4-N(CO)_2C_6H_2(CO)_2N-\right]_n$	2.0	GOOD
PI-2	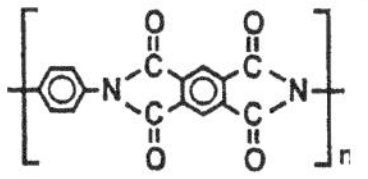 $\left[-C_6H_4-N(CO)_2C_6H_2(CO)_2N-\right]_n$	2.1	GOOD
PI-3	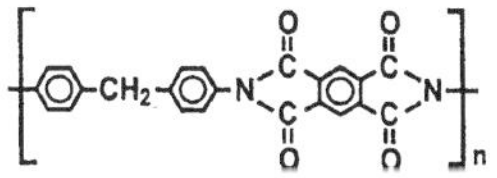 $\left[-C_6H_4-CH_2-C_6H_4-N(CO)_2C_6H_2(CO)_2N-\right]_n$	0.95	GOOD
PI-4	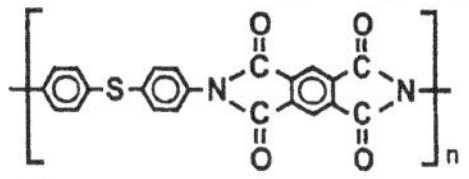 $\left[-C_6H_4-S-C_6H_4-N(CO)_2C_6H_2(CO)_2N-\right]_n$	1.60	GOOD
PI-5	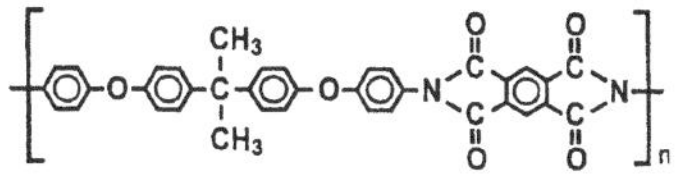 $\left[-C_6H_4-O-C_6H_4-C(CH_3)_2-C_6H_4-O-C_6H_4-N(CO)_2C_6H_2(CO)_2N-\right]_n$	0.23	GOOD
PIF-1	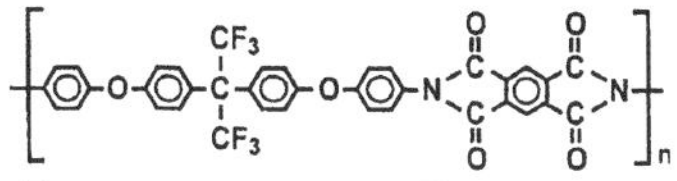 $\left[-C_6H_4-O-C_6H_4-C(CF_3)_2-C_6H_4-O-C_6H_4-N(CO)_2C_6H_2(CO)_2N-\right]_n$	0.21	GOOD
PA-1	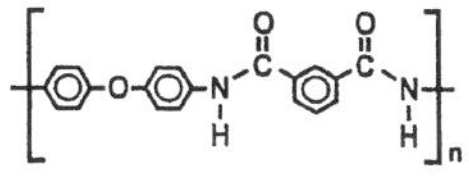 $\left[-C_6H_4-O-C_6H_4-NH-CO-C_6H_4-CO-NH-\right]_n$	—	RANDOM

Table 6.2.3 (Continued)

Polymer code	Polymer structure	Driving margin	Alignment with AFLC
PA-2	$\left[-C_6H_4-NH-CO-C_6H_4-CO-NH- \right]_n$	1.68	GOOD
PAF-1	$\left[-C_6H_4-O-C_6H_4-C(CH_3)_2-C_6H_4-O-C_6H_4-NH-CO-C_6H_4-CO-NH- \right]_n$	—	RANDOM
PAF-2	$\left[-C_6H_4-O-C_6H_4-C(CF_3)_2-C_6H_4-O-C_6H_4-NH-CO-C_6H_4-CO-NH- \right]_n$	0.44	GOOD
PAI-1	$\left[-C_6H_4-O-C_6H_4-N(CO)_2C_6H_3-CO-NH- \right]_n$	2.0	GOOD
PAI-2	$\left[-C_6H_4-N(CO)_2C_6H_3-CO-NH- \right]_n$	2.1	GOOD
PAIF-1	$\left[-C_6H_4-O-C_6H_4-C(CF_3)_2-C_6H_4-O-C_6H_4-N(CO)_2C_6H_3-CO-NH- \right]_n$	0.43	GOOD

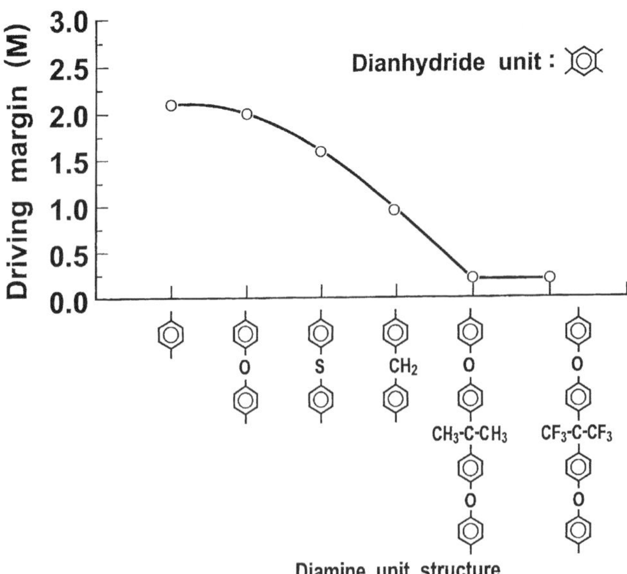

Fig. 6.2.7 Effect of polyimide molecular structure on the driving margin (M).

To obtain a large driving margin, M, or to realize high contrast of an AFLCD of high duty ratio, the selection of highly symmetric polymer materials with symmetrical monomeric units is important.

6.2.3.2 Polarity of alignment layer surface

Together with the symmetry of alignment layer polymers, the polarity of the alignment surface may have an important influence on the driving margins of AFLCDs [10]. The larger the dipole moment in the alignment layer surface, the larger the AFLCD driving margin can become. For polyamide, because of the *meta*-orientation of the diacid part, the structural symmetry is lower and the dipole moment is expected to be larger. However, hydrogen bonding with each neighbouring amide group should effectively reduce the net dipole moment.

6.3 Application of FLC/AFLC Materials to Active matrix Devices

Kohki Takatoh

6.3.1 Introduction

Since the mid-80s, FLC and AFLC materials have been investigated for application to passive matrix LCDs. Using the memory properties or large hysteresis, the intention was to realize high duty-ratio displays without active matrix devices. In the case of passive matrix FLC and AFLC displays, difficulties were experienced with respect to grey scale expression and the improvement of contrast ratio. In the case of SSFLC displays, the grey scale expression has been a serious problem, although it has been partially solved [1, 2]. At FLC '97, Mizutani *et al.* demonstrated that even in the case of SSFLC, digital driving technologies could realize not only grey scale, but also full colour moving pictures, although within a limited area. In the case of AFLC displays, the partial response of liquid crystalline molecules at the bias voltage indispensable for simple matrix driving, causes an imperfect black state and low contrast ratio. It has been improved to 30 by the enlargement of the driving margin of AFLC materials [3]. However, it is still low compared to TN contrast ratio, which is larger than than 200.

On the other hand, since 1995, active matrix displays have been predominant over passive matrix displays in the market. Notebook computers with displays from 10 to 13 inches have been the main application for active matrix LCDs. For this application, TN mode displays have been used and consequently demand for other modes has been limited. FLCs or AFLCs have little chance of being used for this application.

LCD monitors are emerging as a promising new application. For this application, high contrast ratio, wide viewing angle and fast switching are required.

Requirements for LCD monitors include the following:

(1) High contrast ratio of more than 200 or 300
(2) Wide viewing angle of over 70° from upper, lower, left and right directions
(3) Switching speed faster than around 10 ms.

Concerning the contrast ratio, the TN mode has already satisfied the high contrast ratio of over 200. Drawbacks of the TN mode have been narrow viewing angle, particularly in the case of grey scale expression, and slow switching speed. Several modes have been proposed to improve the viewing angle dependences. The most convenient method has been the application of compensation films to the TN mode. However, compensation films are expensive and their application increases costs. Moreover, it is difficult to realize fast response. Secondly, the In-Plane Switching mode (IPS mode) has been attracting much interest because of its advantage in terms of wide viewing angle [4]. However, the improvement of response time is expected to be difficult and the low aperture ratio is an inherent disadvantage. The Multidomain Vertical Alignment (MVA) mode has also been highly evaluated because of its advantage in terms of wide viewing angle and fast response [5]. However, the improvement of switching speed is limited. The Optically Compensated Birefringence (OCB) mode [6] is expected to be the fastest mode in the case of using nematic liquid crystalline materials, and is being investigated with a view to practical usage. The response time is less than 10 ms for all grey scale levels and should be sufficiently fast even for moving picture requirement. However, the problem is that the initial bend state is unstable without bias voltage. The transition from stable splay state to bend state is necessary for every usage, which is a disadvantage for practical products.

To solve these problems, the application of FLC or AFLC materials to TFT-LCD has been proposed. For passive matrix displays, the memory properties of SSFLC or the large hysteresis of AFLC are the most important features in realizing simple matrix driving with high duty ratio. For active matrix driving, these properties are not necessary. However, even without these properties, the application of FLC or AFLC materials is quite attractive because of the wide viewing angle and fast response. In paticular, the fast switching of FLC or AFLC materials could be indispensable when clear moving images such as CRT are required for LCDs. In this section, the application of FLC or AFLC materials to TFT devices is discussed.

6.3.2 Specifications of liquid crystalline materials with spontaneous polarization for TFT driving

The characteristics required for LCDs differ greatly depending on the usages which are, for example, large size LCDs for monitors, small or middle size LCDs for car navigation systems, and so on. Moreover, even if the usage were restricted, the characteristics which depend only on liquid crystalline material properties are few. Almost all the characteristics are decided by the combinations of TFT device properties, such as duty ratio, aperture ratio, storage capacitance and the components like driver ICs, backlights, colour filter. The fabrication process is also the decisive factor. For example, the transmittance usually depends on rubbing process, cell gap and so on. Here, the way to decide two important parameters, P_s and V_{sat} will be explained. Furthermore, the problem of "domain formation" will also be discussed.

6.3.2.1 The relationship between spontaneous polarization and saturated voltage

Figure 6.3.1 shows the equivalent circuit for TN TFT-LCD. In the case of conventional LCD, 60 pictures are formed every second (60 Hz). This means one picture is formed every 16 ms (1 s/60). Gate signals are applied to each gate line (Gm) at one time, from the top of the gate line to the bottom of the gate line. This method is called the line sequential method.

In every 16 ms, all gate lines (G_m) are written out. By applying gate voltage (V_G) into a gate line every 20 µs, (write time t_w), V_G is applied to each gate electrode of the TFT switching device in each pixel. Then, the gate opens or the TFT switch is "on" during t_w, when the signal voltage (V_{ex}) is applied to an ITO electrode in each pixel from the signal line. The charge depending on the signal voltage is applied to the liquid crystalline material (C_g) and the storage capacitance (C_s) of the TFT. After the write time, t_w, the voltage, V_G, is switched off and the gate is closed or the switch is turned off, and the system consisting of C_s and C_g

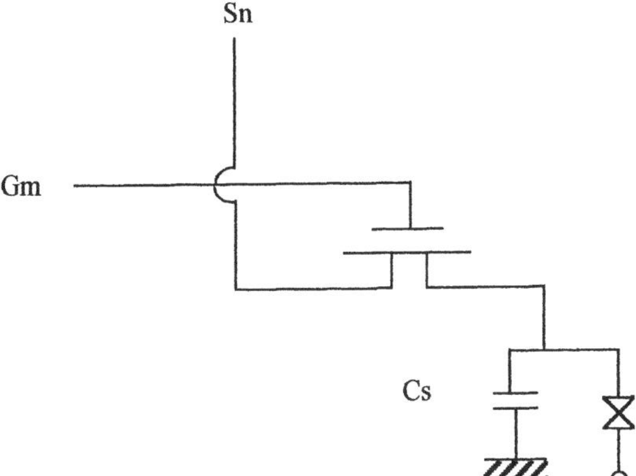

Fig. 6.3.1 Equivalent circuit model of a TFT-LCD.

becomes independent. The voltage on the liquid crystalline material is maintained by C_s and C_g during the 16 ms of hold time. For example, in the case of an XGA LCD (signal line number: gate line number = 1024:768), 768 lines are written out in 16 ms. As a result, signal voltage is applied during 20 μs (16 ms/768) in each pixel from signal line. The period is 60 μs, 30 μs, 27 μs, 20 μs and 16 μs in the case of quarter VGA (320*240), VGA (640*480), SVGA (800*600), XGA (1024*768), SXGA (1280*1024) and UXGA (1600*1200), respectively. The term VGA etc, shows the resolution of the displays. The numbers of signal lines and gate lines are shown in parentheses. In the case of the TN mode, the response time in the slowest grey level is about 100 ms. Therefore the response continues during several frames. This means that the picture does not change during 100 ms, even if the signal changes every 16 ms.

In the case of TN TFT-LCD, the specification of saturated voltage V_{sat} for the liquid crystalline material can be equal to the maximum value of the signal voltage V_{ex} decided by the driver ICs. However, in the case of liquid crystalline materials with spontaneous polarization (P_s), the specification of V_{sat} is not equal to the maximum value of the signal voltage, V_{ex}.

$$V_{sat} = V_{pix} \text{ in the case of TN, IPS, MVA, OCB without} P_s$$

$$V_{sat} < V_{pix} \text{ in the case of (A)FLC mode driven by a TFT device}$$

Figure 6.3.2 shows the equivalent circuit model of a TFT-LCD using liquid crystalline materials with spontaneous polarization. When the reset signal is applied, addi-

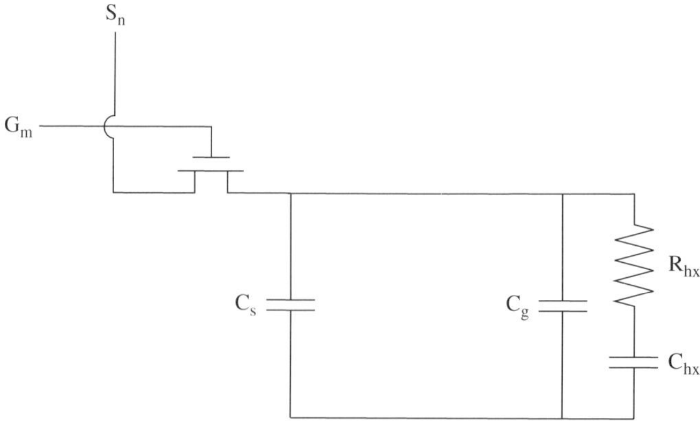

Fig. 6.3.2 Equivalent circuit of FLC/AFLC TFT-LCDs.

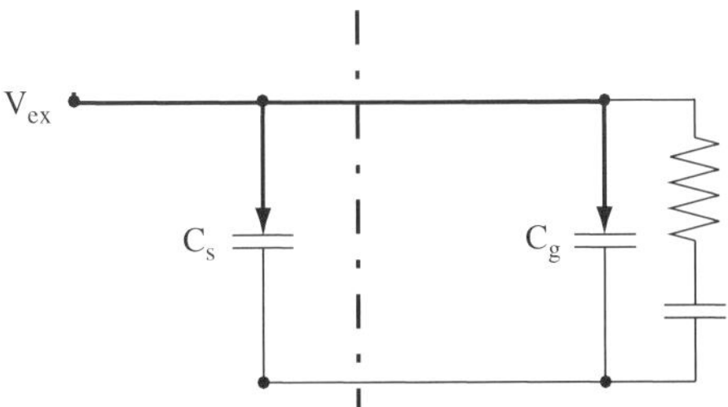

Fig. 6.3.3 Charge is injected into C_{pix}, the sum of C_s and C_g, depending on the magnitude of V_{ex} during write time, t_w, in which the switch of the TFT is on (for example, 15 μs in the case of SXGA).

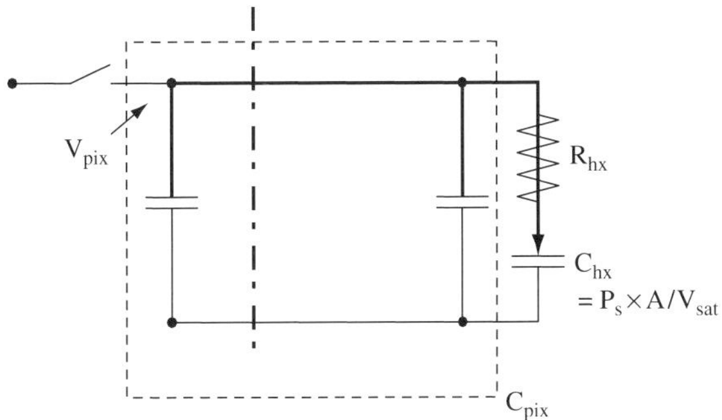

Fig. 6.3.4 Charge is redistributed during the FLC/AFLC switching process in the hold time (ca. 1 ms).

tional structure for applying a reset pulse is necessary. However, in this figure, the additional part for applying the reset signal is omitted.

During the write time, t_w, when the gate is opened, the charge depending on the signal voltage V_{ex} is injected into both C_s and C_g as shown in Fig. 6.3.3. The injected charge is stored in both C_s and C_g.

In the hold time, when the switch of the TFT (gate) is closed after the write time, the charge on C_s and C_g is redistributed into the C_{hx}, the capacitance based on the ferroelectric property as shown in Fig. 6.3.4. In other words, by the rotation of P_s, the charge on C_s and C_g and, as the result, the electric field on the liquid crystalline material is reduced. The resultant voltage on the liquid crystalline material, V_{pix}, can be expressed by equation 1, if the response during t_w can be neglected [18]. For example, in the case of XGA, t_w is 20 μs and the typical response time by a 5 V signal voltage is several hundred μs. Therefore, usually the switching during t_w can be neglected.

$$V_{pix} = \frac{V_{ex}}{(2K+1)} \qquad K = C_{hx}/C_{pix} \tag{1}$$

V_{pix} is the resultant voltage on a pixel electrode. V_{ex} is the signal line voltage. Roughly speaking, in the case of FLC/AFLC materials with a P_s of several $10\,nCcm^{-2}$, K is about 10. So, V_{pix} becomes quite low. To obtain high contrast ratio by AC driving, the development of small P_s materials and the realization of a large storage capacitance C_s is necessary. For example, Takatori *et al.* reported that [50] P_s should be lower than $16\,nCcm^{-2}$ as a typical condition.

In order to determine the specification of P_s and V_{sat} for FLC or AFLC materials, this calculation is not sufficient and it is necessary to raise the accuracy. However, in order to simplify the argument here, the equation for the rough approximation will be used. To obtain a complete response, the resultant pixel voltage V_{pix} should be larger than the saturated voltage V_{sat} of the FLC or AFLC materials.

$$V_{sat} < V_{pix} \tag{2}$$

C_{hx}, P_s and V_{sat} possess the relationship shown in equation 3.

$$P_s = C_{hx} \times V_{sat}/A \tag{3}$$

From equations 1–3, the relationship shown in equation 4 can be obtained.

$$P_s < -\frac{C_{pix}}{2A}(V_{sat} - V_{ex}) \tag{4}$$

Figure 6.3.5 shows the relationship of equation 4. The FLC or AFLC materials with P_s and V_{sat} shown in the shadowed part can attain complete response.

Equation 4 and Fig. 6.3.5 show that the specifications for P_s and V_{sat} can be determined by C_{pix} and V_{ex}. When the values of C_{pix} and V_{ex} of TFT devices are large, the values of P_s and V_{sat} of the liquid crystalline materials can be selected more freely. The magnitude of C_{pix} is mainly determined by the magnitude of C_s. However, because the large C_s reduces the aperture ratio, its value is limited. V_{ex} is determined by driver ICs and usually is less than 7 V.

For the driving of (A)FLC materials of large P_s and V_{sat}, a quasi-DC driving method was proposed. In this method, the signal voltage of the same polarity is applied in a certain term (several seconds) to obtain the complete response [34, 35, 60, 61]. By this method, the response time in the slowest grey level was 50 ms. This response is faster than that of TN, IPS or MVA LCDs. However, for clearer moving pictures, a faster response by AC driving should be realized.

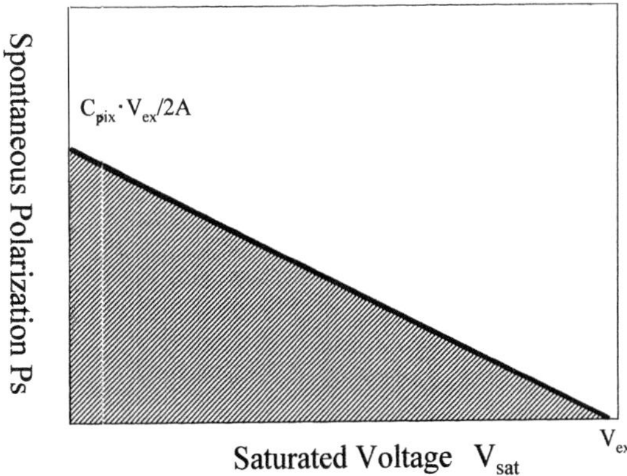

Fig. 6.3.5 Spontaneous polarizations and saturated voltages for complete responses.

6.3.2.2 Domain formation

SSFLC and AFLC show clear domain formation during the switching. The modes for active matrix driving discussed in this chapter, do not show clear, large domain formation like SSFLCs or AFLCs. However, some of these modes show fine domain formation during switching. The formation of domains could obstruct stable grey scale expression [61]. Switching of each mode has not been discussed completely from the aspect of domain formation. However, whether domains are formed during the switching or not could be one of the important specifications for grey scale expression.

6.3.3 Active matrix driving for FLC or AFLC materials [7]

6.3.3.1 Active matrix driving for SSFLC

FLC and AFLC materials have been studied with a view to realizing high duty-ratio displays or high-resolution displays by using the memory properties or large hysteresis. These properties are unnecessary for active matrix driving. However, even without these properties, wide viewing angle and fast response are attractive since they contribute to the high performance of LCDs.

The application of active matrix driving has been considered from the early stage of SSFLC studies. Hartmann showed that [8] grey scale expression could be realized by the application of SSFLC to active matrix driving. Figure 6.3.6 shows the relationship between the applied voltage (V_{ex}) and resultant transmittance. The smooth curve suggests that grey scale can be realized.

This phenomenon can be explained as follows (see Fig. 6.3.7). The charge injected during the write time of the TFT, t_w, is compensated by the rotation of

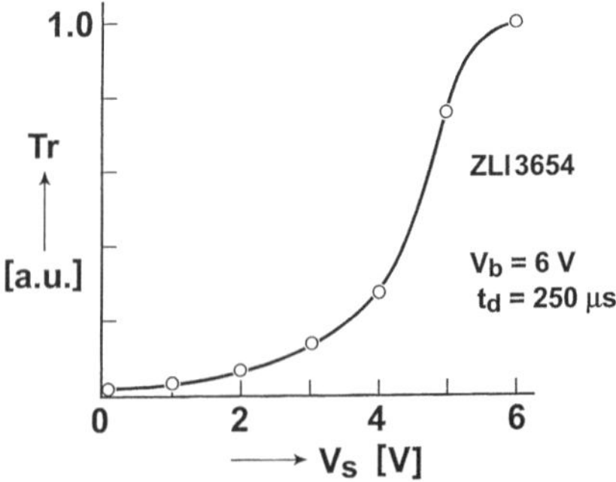

Fig. 6.3.6 Relationship between applied voltage and transmittance for SSFLC active matrix driving.

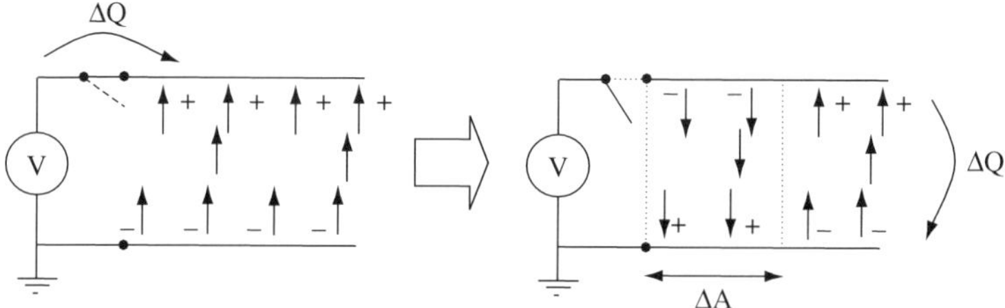

Fig. 6.3.7 Switching of a SSFLC by active matrix driving.

the spontaneous polarization in both write time and hold time. This process determines the area of the switched portion, ΔA, according to equation (5)

$$\Delta Q = P_s \cdot \Delta A \tag{5}$$

and the resultant transmission. ΔQ is the amount of the injected charge. P_s is the spontaneous polarization of the FLC material. It is well known that switching for SSFLC proceeds through domain nucleation and growth kinetics because of its bistable properties. Microscopic observation of the grey scale of a SSFLC driven by a TFT device shows the innumerable minute domains. This switching is called "multidomain switching" (see Fig. 6.3.8).

However, it is considered to be difficult to realize grey scale expression by using this type of switching. The switching only by the change of area of one F-state should be unstable, because the boundary between the two kinds of domain moves easily, and so the desired state cannot be realized accurately in the case of grey scale; sometimes it is excessive and sometimes insufficient. The switching process based on the domain nucleation and growth kinetics is unsuitable for grey scale expression. In contrast to the bistable SSFLC, methods employing the analogue change of the optical axis have been proposed, as listed below. The basic concepts of the Twisted FLC

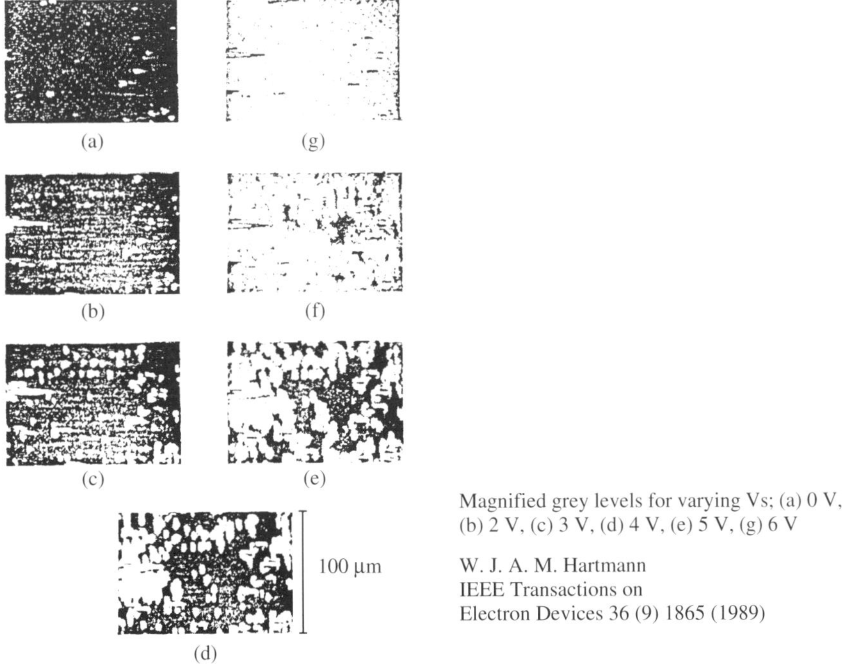

Fig. 6.3.8 Photograph of Multidomain Switching.

mode (3) and the Monostable FLC mode (4) were discussed in the early stage of SSFLC studies [51]. These methods, from (1) to (9), are more appropriate for active matrix driving in view of their advantages with respect to stable grey scale expression.

(1) The application of the electroclinic effect [9]
(2) Deformed Helix FLC (DHF) mode [10–20, 53]
(3) Twisted FLC mode [21–23]
(4) Monostable FLC mode [24–25]
(5) Application of Parallel Stripe Texture and Alternating Polarized Domains (APD) mode [26]
(6) Application of ultra-fine particles [27]
(7) Application of liquid crystalline polymers [28]
(8) Continuous Director Rotation (CDR) mode [63, 64]
(9) Application of the frustoelectric liquid crystalline phase [29–50, 67].

In Fig. 6.3.9, the relationship between applied voltage and transmittance in the case of SSFLCs, AFLCs and modes (1)–(6) and (9) are shown. Theses modes (1)–(6) and (9) show the V-shaped switching property which is appropriate for TFT-driving. Mode (8) shows a unique switching property and requires a special driving scheme.

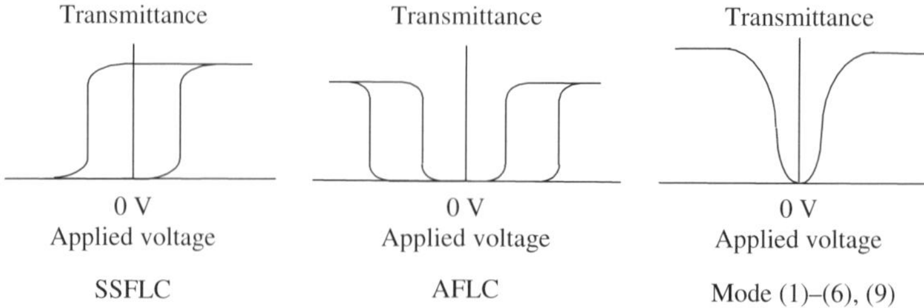

Fig. 6.3.9 Relationship between applied voltage and transmittance in the case of SSFLCs, AFLCs and modes (1)–(6) and (9).

Among these modes, (2) DHF mode, (3) twisted FLC mode, (4) monostable FLC mode, (8) CDR mode, (5) the application of Parallel Stripe Texture and (9) the application of the frustoelectric liquid crystalline phase are reviewed below.

6.3.3.2 Deformed-Helix FLC (DHF) mode [10–20, 53]

When a liquid crystalline material showing a chiral smectic C phase is injected into a cell of the thickness d,

(1) in the case of d much larger than the helical pitch length of the chiral smectic C phase, the chiral structure is maintained and pitch bands or dechiralization lines are observed;
(2) in the case of d shorter than the pitch length, the surfae stabilized ferroelectric liquid crystalline (SSFLC) state is realized.
(3) in the case that d and the pitch length are of similar values, the twisted state is observed.

When liquid crystalline material having a helical pitch much shorter than the thickness d is injected, the helical structure can be maintained even in cells of 1–2 μm thickness. Both directions of the optical and helical axis become parallel to the rubbing direction. By applying an electric field, the helical structure is deformed and the optical axis rotates continuously (Fig. 6.3.10). As a result, a V-shaped relationship between applied voltage and transmittance without threshold and hysteresis properties can be realized.

In the case of a helical pitch shorter than the wavelength of visible light, colouration due to selective reflection from the helical structure disappears and the rotation of the optical axis by deformation of the helical structure can be used for optical switching. This mode is called the "Deformed-Helix Ferroelectric" (DHF) liquid crystal mode. The DHF mode can realize stable continuous grey scale. Moreover, the viewing angle dependence of the contrast is small even in

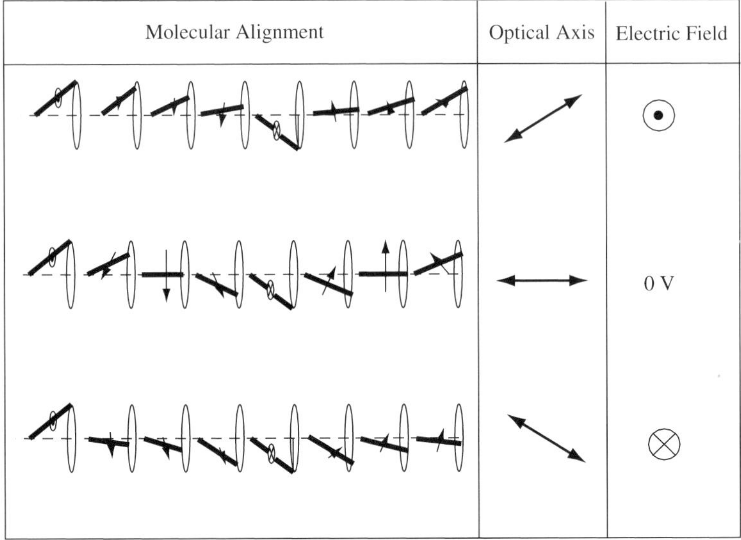

Fig. 6.3.10 Structure and Switching of the DHF Mode.

the case of grey scale, because rotation of the optical axis occurs in the plane parallel to the substrate.

The alignment of the DHF mode was reported to be less sensitive than that of the SSFLC mode to surface treatment and cell thickness variation [10]. In the case of the SSFLC mode, these factors are critical for display performance, especially so for memory properties. From the viewpoint of liquid crystalline materials, improving alignment by introducing a chiral nematic phase with a long pitch and maintaining a tight pitch of the SmC* phase has been reported [19]. In DHF cells with alignment layers rubbed in parallel or antiparallel directions, defects which look like strings are observed as shown in Fig. 6.3.11.

These string-like defects are considered to be a kind of zigzag defect. In each string defect, helical patterns can be observed. Moreover, in each adjacent string, the directions of the helical patterns are opposite.

In the case of cells with parallel or antiparallel rubbing directions, the string-like defects do not align accurately in the parallel direction. The directions are disturbed over several degrees. By twisting the rubbing directions on the substrates at appropriate angles, the directions of the strings become parallel. Furthermore, by applying an electric field, the appearance of the defects becomes dim. In the case of alignment layers rubbed in parallel/antiparallel directions, the defects do not effectively disappear. By twisting the rubbing directions at appropriate angles, the homogeneous alignment without string-like defects can be realized [20]. This string-like zigzag type of defect can also be avoided by using high pretilt angle polyimide without applying an electric field [16]. However,

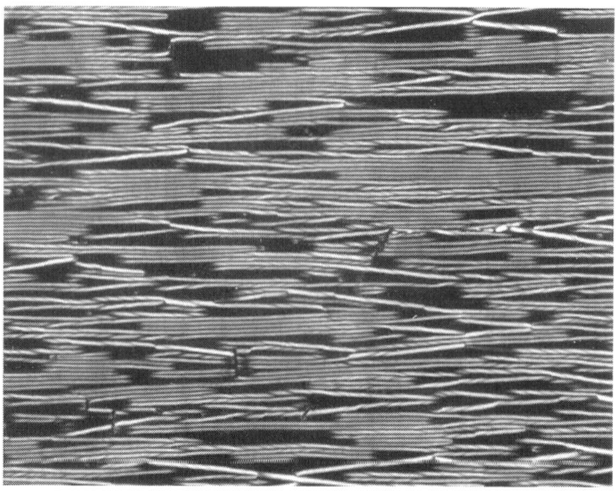

Fig. 6.3.11 Photographs of String Defects.

focal-conic types of defect occur even if high pretilt angle polyimides are applied, although they can be avoided by applying an electric field.

Commercially, the application of an electric field for LC alignment is undesirable. These focal-conic defects are attributed to the chevron structure [53]. Cnossen reported the method of forming the tilted bookshelf structure by selecting alignment layer materials and preventing the formation of the chevron structure. In the DHF mode with the helical structure, the excess free energy to form the chevron structure should be $8 \times 10^{-6}\,\mathrm{Jm}^{-2}$, which is two orders of magnitude larger than that of SSFLC mode. The large excess free energy can be explained as follows. Generally speaking, across the chevron tip, the director profile must be continuous and symmetric. Therefore, the liquid crystalline molecules must be parallel to the substrate near the tip. In the case of the DHF mode, this implies that the helix must be locally unwound. Therefore, the chevron structure of the DHF mode can be more easily prevented from forming than that of the SSFLC mode. The stress in the chevron tip is of the same order as the anchoring energy on alignment layers, and the stress could be the driving force for the deformation from chevron to tilted bookshelf structure. By making the interaction energy between the liquid crystalline molecules and the alignment layer small on one substrate, the tilted bookshelf structure can be realized. For this purpose, unrubbed polymethylsilsesquioxane (PMSQ) was selected for one alignment layer. The surface energy on PMSQ is low and the anchoring energy of the layer is reduced to a minimum by leaving it unrubbed. Under these conditions, the tilted bookshelf smectic layer was confirmed by the single X-ray diffraction peak. The width of the defect lines perpendicular to the layers was observed to be equal to twice the cell gap as is expected for the tilted bookshelf structure.

The most serious problem with respect to the DHF mode is that the helical structure is unwound and the alignment is deteriorated by applying an electric voltage larger than the saturated voltage. This is different from the cholesteric helical structure which is restored by cutting off the electric field. This phenomenon restricts applications, and, furthermore, is a critical problem from the viewpoint of product quality assurance.

6.3.3.3 Twisted FLC mode [21–23]

Even in the case of an FLC cell in which the alignment layers are rubbed in mutually parallel directions, a twisted state is stabilized depending on the cell gap, the polarity of the alignment surface and the FLC material properties. The twisted state is stabilized by the same polarities on both alignment layer surfaces.

In 1992, the Twisted FLC mode was proposed by Patel [21]. The structure of this mode is shown in Fig. 6.3.12.

In this mode, the rubbing directions on the two substrates are perpendicular to each other. The direction of liquid crystalline molecules is twisted at right angle through cell gap, as in the Twisted Nematic (TN) mode. Therefore, without

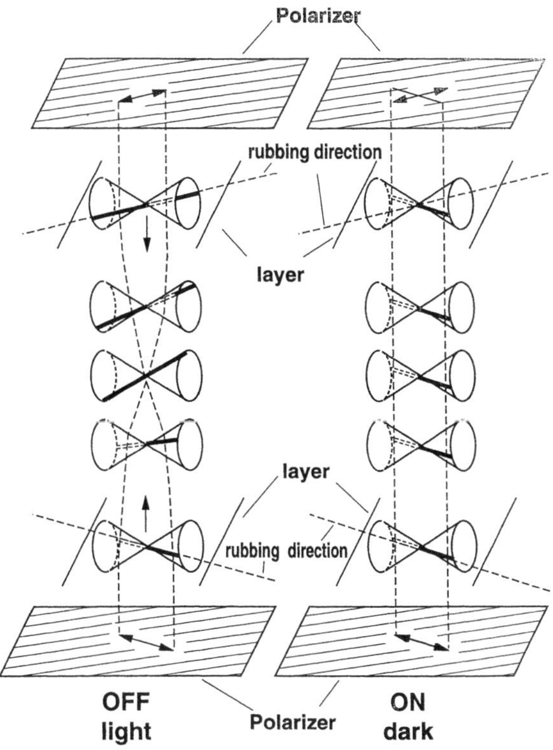

Fig. 6.3.12 Structure of Twisted FLC Mode.

an applied voltage, the incident polarized light rotates through 90°. By setting the polarizers crossed, the rotated polarized light passes through the analyzer to show the bright state. By applying an electric field, the molecular orientation continuously changes into the two alignments of the SSFLC state depending on the polarity of the applied field. These states look dark. The grey scale can be realized depending on the magnitude of the applied voltage. The relationship between applied voltage and transmittance can be expressed by V-shaped curve. This mode is unique among the modes using FLC or AFLC materials in that the change of optical rotatory power is used for light switching, as in the TN mode. If typical ferroelectric liquid crystalline materials with phase sequence Cr-SmC*-SmA-N-I or Cr-SmC*-SmA-I are injected into this type of cell with twisted rubbing directions, two kinds of domains with layers perpendicular to each other are formed. These two domains are formed during the transition from nematic phase to smectic A phase. Therefore, for the Twisted FLC mode, FLC materials which possess no smectic A phase are used.

The direction of layer formation depends on the sign of the spontaneous polarization, the direction of rubbing treatment and the polarity of the alignment surfaces. Figure 6.3.13 shows the directions of smectic layer formation and the rubbing treatment when the sign of the spontaneous polarization is negative and the dipole moment of the liquid crystalline molecules points from alignment layer surface to liquid crystalline material. Figure 6.3.14 shows the relationship between the direction of the smectic layers, the FLC molecules and the dipole moment of the FLC molecules when the sign of the spontaneous polarization is negative. A card in which these two figures are drawn on both sides is very useful to understand this relationship.

In Table 6.3.1, four kinds Twisted FLC mode cells are listed with rubbing directions from Fig. 6.3.13. In cells II and IV, two kinds of domains can be observed in the same ratio and defects are observed between these domains. Among the characteristics observed with respect to the defects between these two domains are:

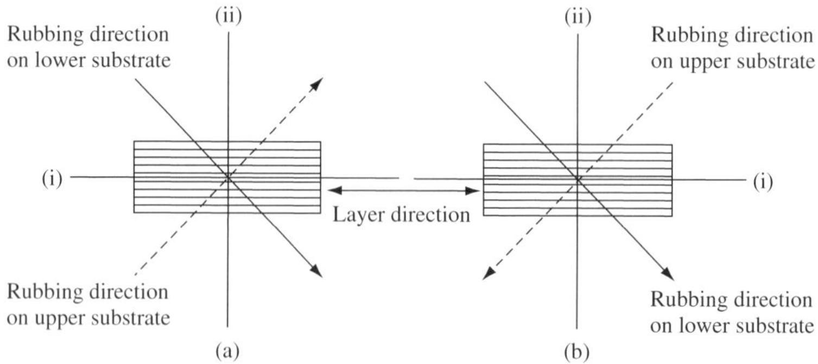

Fig. 6.3.13 Directions of rubbing treatments on upper and lower substrates, smectic layer formation and measurements for viewing angle dependence (i) and (ii).

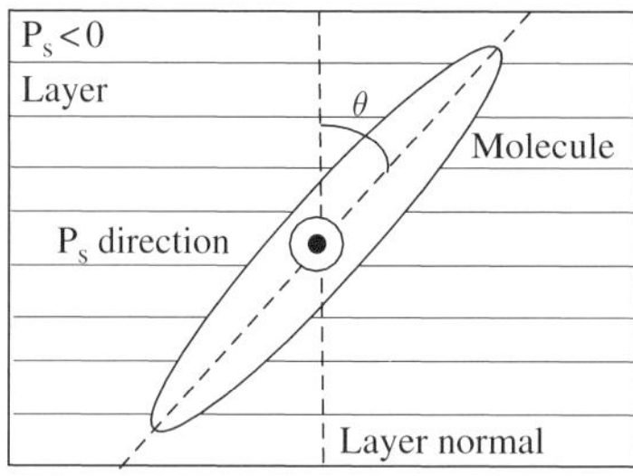

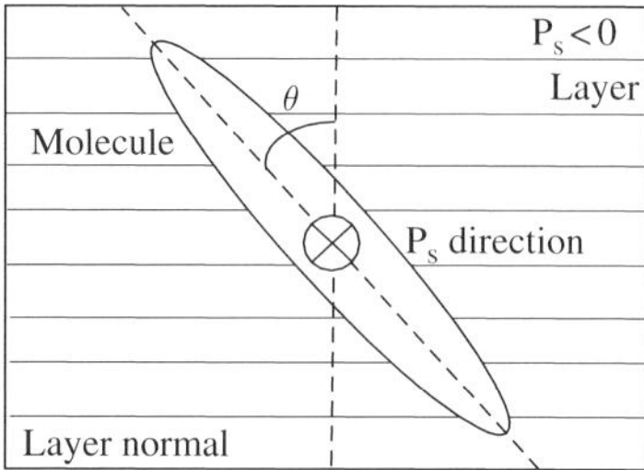

Fig. 6.3.14 The relationship between sign of spontaneous polarization, electric field direction and direction of layer formation.

Table 6.3.1 Four kinds of Twisted FLC mode cells

Cell number	Type of polyimide	Rubbing direction
I	Low pretilt	(a)
II	Low pretilt	(b)
III	High pretilt	(a)
IV	High pretilt	(b)

(1) a zigzag shape in the direction parallel to the layer normal and a smooth shape in the direction parallel to the layers. Therefore, the defects are observed as though zigzag defects had rotated at right angles; (2) on applying voltage, the defects disappeared and on turning off the voltage, the defects reappeared in different patterns.

On the other hand, in cells I and III, one specific domain predominated. This was especially marked in cell III with high pretilt alignment layers where uniform alignment of one domain without any other was observed. Figures 6.3.15 and 6.3.16 show the two possible orientations with the directions of rubbing treatment as in cases (a) and (b), respectively. In the case of (a) in Fig. 6.3.15 orientation (2) is more twisted or unstable than orientation (1). For this reason, a single domain with orientation (1) was observed predominantly in cell I and III. A larger pretilt angle causes a more significant energy difference between orientation (1) and (2). This explains the uniform alignment, without another domain, in cell III. In the case of rubbing direction (b), there is no energy difference between the two possible orientations (1) and (2), shown in Fig. 6.3.16 and two kinds of domain were observed in the same ratio, without any relationship with the values of the pretilt angle.

This mode shows the characteristic viewing angle dependence on the direction of rubbing treatment, which is not observed in other modes using FLC or AFLC materials. In Fig. 6.3.17, the viewing angle dependence of transmittance for cell I is shown for various applied voltages and in Fig. 6.3.18 the direction of strong viewing angle dependence is shown.

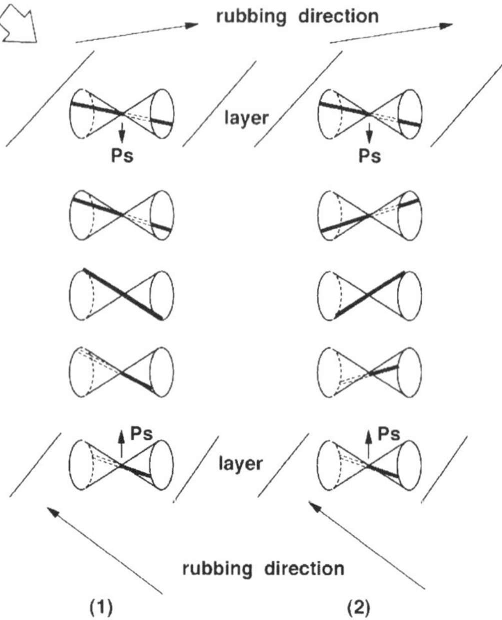

Fig. 6.3.15 The relationships among sign of P_s, polarity of alignment layer surface and direction of molecular orientation in the case of (a).

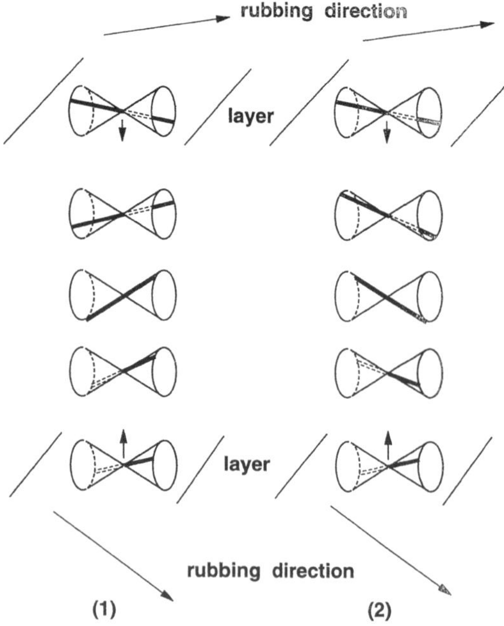

Fig. 6.3.16 The relationships among sign of P_s, polarity of alignment layer surface and direction of molecular orientation in the case of (b).

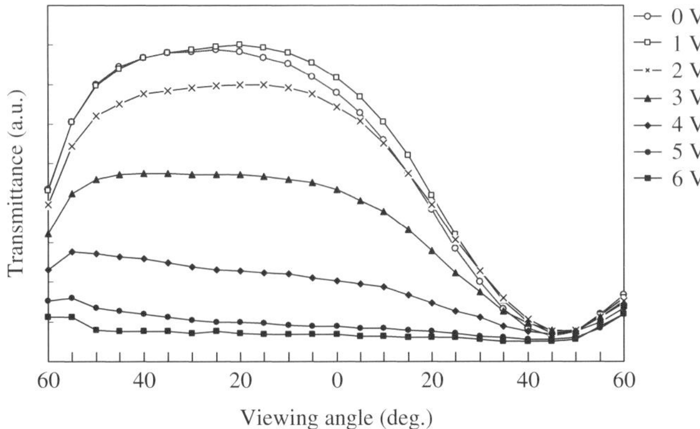

Fig. 6.3.17 Viewing Angle Dependence of Transmittance for Cell I in direction (i).

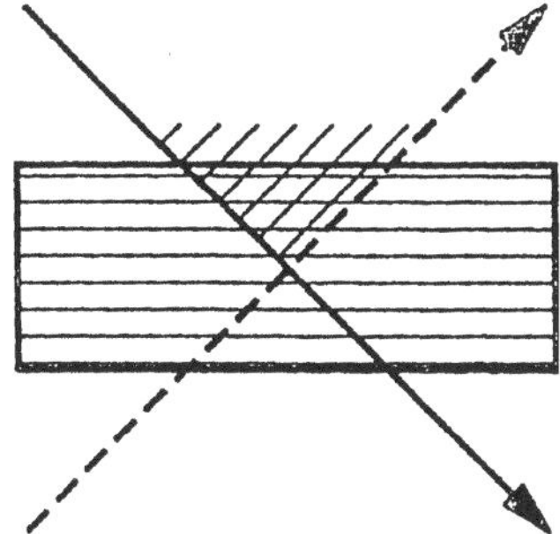

Fig. 6.3.18 Direction of Strong Viewing Angle Dependence.

The shadowed part in Fig. 6.3.18 shows a direction in which strong viewing angle dependence of transmittance is observed. The viewing angle dependence becomes strongest at ca. 45° which is the same as the tilt angle of this LC compound. Exactly speaking, it is the same as (π/2-tilt angle). From this viewing angle, the transmittance becomes constant independent of applied voltage. The arrow in Fig. 6.3.15 shows the direction in which the strongest viewing angle dependence can be expected. In terms of directions of rubbing treatment and layer formation, this direction agrees with the shadowed part in Fig. 6.3.18. Figure 6.3.17 also shows that viewing angle dependence is strongest without an applied voltage and decreases with increase in applied voltage. This is because the stronger the applied voltage, the smaller the averaged declination of the liquid crystal molecules becomes. On the other hand, in the case of a TN-LCD, the strongest viewing angle dependence is observed when grey scale is displayed. Because of this strong viewing angle dependence, the twisted FLC mode is considered to be unattractive for usage in large displays. The viewing angle dependence of cell

II is reduced in comparison with that of cell I. In cell II, two kinds of infinite domains are formed. When viewed from one side of direction (ii) in Fig. 6.3.13, numerous, infinite black and white spots are observed. When viewed from the other side, these spots are changed from black to white and vice versa. This is because the domains with different orientation shown in Fig. 6.3.16 possess the opposite viewing angle dependence. In cell II, there are two kinds of domains in the same ratio which show the opposite viewing angle dependence, and the viewing angle dependence of transmittance for the whole area of the electrode is reduced.

Parallel stripe textures

Depending on the production condition, the stripe-shaped texture parallel to the rubbing direction or perpendicular to the layer direction can be observed in SSFLC. In some materials showing no SmA phase, this stripe-shaped texture can be clearly observed when the smectic C phase is formed from the isotropic liquid or cholesteric phase directly without applied voltage. The stripe-shaped texture can also be observed in the texture formed by applying a low frequency AC or DC voltage to the smectic C phase. The characteristic electro-optical properties can be observed in LCDs with the stripe-shaped texture. Some attempts were made to apply the stripe-shaped texture to both simple matrix and active matrix LCDs. Here, the properties and the structure of SSFLCD with the stripe-shaped texture are explained and the applications to displays are introduced. In this section, the applications of FLC and AFLC materials to active matrix driving are discussed. However, here, the applications to both simple matrix driving (SBF mode and "texture mode") and active matrix driving (APD mode) are discussed.

The stripe-shaped texture parallel to the rubbing direction is observed due to the formation of periodic undulation of the layer structure (Fig. 6.3.19). In the case that a strong AC voltage is applied, the chevron structure is deformed into the bookshelf structure and the strain occurs. The strain is relaxed by the rotation of the layer direction [55] which forms the periodic layer undulation.

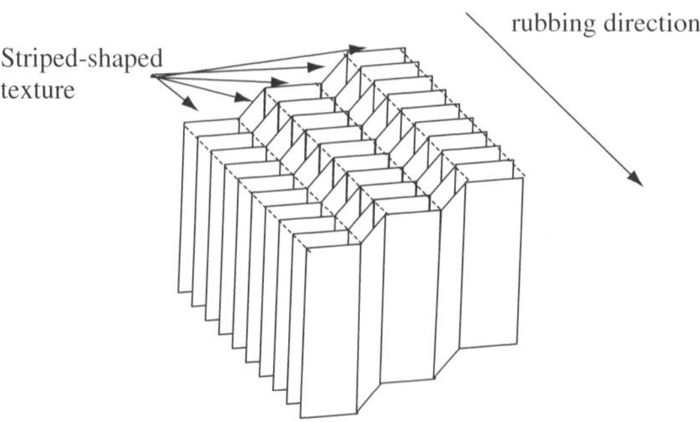

Fig. 6.3.19 Structure of the stripe-shaped texture parallel to the rubbing direction.

In Fig. 6.3.19, the structure of the stripe-shaped texture is shown. The width of the stripe becomes equal to the cell gap and the layer rotation angle becomes equal to chevron angle [68]. In certain conditions the tilted bookshelf structure can also be realized; in this the width of the stripe becomes twice that of the cell gap [54].

On the other hand, liquid crystalline materials showing the N* (Ch)-SmC* phase transition form two kinds of domain on transition into the smectic C phase. In each domain, the layer normal rotates at tilt angle θ, from the rubbing direction and the liquid crystalline molecules align parallel to the rubbing direction (Fig. 6.3.19). In most cases, these two kinds of domains are formed at random. However, depending on the liquid crystalline materials, these two kinds of domains are formed in a striped pattern. The size of P_s does not affect the formation of the stripe-shaped texture and the even FLC materials with a small P_s can form the texture.

Three kinds of LCDs with textures of stripe-shaped patterns have been proposed. Of these three, the two modes using memory properties are studied with respect to simple matrix driving.

6.3.3.4 The "Texture Mode" for simple matrix driving [56–58]

It was observed that by applying an AC field, for example 25 V, 25 Hz, to an SSFLC cell with a chevron structure (Texture I), the texture with the stripe-shaped pattern parallel to the rubbing direction (Texture III) is formed. During the change from Texture I to Texture III, a rather inhomogeneous roof-type texture (Texture II) is also observed (see 6.1.4). Because of the change from chevron structure to bookshelf structure with accompanying stripe-shaped pattern formation, the memory angle increases, for example, from 12° to 32–45° [56]. In Fig. 6.3.20, the relationship between applied voltage and the change of transmittance is shown for textures I, II and III.

Concerning texture II, the value of V_{th} does not change from that of texture I, but the slope of the curve becomes small. Moreover, the transmittance at V_{sat}

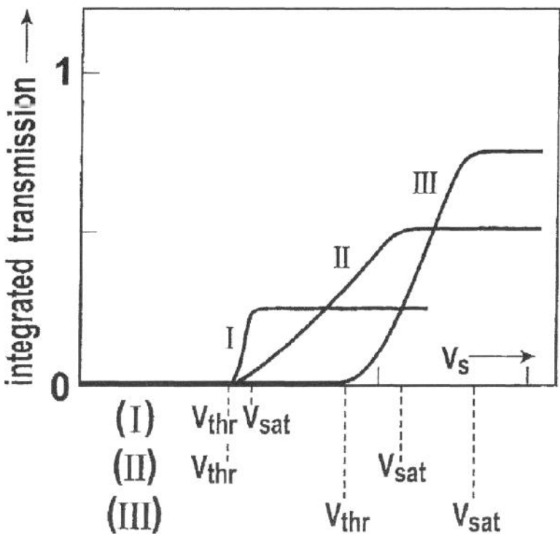

Fig. 6.3.20 The relationship between applied voltage and transmittance for texture I, II and III (see also the photographs in Fig. 6.1.4).

becomes larger. These phenomena imply that in the case of texture II, the magnitude of the chevron angle not only decreases, but also is distributed. In the case of texture III, both V_{th} and the saturated transmittance increase, which can be explained by the change from chevron structure to bookshelf structure. The slope of the curve becomes steeper than that for texture II. However, the slope is smaller than that for texture I, which suggests that the structure is not completely homogeneous.

In these LCDs, the switching occurs in each stripe-shaped domain, which is different from the switching of domains nucleation and growth kinetics through the formation of boat form domains in the case of SSFLC. The grey scale expression is considered to be stabilized by this kind of switching. In fact, a simple matrix 140×170 LCD having texture III, could realize grey scale expression by using a column pulse of variable amplitude [58]. The amplitude of the pixel voltage determines the resultant switched area and grey level. Grey level is not expressed by continuous director rotation, but by the portion of switched area. In this sense, "texture mode" is different from the twisted FLC mode, the DHF mode and so on.

6.3.3.5 Short pitch bistable ferroelectric (SBF) liquid crystal [59] for simple matrix driving

The larger the spontaneous polarization of the SSFLC, the faster the switching speed becomes. Liquid crystalline materials of large spontaneous polarization with a high concentration of chiral dopant usually possess a short pitch length. To form a SSFLCD, the cell gap should be sufficiently narrow compared to the chiral pitch length. Therefore, the pitch lengths applicable to SSFLCD are limited. Moreover, the bistability decreases in the case of FLC materials of large spontaneous polarization because of the depolarization field originating from the existence of ionic impurities. As a result, the magnitude of spontaneous polarization applicable for SSFLCDs is limited.

On the other hand, in the cell with a stripe-shaped texture formed by applying a $15\,V\mu m^{-1}$, 10 Hz AC voltage at the transition from SmA phase to SmC* phase, the realization of excellent bistability was confirmed even with a short pitch length and a large spontaneous polarization. For this kind of cell, the structure of horizontal chevrons as shown in Fig. 6.3.21 is proposed.

In each stripe-shaped domain, the bookshelf structure with layers perpendicular to the substrate is realized. As a result, both memory angle and contrast ratio can be improved. In the stripe-shaped domain, the formation of the helical structure can be prevented. Therefore, even liquid crystalline materials with a large spontaneous polarization and a short pitch length can realize stable bistability. Usage of this mode offers the following advantages: liquid crystalline materials with large spontaneous polarization can be used for fast response; stable memory properties can be realized; the memory angle can be increased because of the bookshelf structure; the structure shows recovery from alignment degradation by applying an electric field.

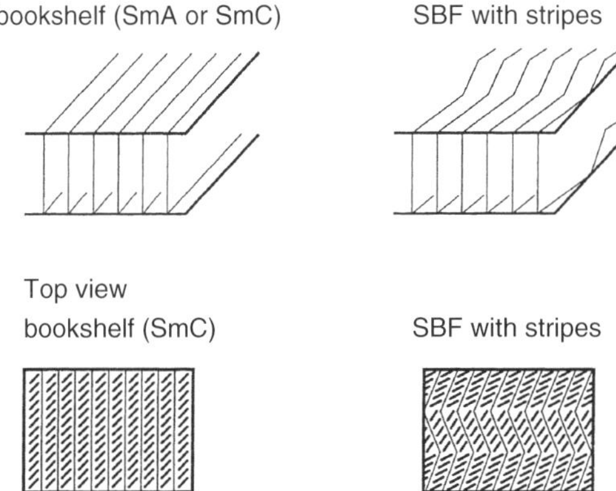

Fig. 6.3.21 Horizontal chevron structure of stripe-shaped texture.

6.3.3.6 Alternating Polarization Domains (APD) mode for active matrix driving

In the alternating polarization domains (APD) mode proposed by Funfschilling and Schadt, the stripe-shaped texture formed during the transition from the cholesteric phase to the smectic C phase is used. The structure of this stripe-shaped texture and the mechanism of the switching with applied voltages of both polarities is shown in the Fig. 6.3.22.

Being different from other FLC and AFLC modes, these modes with stripe-shaped textures show the largest birefringence when the electric field is not applied. In one domain without applied voltage, the direction of P_s is opposite to

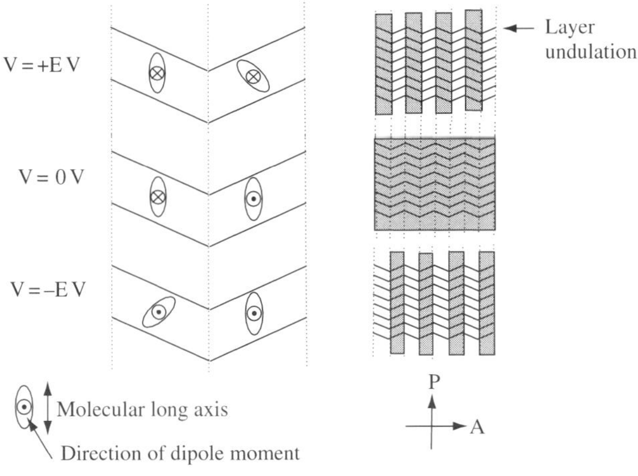

Fig. 6.3.22 Structure and switching mechanism of APD mode.

that in the adjacent two domains. Therefore, the sum of P_s over the whole area becomes 0. In the case that an electric field of one polarity is applied, the liquid crystalline molecules in only one kind of domain rotate at twice the tilt angle and the direction of P_s is reversed. Liquid crystalline molecules in the other kind of domain hardly move. The switching is not a homogenuous change of texture, but the grey scale is determined by the area of the switched portion. This switching is not desirable from the viewpoint of grey level stability, although the size of the domain is limited. Moreover, even in the case that the optical axis rotates at 45° in one kind of domain, half of the area is left without switching. Therefore, the transmittance of the white state must be half of the other modes in which the whole area responds. This has to be the largest drawback of this mode. Liquid crystalline molecules in adjacent domains respond under the application of an electric field of reverse polarity, and without an applied electric field, the liquid crystalline molecules in both kinds of domains revert to the rubbing direction.

6.3.3.7 Monostable SSFLC mode with tilted bookshelf structure

In Fig. 6.3.23, the structure of the monostable SSFLC mode with tilted bookshelf layer structure is shown. The angle δ must be equal to the tilt angle θ to obtain this structure.

The alignment of the liquid crystalline molecules is most stabilized in the direction parallel to the rubbing direction. The P_s direction becomes parallel to

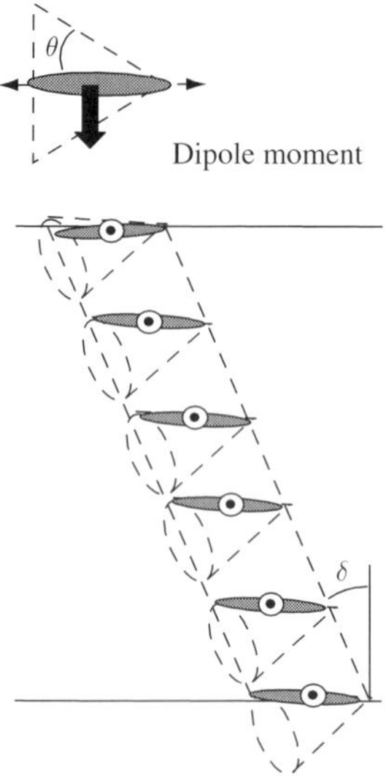

Fig. 6.3.23 Structure of the monostable SSFLC mode with tilted bookshelf layer structure.

the substrates and perpendicular to the rubbing direction. Without applied voltage, the optical axis is parallel to the rubbing direction. When the electric field is applied, the molecules rotate continuously depending on the polarity and the magnitude of the electric field. P_s is directed parallel to the applied electric field when the electric field applied is larger than the saturated voltage. Based on this process, grey scale expression can be realized. The LCD with ferroelectric liquid crystalline material reported by Nito *et al.* was explained by using this structure [24–25]. In this LCD, a smectic C phase of long chiral pitch formed through the SmA-SmC* phase transition was used. Normal SSFLC characteristics with bistablity was observed when the alignment layers on both substrates were rubbed in the same direction (parallel rubbing). On the other hand, when the alignment layers were rubbed in opposite directions (antiparallel rubbing), the LCD with the structure shown in Fig. 6.3.23 could be obtained [24–25]. This mode can be realized only when the angle δ becomes equal to the tilt angle θ, and so it is expected to be used with only a limited variety of materials. Furthermore, it is expected to be difficult to realize this relationship over a wide temperature range.

6.3.3.8 Continuous Director Rotation (CDR) mode [63, 64]

Like the APD mode, depending on the materials, the homogeneous texture of the smectic C phase with the stripe-shaped pattern can be formed by direct transition from the cholesteric phase or isotropic liquid phase without an applied electric field. However, most materials having the transition without a smectic A phase, do not exhibit such a homogeneous texture, but forms two kinds of layers at random.

Patel and Goodby reported a method whereby homogeneous domains can be formed by applying a DC electric field during the transition from the cholesteric phase or isotropic liquid phase to the smectic C phase [65]. During layer formation, according to the direction of the DC electric field, only one of two kinds of domain is selected. By using Fig. 6.3.14, the way in which the layers are formed depending on the sign of P_s, the polarity of the DC field and the size of the tilt angle can be understood.

In the CDR mode (Continuous Director Rotation mode), such a texture has been proposed for use with active matrix driving. The structure and the switching mechanism of the CDR mode are shown in Fig. 6.3.24. Without applied voltage, the liquid crystalline molecules align in parallel to the direction of rubbing. By applying an electric field, the molecules rotate through twice the tilt angle from the direction of rubbing. By turning off the electric field, the molecules return to the initial alignment state. The molecules rotate according to the magnitude of the electric field, and the rotation angle changes continuously. In this mode, the response to the signal of positive/negative polarity becomes asymmetrical. For example, in Fig. 6.3.24(a), when an electric field is applied in the upward direction, the liquid crystalline molecule rotates through twice the tilt angle. However, the molecule hardly rotates with the reverse electric field. Figure 6.3.24(b) also shows

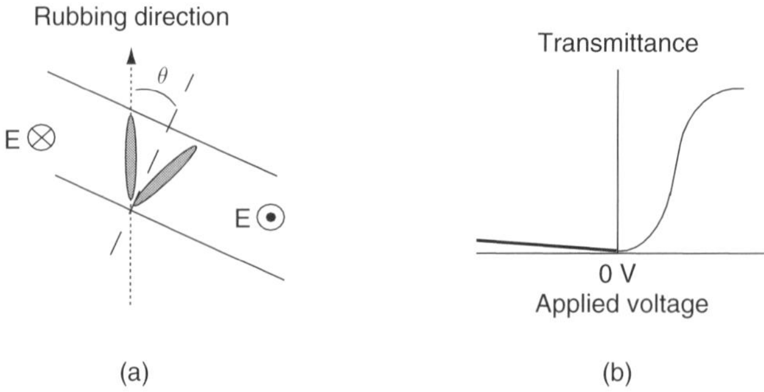

Fig. 6.3.24 Structure, switching mechanism and relationship between applied voltage and transmittance for the CDR mode.

the relationship between the applied voltage and the transmittance. The name "half V-shaped switching FLC mode" is proposed for this characteristic [64].

Generally, in order to prevent disturbance by ionic impurity, it is necessary to apply the signal voltage of different polarity in turns. In the case of the CDR mode, the required transmittance is realized by applying a positive or negative signal, and a black display will be formed by applying a negative or positive signal of the other polarity. Therefore, the CDR mode cannot be used with the usual driving methods employed today. For example, when it is driven by 60 Hz, pictures will be observed from 30 Hz, and flickering is recognized. Also, it is reported that a clear moving picture equal in quality to CRT displays is not obtained even by liquid crystal materials responding at high speed [66]. Even if liquid crystalline molecules respond at infinitely high speed, our eyes feel blur when each picture is exhibited continuously for one frame (6).

--- picture	--- picture	--- picture	--- picture	--- picture	(6)
16 ms	16 ms	16 ms	16 ms	16 ms	

--- picture	--- black state	--- picture	--- black state	--- picture	--- black state ---	(7)
8 ms	8 ms	8 ms	8 ms	8 ms	8 ms	

In the case of CRT displays, one pixel emits over only a short period during one frame. It is reported that this blur can be avoided by displaying a black state in the second half portion of one frame (7). This driving scheme is called "In-pulse driving". For example, when, 120 Hz AC driving is used and the same signal except for polarity is applied in each successive frame, the display shows the transmittance according to the sequence of 8 ms in which the positive signal is applied and 8 ms in which black will be displayed by the negative signal. This is equal to in-pulse driving which displays black in 50% of the frame period. For this reason, this CDR mode has been investigated for in-pulse driving [64], especially for displays which need a high-speed performance.

6.3.3.9 The application of the frustoelectric phase [29–49, 67]

In 1995, Inui *et al.* reported that the liquid crystalline mixture (I) shown below gives switching without hysteresis or threshold properties [29].

F F

$C_{11}H_{23}O$–⟨ring⟩–⟨ring⟩–COO–⟨ring⟩–$CO_2 - \overset{*}{C}H(CF_3)C_4H_8OCH_3$ 40%

F

$C_{11}H_{23}O$–⟨ring⟩–⟨ring⟩–COO–⟨ring⟩–$CO_2 - \overset{*}{C}H(CF_3)C_5H_{10}OC_2H_5$ 40%

F

$C_{11}H_{23}O$–⟨ring⟩–⟨ring⟩–COO–⟨ring⟩–$CO_2 - \overset{*}{C}H(CF_3)C_4H_8OC_2H_5$ 20%

Mixture I

Liquid Crystalline Mixture (I)

Figure 6.3.25 shows the relationship between applied voltage and transmittance of the frustoelectric phase measured by applying a triangular wave electric field. The V-shaped relationship can be observed. Without applied voltage, the optical axis is parallel to the layer normal.

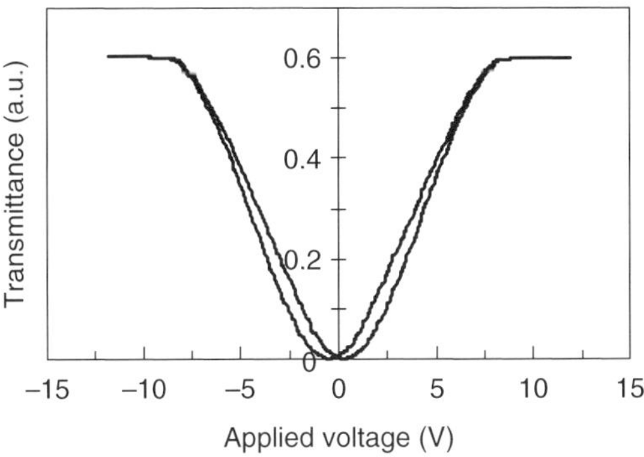

Fig. 6.3.25 Relationship between applied voltage and transmittance of the frustoelectric phase.

The alignment of the frustrated phase has been investigated in terms of the thickness, the surface polarity and the chemical structures of the polyimide alignment layer [42]. Concerning the cell thickness, in the case of thin alignment layers, V-shaped switching could not be observed, but W-shaped switching was obtained. However, it was also observed that by using SiO (thickness 1000Å evaporated at normal incidence) as insulating layers, good V-shaped switching could be observed even with thin nylon alignment layers [49]. In the case of alignment directly on ITO electrode without an alignment layer using the temperature gradient method, typical ferroelectric switching was observed [42].

The effect of surface polarity has also been discussed. Rudquist *et al.* [49] expressed the view that strong polar interaction between the liquid crystalline materials and the alignment layer surfaces forming the twisted state and a sufficiently large spontaneous polarization are indispensable for realizing V-shaped switching based on the collective model described below. On the other hand, Chandani *et al.* compared the alignment on several alignment layers of different polarities and expressed the view that no relationship between the alignments and the polarities was observed. At lower temperature, V-shaped switching was more easily realized and this phenomenon was explained in terms of the change in the mobility of ionic impurities.

In the case of SSFLCs, thin alignment layers with high dielectric constant, high ionic mobility and low spontaneous polarization alleviate the reverse field and enhance the bistability. For V-shaped switching, the situation is opposite with respect to alignment layer thickness, ionic mobility and the magnitude of spontaneous polarization. A thick alignment layer, low ionic mobility and high spontaneous polarization realize V-shaped switching.

Regarding switching appearance, two different observations have been reported. In one observation, a continuous change of texture without domain boundary movement during increase and decrease of the voltage was observed [37, 52]. In the other observation, it was reported that during switching, the occurrence of fine stripe-shaped domains was observed [42]. The appearance of the stripe-shaped domains looks faint and is different from that of the pronounced stripe-shaped domains observed during the switching of tristable AFLC materials. In practical usage, when the stripe-shaped domain appears clearly, it affects the stability of grey scale expression. However, the faint appearance of the texture is not expected to play an important role in the display behaviour.

Proposed structures and switching mechanisms

For this liquid crystalline state, two models of the structure and switching mechanism have been proposed. They are called the "random model" and the "collective model".

Random model [29–31, 36–44]

This model was proposed and has been developed by Fukuda *et al.* In this model, liquid crystalline molecules align in the same direction in each layer, but do not

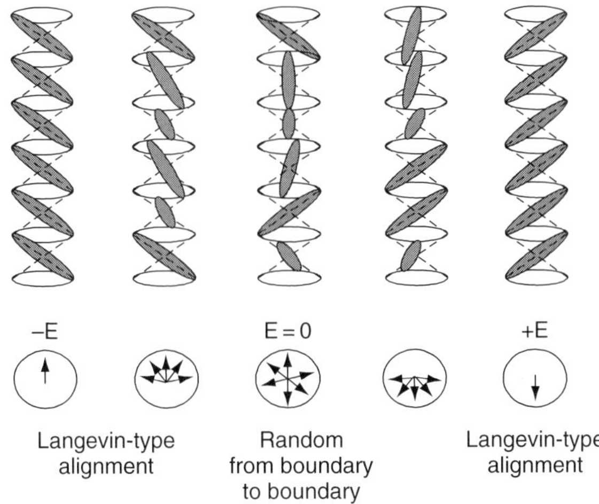

Fig. 6.3.26 Structure and switching mechanism of the random model.

possess any orientational relationship between each layer. Molecular orientation or the C-director orientation in the direction parallel to the layer normal is random (Fig. 6.3.26).

The molecular interaction between each layer is weak and the molecular directors in one layer are not strongly bound by molecular directors in the adjacent layers. As a result, neither a ferroelectric liquid crystal state nor an antiferroelectric liquid crystal state can be realized in forming this random orientation. In the early stages, these materials were called "thresholdless antiferroelectric liquid crystals (TLAF)". However, for these materials antiferroelectric order or antiferroelectricity cannot be observed. Recently, the terms "frustoelectricity" and "frustrated electricity" have been proposed [36]. These terms imply that the formation of both ferroelectricity and antiferroelectricity are prohibited and neither phase can be formed.

The liquid crystalline material of the frustoelectric liquid crystal phase itself exhibits a helical structure. When it is in contact with polymer surfaces, the random molecular orientation must be induced [37]. By contact with the alignment layer surfaces, the liquid crystalline molecules are forced to tilt randomly on a microscopic scale, not only in the sense of right and left, but also with respect to the azimuthal angle distribution, because the interface consists of a rough polymer surface. As a result, the polymer surfaces induce randomization by breaking the intrinsic tilting correlation from layer to layer. In this alignment, the optical axis is parallel to the layer normal. By applying an electric field, Langevin-type alignment and finally ferroelectric orientation are induced (Fig. 6.3.26). This process is analog response and realizes V-shaped switching without threshold properties and hysteresis.

Collective model [45–49]
This model has been proposed by Park *et al.* [45–47] and Rudquist *et al.* [48, 49], independently. Rudquist *et al.* expressed the view that this phase can be identified as a chiral smectic C phase.

In Figs 6.3.27 and 6.3.28, the models proposed by Rudquist *et al.* and Park *et al.* are shown, respectively. Rudquist *et al.* assumed a bookshelf structure. Because of the polarity on both alignment layer surfaces, the liquid crystalline materials have a tendency to form the twisted state. In the twisted state, the directions of the molecular dipole moments are splayed between the substrates (Fig. 6.3.27(a)) and the polarization charge density $\rho_p = -\nabla \cdot \mathbf{P}$ appears. In the case of a large P_s, the electrostatic energy due to the charge density cannot be ignored. As a result, the region of uniformly aligned structure appears (Fig. 6.3.27(b)) or the thickness of the twisted orientation, ζ_p, is reduced into the narrow regions near the surfaces. ζ_p can be expressed by $(K\varepsilon/P_s^{\,2})^{1/2}$ where K is the elastic constant and ε is the dielectric constant. When P_s, K, ε_r are $1\,\mathrm{nCcm^{-2}}$, 5×10^{-12} N, 9, respectively, ζ_p is $2\,\mu\mathrm{m}$. In this case, ζ_p is equal to the cell gap d and the liquid crystalline material forms the twisted state. When P_s is $10\,\mathrm{nCcm^{-2}}$, ζ_p becomes d/10. This result suggests that to form the structure shown in Fig. 6.3.27, the magnitude of P_s should be more than a certain value which is of the same order as the maximum value required from the driving condition of TFT devices.

Park *et al.* assumed the chevron structure as shown in Fig. 6.3.28 based on the results of X-ray measurements. They argued that the effective birefrengence, switching current, second harmonic generation and IR absorption anisotropy measured for this phase could be explained by the model as shown in Fig. 6.3.28.

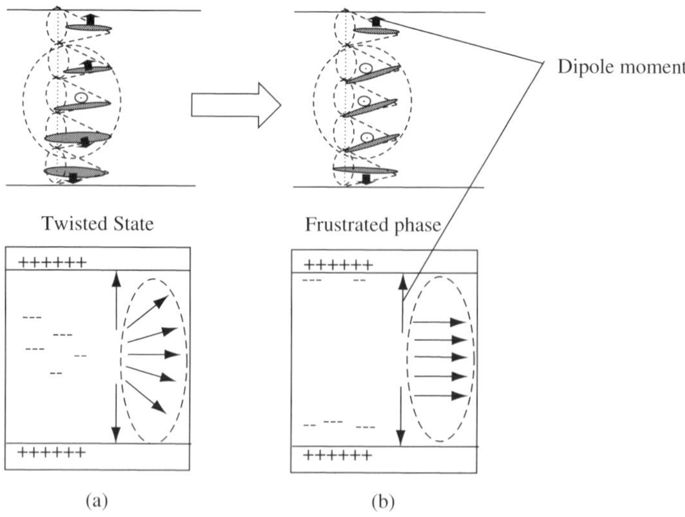

Fig. 6.3.27 Structures of the "collective model" by Rudquist *et al.*

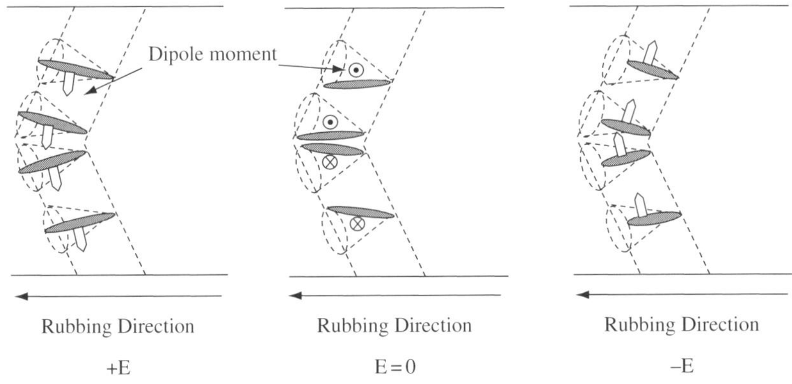

Fig. 6.3.28 Structures of the "collective model" by Park *et al.*

Fig. 6.3.29 15″ TFT-LCD [7, 34, 35, 60–62] using materials showing the frusto-electric phase.

IR absorption anisotropies of the phenyl ring vibration mode at 1605 cm^{-1} for the SmA phase and the frustoelectric phase without applied voltage show that the molecular ordering in this phase at zero field is almost the same as that in SmA phase. Furthermore, the maxima and minima of the IR absorption anisotropies under the electric field do not change from those at zero field. These phenomena mean that the molecular distribution at zero voltage is the same as that under the electric field. These phenomena including the chevron structure can only be explained by the model shown in Fig. 6.3.28 [46].

In the case of SSFLC switching, no signal can be observed by SHG interferometry. However, during the V-shaped switching of this phase, a change in the SHG intensity can be observed. The results can be well explained by the model in which dipole orientation at zero field is parallel to the substrate with an azimuthal angle of 90°. Furthermore, the results indicate that azimuthal rotation of the dipole moments is limited from 0° to 180° (half a cone), but not from 0° to 360° during the switching.

These results can be explained by the model shown in Fig. 6.3.28 in which the molecules have a tendency to align themselves parallel to a substrate surface. It is noted that in this model, if the chevron cusp is at the centre of the cell thickness, no SHG would be observed. So, the SHG observation at zero field is considered to be due to the asymmetric chevron structure, because of the asymmetric rubbing treatment. The authors fabricated a 15″ TFT-LCD [7, 34, 35, 60–62] using the materials showing the frustoelectric phase. Full colour moving images with high contrast ratio and wide viewing angle could be realized by using quasi-DC driving. Figure 6.3.29 shows the displayed images at various angles. The colour shift of the display caused by changing of the viewing angles was slight compared with that for TN-LCDs. The viewing angle was more than ±70° in the horizontal and vertical directions.

References

6.1 Molecular Orientations and Display Performance in FLC Displays

[1] N. A. Clark and S. T. Lagerwall, Appl. Phys. Lett., **36**, 899 (1980).

[2] M. Koden, Ferroelectrics, **179**, 121 (1996).

[3] L. Beresnev, V. Chriginov, E. Pozhidayev, D. Dergachev, M. Schadt and J. Funfschilling, Liq. Cryst., **5**, 1171 (1989).

[4] T. Tanaka, K. Sakamoto, K. Tada and J. Ogura, SID '94 Digest, 430 (1994).

[5] A. G. H. Verhulst, G. Cnossen, J. Funfschilling and M. Schadt, J. SID, **3**(3), 133 (1995).

[6] W. J. A. M. Hartmann, Ferroelectrics, **122**, 1 (1991).

[7] Y. Sato, T. Tanaka, H. Kobayashi, K. Aoki, H. Watanabe, T. Takeshita, Y. Ouchi, H. Takezoe and A. Fukuda, Jpn. J. Appl. Phys., **28**, L483 (1989).

[8] H. Rieger, C. Escher, G. Illian, H. Jahn, A. Kaltbeitzel, T. Harada, A. Weippert and E. Luder, SID '91 Digest, 396 (1991).

[9] T. Kitamura, M. Isogai, Y. Kawabata, Y. Hanawa, Y. Uesugi, K. Iida and K. Matsumura, Proc. of Japanese Liquid Crystal Symposium, 410 (1991).

[10] A. Mochizuki, K. Motoyoshi and M. Nakatsuka, Ferroelectrics, **122**, 37 (1991).

[11] T. Nonaka, J. Li, A. Ogawa, B. Hornung, W. Schmidt, R. Wingen and H. R. Duebal, Liq. Cryst., **26**, 1599 (1999).

[12] Y. Asao, T. Togano, M. Terada, T. Moriyama, S. Namamura and J. Iba, Jpn. J. Appl. Phys., **38**, 5977 (1999).

[13] T. Furukawa, M. Shigeta, H. Uchida and M. Koden, Proc. IDW '00, 251 (2000).

[14] J. Kanbe, H. Inoue, A. Mizutome, Y. Hanyuu, K. Katagiri and S. Yoshihara, Ferroelectrics, **114**, 3 (1991).

[15] M. Koden, T. Shinomiya, N. Itoh, T. Kuratate, T. Taniguchi, K. Awane and T. Wada, Jpn. J. Appl. Phys., **30**, L1823 (1991).

[16] M. Koden, T. Numao, N. Itoh, M. Shiomi, S. Miyoshi and T. Wada, Proc. Japan Display '92, 579 (1992).

[17] A. Tsuboyama, Y. Hanyuu, S. Yoshihara and J. Kanbe, Proc. Japan Display '92, 53 (1992).

[18] M. Terada, S. Yamada, K. Katagiri, S. Yoshihara and J. Kanbe, Ferroelectrics, **149**, 283 (1993).

[19] Y. Hanyu, K. Nakamura, Y. Hotta, S. Yoshihara and J. Kanbe, SID '93 Digest, 364 (1993).

[20] M. Koden, H. Katsuse, A. Tagawa, K. Tamai, N. Itoh, S. Miyoshi and T. Wada, Jpn. J. Appl. Phys., **31**, 3632 (1992).

[21] P. W. H. Surguy, P. J. Ayliffe, M. J. Birch, M. F. Bone, I. Coulson, W. A. Crossland, J. R. Hughes, P. W. Ross, F. C. Saunders and M. J. Towler, Ferroelectrics, **122**, 63 (1991).

[22] M. J. Towler, J. C. Jones and E. P. Raynes, Liq. Cryst., **11**, 365 (1992).

[23] M. Koden, H. Katsuse, N. Itoh, T. Kaneko, K. Tamai, H. Takeda, M. Kido, M. Matsuki, S. Miyoshi and T. Wada, Ferroelectrics, **149**, 183 (1993).

[24] M. Koden, T. Furukawa, M. Kabe, S. Okamoto, A. Sakaigawa, T. Sako, M. Sugino and A. Tagawa, J. C. Jones, M. H. Anderson, P. E. Dunn, J. R. Hughes, K. P. Lymer, V. Minter, K. G. Russell and A. J. Slaney, SID 98 Digest, 778 (1998).

[25] R. Hasegawa, Y. Mori, H. Sasaki and M. Ishibashi, Mol. Cryst. Liq. Cryst., **262**, 77 (1995).

[26] M. H. Anderson, J. R. Hughes, J. C. Jones and K. G. Russell, Proc. 14th International Display Research Conference (IDRC '97), L-40 (1997).

[27] M. Koden, Ferroelectrics, **246**, 87 (2000).

[28] A. Sakaigawa, T. Sako, T. Furukawa, M. Kabe, A. Tagawa, S. Okazaki and M. Koden, Proc., IDW '99, 293 (1999).

[29] H. Furue, Y. Iimura, H. Hasebe, H. Takatsu and S. Kobayashi, Proc. IDW '98, 209 (1998).

[30] P. A. Gass, M. J. Towler, M. Shigeta, K. Tamai, H. Uchida, P. E. Dunn, S. D. Haslam and J. C. Jones, Proc. 14th International Display Research Conference (IDRC '97), L-28 (1997).

[31] P. A. Gass, P. E. Dunn, S. D. Haslam and J. C. Jones, Proc. IWD '98, 197 (1998).

[32] N. Wakita, T. Uemura, H. Ohnishi, H. Mizuno and H. Yamazoe, Ferroelectrics, **149**, 229 (1993).

[33] P. W. Ross, SID '92 Digest, 217 (1992).

[34] H. Shiroto, Y. Kiyota, T. Makino, S. Kasahara, T. Yoshihara and A. Mochizuki, Proc. Asia Display '95, 417 (1995).

[35] S. R. Lee, O. K. Kwon, S. H. Kim and S. J. Choi, SID 97 Digest, 1051 (1997).

[36] M. Koden, S. Miyoshi, M. Shigeta, M. Sugino, K. Nonomura, T. Numao, H. Katsuse, A. Tagawa, J. C. Jones, C. V. Brown, J. R. Hughes, A. Graham, M. J. Bradshaw, D. G. McDonnell, P. Gass, M. J. Towler and E. P. Raynes, Proc. of 4th International Display Workshops (IDW '97), 269 (1997).

[37] T. Minato and K. Suzuki, Liq. Cryst., **23**, 43 (1997).

[38] K. Nakagawa, T. Shinomiya, M. Koden, K. Tsubota, T. Kuratate, Y. Ishii, F. Funada, M. Matsuura and K. Awane, Ferroelectrics, **85**, 39 (1988).

[39] S. Garoff and R. B. Mayer, Phys. Rev., **A19**, 338 (1979).

[40] Y. Hattori, M. Nagai and H. Hama, Proc. 20th Japanese Liquid Crystal Symposium, 270 (1994).

[41] D. C. Ulrich, B. Henley, C. Tombling, M. D. Tillin, D. Smith and N. Dodgson, Proc. IDW '00, 293 (2000).

[42] M. D. Wand, W. N. Thurmes, R. T. Vohra and K. M. More, Proc. of IDW '97, 89 (1997).

[43] H. Ohnishi, T. Uemura, I. Ota, T. Sakurai, K. Takeuchi and K. Yoshino, Proc. Japan Display '86, 352 (1986).

[44] J. S. Patel, Appl. Phys. Lett., **60**, 280 (1992).

[45] K. Nito, E. Matsui, M. Miyashita, S. Arakawa and A. Yasuda, J. SID, **1(2)**, 163 (1993).

6.2 Alignment and Performance of AFLCD

[1] A. D. L. Chandani, T. Hagiwara, Y. Suzuki, Y. Ouchi, H. Takezoe and A. Fukuda, Jpn. J. Appl. Phys., **27**, L729 (1988).

[2] K. Itoh, M. Johno, Y. Takanishi, Y. Ouchi, H. Takezoe and A. Fukuda, Jpn. J. Appl. Phys., **30** (4), 735 (1991).

[3] Y. Yamada, N. Yamamoto, K. Mori, N. Nakamura, T. Hagiwara, Y. Suzuki, I. Kawamura, H. Orihara and Y. Ishibashi, Jpn. J. Appl. Phys., **29** (9), 1757 (1990).

[4] Y. Yamada, N. Yamamoto, M. Yamawaki, I. Kawamura and Y. Suzuki, Proceedings of Japan Display 92, 57 (1992).

[5] N. Yamamoto, N. Koshoubu, K. Mori, K. Nakamura and Y. Yamada, Ferroelectrics, **149**, 295 (1993).

[6] E. Tajima, S. Kondoh and Y. Suzuki, Ferroelectrics, **149**, 255 (1993).

[7] N. Koshoubu, K. Nakamura, N. Yamamoto, Y. Yamada, N. Okabe and Y. Suzuki, Proceedings of Asia Display '95 (1995).

[8] Y. Yamada, N. Yamamoto, K. Nakamura, N. Koshoubu, S. Ohmi, R. Sato, K. Aoki and S. Imai, Proceedings of SID 95, 789 (1995).

[9] K. Nakamura, A. Takeuchi, N. Yamamoto, Y. Yamada, Y. Suzuki and I. Kawamura, Ferroelectrics, **179**, 131 (1996).

[10] Y. S. Negi, I. Kawamura, Y. Suzuki, N. Yamamoto, Y. Yamada, M. Kakimoto and Y. Imai, Mol. Cryst. Liq. Cryst., **239**, 11 (1994).

[11] Y. S. Negi, N. Yamamoto, Y. Suzuki, I. Kawamura, Y. Yamada, M. Kakimoto and Y. Imai, Jpn. J. Phys. **31**, Part 1, 12A, 3934 (1992).

6.3 Application of FLC/AFLC Materials to Active matrix Devices

[1] H. Mizutani, A. Tsuboyama, Y. Hanyu, S. Okada, M. Terada and K. Katagiri, Ferroelectrics, **213**, 573 (1998).

[2] M. Koden, T. Furukawa, M. Kabe, S. Okamoto, A. Sakaigawa, T. Sako, M. Sugino, A. Tagawa, J. C. Jones, M. A. Anderson, P. E. Dunn, J. R. Hughes, K. P. Lymer, V. Minter, K. G. Russell and A. J. Slaney, SID 98 DIGEST, 778 (1998)/ N. Itoh, H. Akiyama, Y. Kawabata, M. Koden, S. Miyoshi, T. Numao, M. Shigeta, M. Sugino, M. J. Bradshaw, C.V. Brown, A. Graham, S. D. Haslam,

J. R. Hughes, J. C. Jones, D. G. McDonnell, A. J. Slaney, P. Bonnett, P. A. Glass, E. P. Raynes and D. Ulrich, Proc. IDW '98 205 (1998).

[3] K. Nakamura, A. Takeuchi, N. Yamamoto, Y. Yamada, Y. Suzuki and I. Kawamura, Ferroelectrics, **179**, 131 (1996).

[4] M. Oh-e, M. Ohta, S. Aratani and K. Kondo, Proceedings of Asia Display '95, 577 (1995)/ M. Ohta, M. Oh-e and K. Kondo, Proceedings of Asia Display '95, 707 (1995)/S. Aratani, H. Klausmann, M. Oh-e, M. Ohta, K. Ashizawa, K. Yanagawa and K. Kondo, Jpn. J. Appl. Phys. Part 2, 36, (1A/B), L27 (1997).

[5] K. Ohmuro, S. Kataoka, T. Sasaki and Y. Koike, *SID '97* Digest of Tech. Papers 845 (1997)/ K. Koike, S. Kataoka, T. Sasaki, H. Chida, H. Tsuda, A. Takeda and K. Ohmuro: Proceedings of AM-LCD '97, 25 (1997)/ S. Yamauchi, M. Aizawa, J. F. Clerc, T. Uchida and J. Duchen, SID '89 Digest of Tech. Papers, 378 (1989)/ A. Takeda, S. Kataoka, T. Sasaki, H. Chida, H. Tsuda, K. Ohmuro and K. Koike, SID '98 Digest of Tech. Paper 1077 (1998).

[6] P. J. Bos and K. R. Koehler/Beran, Mol. Cryst. Liq. Cryst., **113**, 329 (1984)/ P. J. Bos, P. A. Johnson and K. R. Koehler/Beran, SID 83 DIGEST, 30 (1983)/ T. Miyashita, Y. Yamaguchi and T. Uchida, Jpn. J. Appl. Phys., **34**, Part 2 (2A), L177, (1995)/ Y. Yamaguchi, T. Miyashita and T. Uchida, SID 93 DIGEST, 277 (1993)/ T. Uchida, T. Ishinabe and M. Suzuki, SID 96 DIGEST, 618 (1996)/ P. L. Bos and J. A. Rahman, SID 93 DIGEST, 273 (1993)/ M. Xu, D.-Y. Yang and P. J. Bos, SID '98 DIGEST, 139 (1998)/ M. Noguchi and H. Nakayama, SID 97 DIGEST, 739 (1997).

[7] K. Takatoh, AM-LCD 97 DIGEST, 29 (1997)/ K. Takatoh, Presented at the FLC 97 Conference in Brest, France, 20–24, July, 1997/ K. Takatoh, EKISHO, **4** (3), 235 (2000) (in Japanese)/ K. Takatoh, H. Yamaguchi, R. Hasegawa, T. Saishu and R. Fukushima, Polym. Adv. Technol. **11**, 413 (2000).

[8] W. J. H. M. Hartmann, IEEE TRANSACTIONS ON ELECTRON DEVICES, **36** (9), 1895 (1989)/ W. J. H. M. Hartmann, *J. Appl. Phys.*, **66** (3), 1132 (1989).

[9] B. R. Ratna, H. Li, K. S. Neison, J. Naciri and P. P. Bey, Jr., SID 97 DIGEST, 207 (1997).

[10] L. A. Beresnev, V. G. Chigrinov, D. I. Dergachev, E. P. Poshidaev, J. Funfschilling and M. Schadt, *Liq. Cryst.*, **5** (4), 1171 (1989).

[11] J. Funfschilling and M. Schadt, SID 96 DIGEST, 691 (1996).

[12] J. Funfschilling and M. Schadt, J. Appl. Phys., **66** (8), 3877 (1989).

[13] A. G. H. Verhulst and G. Cnossen, Proc. IDRC, 377 (1994).

[14] J. Funfschilling and M. Schadt, Jpn. J. Appl. Phys. Part 1, **33** (9A), 4950 (1994).

[15] A. G. H. Verhulst, G. Cnossen, J. Funfschilling and M. Schadt, SID 95 DIGEST, 133 (1995).

[16] A. G. H. Verhulst and G. Cnossen, *Ferroelectrics*, **179**, 141 (1996).

[17] J. Funfschilling and M. Schadt, *Jpn. J. Appl. Phys.*, Part 1, **35** (11), 5765 (1996).

[18] T. Verhulst, *Jpn. J. Appl. Phys.*, Part 1, **36** (2), 720 (1997).

[19] M. D. Wand, R. Vohra, M. O'Callaghan, B. Roberts and C. Escher, *Proc. SPIE*, **1665**, 176 (1992).

[20] H. Nagata, K. Takatoh and T. Saishu, Proceedings of Annual Meeting for Japanese Liquid Crystal Society, 176 (1995).

[21] J. S. Patel, *Appl. Phys. Lett.*, **60** (3), 280 (1992).

[22] V. Pertuis and J. S. Patel, *Ferroelectrics*, **149**, 193 (1993).

[23] K. Takatoh, H. Nagata and T. Saishu, *Ferroelectrics*, **179**, 173 (1996).

[24] K. Nito, E. Matsui, M. Miyashita, S. Arakawa and A. Yasuda, *SID 93 DIGEST*, 163 (1993).

[25] K. Nito, T. Fujioka, N. Kataoka, A. Yasuda, *AM-LCD 94 DIGEST*, 48 (1994), K. Nito, Y. Imanishi, T. Fujioka, N. Kataoka and A. Yasuda, *Technical Report of IEICE*, 1 (1997) (in Japanese).

[26] J. Funfschilling and M. Schadt, *Ferroelectrics*, **213**, 195 (1998).

[27] A. Yasuda, Y. B. Yang, K. Nito and H. Takanashi, *Eurodisplay 93 DIGEST*, 53 (1993), A. Yasuda, H. Takanishi, K. Nito and E. Mastui, *Jpn. J. Appl. Phys.*, **36**, 228 (1997).

[28] S. Kataoka, Y. Iimura, S. Kobayashi, H. Hasebe and H. Takatsu, *SID 96 DIGEST*, 699 (1996)/ S. Kataoka, Y. Taguchi, Y. Iimura, S. Kobayashi, H. Hasebe and H. Takastu, *Mol. Cryst. Liq. Cryst.*, **292**, 333 (1997)/ H. Furue, T. Miyama, Y. Iimura, H. Hasebe, H. Takatsu and S. Kobayashi, *Jpn. J. Appl. Phys.*, **36**, Part 2, L1517 (1997).

[29] S. Inui, N. Iimura, T. Suzuki, H. Wane, K. Miyachi, Y. Takanishi and A. Fukuda, *J. Mater. Chem.*, **6** (4), 671 (1996).

[30] A. Fukuda, *Proc. Asia Display 95*, 61 (1995).

[31] S. S. Seomun, Y. Takanishi, K. Ishikawa, H. Takezoe and A. Fukuda, *Jpn. J. Appl. Phys.*, **36**, 3586 (1997).

[32] T. Saishu, K. Takatoh, R. Iida, H. Nagata and Y. Mori, *SID 96 DIGEST*, 703 (1996).

[33] T. Yoshida, T. Tanaka, J. Ogura, H. Wakai and H. Aoki, *SID 97 DIGEST*, 841 (1997).

[34] R. Hasegawa, H. Fujiwara, H. Nagata, T. Saishu, R. Iida, Y. Hara, H. Akiyama, H. Okumura and K. Takatoh, *AM-LCD 97 DIGEST*, 119 (1997).

[35] H. Okumura, M. Akiyama, K. Takatoh and Y. Uematsu, *SID 98 DIGEST*, 1171 (1998).

[36] T. Matsumoto, A. Fukuda, M. Johno, Y. Motoyama, T. Yui, S. S. Seomun and M. Yamashita, *J. Mater. Chem.* **9**, 2051 (1999), Addition and correction, *J. Mater. Chem.*, **9**, 3179 (1999).

[37] S.-S. Seomun, Y. Takanishi, K. Ishikawa, H. Takezoe and A. Fukuda, *Jpn. J. Appl. Phys.*, **36**, Part 1, (6A) 3586 (1997).

[38] A. Fukuda and T. Matsumoto, *Proc. IDW '97 (Nagoya)*, 3555 (1997).

[39] A. Fukuda, Y. Takanishi, T. Isozaki, K. Ishikawa and H. Takezoe, *J. Mater. Chem.*, **4**, 997 (1997).

[40] A. Fukuda, S.-S. Seomon, T. Takahashi, Y. Takanishi and K. Ishikawa, *Mol. Cryst. Liq. Cryst.*, **303** , 379 (1997).

[41] S.-S. Seomon, T. Gouda, Y. Takanishi, K. Ishikawa, H. Takezoe and A. Fukuda, *Liq. Cryst.*, **26**, 151 (1999).

[42] A. D. L. Chandani, Y. Cui, S.-S. Seomon, Y. Takanishi, K. Ishikawa, H. Takezoe and A. Fukuda, *Liq. Cryst.*, **26**, 167 (1999).

[43] S.-S. Seomon, Y. Takanishi, K. Ishikawa, H. Takezoe, A. Fukuda, T. Fujiyama, T. Maruyama and S. Nishiyama, *Mol. Cryst. Liq. Cryst.*, **303**, 181 (1997).

[44] S.-S. Seomon, B. Park, A. D. L. Chandani, D. S. Herman, Y. Takanishi, K. Ishikawa, H. Takezoe and A. Fukuda, *Jpn. J. Appl. Phys.*, **37**, L691 (1998).

[45] B. Park, Nakata, T. Ogasawara, Y. Takanishi, K. Ishikawa and H. Takezoe, *SID '99 Digest*, 404 (1999).

[46] B. Park, S.-S. Seomon, M. Nakata, M. Takahashi, Y. Takanishi, K. Ishikawa and H. Takezoe, *Jpn. J. Appl. Phys. 1, Regul. Pap. Short Notes Rev. Pap.*, **38** (3A) 1474 (1999).

[47] B. Park, M. Nakata, M. Takahashi, S.-S. Seomon, Y. Takanishi, K. Ishikawa and H. Takezoe, *Phys. Rev., E***59** (4), 3815 (1999).

[48] P. Rudquist, J. P. F. Lagerwall, M. Buivydas, F. Gouda, S. T. Lagerwall, R. F. Shao, D. Coleman, S. Bardon, D. R. Link, T. Bellini, J. E. Maclennan, D. M. Walba, N. A. Clark and X. H. Chen, *SID '99 Digest*, 409 (1999).

[49] P. Rudquist, J. P. F. Lagerwall, M. Buivydas, F. Gouda, S. T. Lagerwall, N. A. Clark, J. E. Maclennan, R. F. Shao, D. A. Coleman, S. Bardon, T. Bellini, D. R. Link, G. Natale, M. A. Glaser, D. M. Walba, M. D. Wand and X.-H. Chen, *J. Mater. Chem.*, **9**, 1257 (1999).

[50] K. Takatori and K. Sumiyoshi, *AM-LCD 99 DIGEST*, 53 (1999).

[51] N. A. Clark and S. T. Lagerwall, Ferroelectrics, **59**, 25 (1984).

[52] T. Yoshida, T. Tanaka, J. Ogura, H. Wakai and H. Aoki, *SID 97 DIGEST*, 841 (1997).

[53] G. Cnossen, *SID 96 DIGEST*, 695 (1996).

[54] R. F. Shao, Z. Zhuang and N. A. Clark, *Liq. Cryst.*, **14**, 1079 (1993).

[55] I. Dierking, F. Giesselmann, J. Schacht and P. Zugenmaier, *Liq. Cryst.*, **19** (2), 179 (1995).

[56] W. J. A. M. Hartmann, *Ferroelectrics*, **122**, 1 (1991).

[57] A. G. H. Verhulst and F. J. Stommels, *Ferroelectrics*, **121**, 79 (1991).

[58] W. J. A. M. Hartmann, A. G. H. Verhulst, A. Luyckx-Smolders and F. Stommels, *Proc. of the SID*, 32/2 (1991).

[59] J. Funfschilling and M. Schadt, *Jpn. J. Appl. Phys.*, **30** (4) 741 (1991).

[60] R. Hasegawa, H. Fujiwara, H. Nagata, Y. Hara, T. Saishu, R. Fukushima, M. Akiyama, H. Okumura and K. Takatoh, *J. SID* in press.

[61] R. Hasegawa, H. Yamaguchi, R. Fukushima and K. Takatoh, *Ferroelectrics*, **246**, 111 (2000).

[62] K. Takatoh, H. Yamaguchi, R. Hasegawa, T. Saishu and R. Fukushima, *Polym. Adv. Technol.*, **11**, 413 (2000).

[63] T. Nonaka, J. Li, A. Ogawa, B. Hornung, W. Schmidt, R. Wingen and H.-R. Duebal, *Liq. Cryst.*, **26** (11), 1599 (1999).

[64] Y. Asao, T. Togano, M. Terada, T. Moriyama, S. Nakamura and J. Iba, *Jpn. J. Appl. Phys.*, **38**, 5977 (1999).

[65] J. S. Patel and J. W. Goodby, *J. Appl. Phys.*, **59** (7), 2355 (1986).

[66] T. Kurita, A. Saito and I. Yuyama, *IDW 98*, 823 (1998).

[67] T. Saishu, H. Yamaguchi, R. Hasegawa, R. Fukushima and K. Takatoh, *AM-LCD 2000*, 101 (2000).

[68] R. F. Shao, P. C. Willis and N. A. Clark, *Ferroelectrics*, **121**, 127 (1991).

Index